Green Energy and Technology

For further volumes:
http://www.springer.com/series/8059

Darren P. Broom

Hydrogen Storage Materials

The Characterisation of Their Storage Properties

Springer

Dr. Darren P. Broom
Hiden Isochema Ltd
231 Europa Boulevard
Warrington
WA5 7TN
UK
e-mail: dbroom@hidenisochema.com;
darrenbroom@inbox.com

ISSN 1865-3529 e-ISSN 1865-3537

ISBN 978-0-85729-220-9 e-ISBN 978-0-85729-221-6

DOI 10.1007/978-0-85729-221-6

Springer London Dordrecht Heidelberg New York

British Library Cataloguing in Publication Data
A catalogue record for this book is available from the British Library

Cover design: eStudio Calamar, Berlin/Figueres

Printed on acid-free paper

Springer is part of Springer Science+Business Media (www.springer.com)

Preface

The amount of research activity in the hydrogen storage field has increased substantially over the last decade or so, primarily due to the practical need for a hydrogen storage method suitable for use in hydrogen fuel cell cars and other hydrogen-based transportation technology. Hydrogen can store a large amount of chemical energy per unit mass but under ambient conditions it exists in its pure form only as a low density gas. As a consequence, a number of studies have identified the problem of storing hydrogen, for use as a fuel, as a major obstacle to the smooth transition from a fossil fuel-based transportation system to one in which hydrogen is the principal energy carrier. This so-called hydrogen energy transition is seen by many as the answer to the numerous problems associated with our current reliance on oil, which include its finite nature and issues such as energy security and climate change. The international effort to accelerate this transition is well underway, with the recent introduction of the Honda FCX Clarity, the first production hydrogen fuel cell car, leading the way towards wider commercialisation of hydrogen fuel cell technology by the automotive industry. However, the practical problems of fuel cell durability and cost, and effective hydrogen storage, still exist. In the case of the latter, the storage of hydrogen in a solid state material is a very promising potential solution, and the discovery or development of a highly efficient reversible hydrogen storage material would therefore mark a step change in the transition to a hydrogen-fuelled future.

Reversible hydrogen storage materials tend to be either hydrides or microporous adsorbents. A number of books are already available on metal hydride materials, as well as others that cover the use of gas adsorption measurement for the characterisation of porous adsorbents. The former tend to focus on metal-hydrogen system properties, whether thermodynamic, magnetic, crystallographic and so forth, and the latter on the characterisation of the porous structure or surface area of porous materials and powders using the adsorption of gases other than hydrogen. In addition, in the last year or so, undoubtedly due to the recent increase in the level of interest in hydrogen energy, a handful of books have been published with solid state hydrogen storage as their main focus; however, none of these have yet dealt specifically, from a practical point of view, with the characterisation of the hydrogen storage, or hydrogen sorption, properties of candidate media. The importance of this is clear because without high accuracy characterisation it is not possible to effectively assess the storage capabilities of a material and hence make a meaningful comparison with other potential candidates.

However, the practical measurement of hydrogen uptake and release, particularly for storage applications, can be technically demanding, due mainly to the physical properties of hydrogen, but also to the high pressure measurement conditions required for hydrogen storage applications and the sensitivity of many of the materials to contamination of one form or another. Although this aspect of hydrogen storage material research has been partially addressed by some of the recent books on the topic, it is not their main focus and measurement accuracy is not treated in detail. This book is therefore an attempt to provide a monograph on this important aspect of hydrogen storage material research. Coverage of the various types of hydrogen storage material is included for completeness, along with some of the common complementary characterisation techniques used in the field. It is hoped that it will be of practical use as an introduction to relative newcomers to the field, while also acting as a useful reference source for experienced hydrogen storage material or hydride researchers. In addition, although the main motivation behind this work is the current search for a material suitable for storing hydrogen for use as an energy carrier, it is also hoped that much of the content will be of interest to those studying hydrogen absorption or adsorption by materials for many other practical applications, as well as for scientific interest.

The basis for much of the content, in particular Chap. 6, which covers error sources and issues affecting measurement accuracy, was prepared during a three year postdoctoral period spent at the European Commission's Institute for Energy in the Netherlands. The remainder has been completed whilst I have been employed by Hiden Isochema Ltd in the UK. However, any views or opinions expressed are entirely my own and I also take full responsibility for any inaccuracies contained herein. I hope that the book is useful for those with an interest in the subject and would very much welcome any feedback or questions regarding any of the topics or issues covered and discussed (darrenbroom@inbox.com).

Warrington, August 2010 Darren P. Broom

Acknowledgements

As the groundwork for much of this book, in particular Chap. 6, was performed during a three year postdoctoral period spent at the European Commission's Institute for Energy in the Netherlands, I would like to express my gratitude for the opportunity given to me by past and present members of the Clean Energies Unit, including Dr. Jean-Bernard Veyret, Dr. Constantina Filiou, Dr. Pietro Moretto and Dr. Marc Steen. Partial funding for the work was gratefully received from the European Commission DG Research under contract SES6-2006-518271/ NESSHY.

The remainder has been carried out since I returned to the UK to work for Hiden Isochema Ltd. I am therefore grateful to the Managing Director of Hiden Isochema, Dr. Mike Benham, and my other colleagues at the company, including Dr. Mark Roper, Dr. Pete Woodhead, Dr. Michelle Mercer, Charlie Cook, John Bullis, Gerry Duffy and Kathryn Gallimore, from whom I have learnt a great deal. I owe a debt of gratitude, in particular, to Mark who kindly proofread the manuscript in its entirety and provided me with valuable feedback. In addition, I would also like to thank Professor Mark Thomas, Dr. Allan Walton, Dr. Jenny Vitillo, Dr. Dan Bull, Dr. Eveline Weidner, Dr. José Francisco Fernández, Steve Tedds, Matt Beckner, and anyone else whose help, of one kind or another, I have been fortunate enough to receive whilst working for Hiden Isochema.

The completion of this book would not have been possible without the electronic journal access that is available to external users of the University of Manchester's Joule and John Rylands University Libraries, and so I would also like to acknowledge the importance of this valuable service. Furthermore, I would like to thank Anthony Doyle and Claire Protherough from Springer for initiating and supporting this project, and for guiding me patiently and professionally through to its completion. I am particularly grateful to Professor Keith Ross of the University of Salford for introducing me to the study of the behaviour of hydrogen in materials and for the supervision and guidance he provided whilst I was a member of his research group at the University of Salford's Institute for Materials Research. I would finally like to thank my parents for their invaluable support over the years.

Contents

Chapter 1
Introduction

The search for new hydrogen storage materials is currently receiving a great deal of attention from the scientific community. The target is a material that can satisfy the tough demands of a practical hydrogen store for use in the automotive transportation sector: one that can store large amounts of hydrogen in a light, compact form, can be refuelled rapidly, and is affordable, safe and easy to produce in large quantities.

The criteria for an onboard hydrogen storage system have been defined by the US Department of Energy (DOE), in close collaboration with the automotive industry. The most frequently quoted target is a gravimetric system storage capacity exceeding 6 wt% (2.0 kWh kg^{-1}). This was the 2010 target [1] until recently but has since been reduced to 4.5 wt% (1.5 kWh kg^{-1}). At the time of writing, the 2015 target is 5.5 wt% (1.8 kWh kg^{-1}). However, in addition to a high gravimetric capacity, the 2015 requirements also include a volumetric energy density of at least 1.3 kWh L^{-1} (0.04 kg L^{-1}), a refuelling time of less than 3.3 min (1.5 kg min^{-1}), a minimum lifetime of 1500 hydrogenation/dehydrogenation cycles, an operating pressure in the range 3–100 atm (0.3–10.1 MPa) and a net storage system cost of less than \$2 kWh^{-1}. These figures are not currently satisfied by any known storage system. The most promising chance of a solution to this problem is thought to be through the use of solid state storage and, as a result, the search for suitable candidate materials has intensified.

Characterisation of the hydrogen sorption properties of potential storage materials is a crucial factor in this search, but it is also a technically demanding process during which mistakes can easily be made. The main subject of this book is therefore the determination of these properties. In this introductory chapter, we will look at some of the background to the topic: the use of hydrogen as an energy carrier, its storage and the use of solid state materials for this purpose. We will also introduce the measurement methods for gaseous hydrogen uptake determination, and briefly discuss the accuracy of these techniques. We will close the chapter by defining some of the terminology used throughout the book.

D. P. Broom, *Hydrogen Storage Materials*, Green Energy and Technology,
DOI: 10.1007/978-0-85729-221-6_1,

1.1 The Hydrogen Energy Transition

The intense interest in potential hydrogen storage materials is primarily due to the current drive towards a hydrogen economy, also known as the *hydrogen energy transition* [2]. The idea is to convert our transportation system from one dominated by oil to one dominated by hydrogen. If the conversion was successful, and the hydrogen predominantly produced using non-carbon emitting and sustainable methods, it would go a long way towards mitigating the negative effects that our current reliance on oil has on the world's climate. It could also substantially reduce the sensitivity of our current global energy supply network to geopolitical instabilities, and provide an answer to the continuing depletion of our fossil fuel reserves.

At the centre of a future hydrogen energy transportation network stands the hydrogen fuel cell [3]. Vehicles powered by these fuel cells produce only water as a waste product, and so a further significant benefit is the elimination of the harmful and polluting effects of the exhaust fumes from petroleum-based Internal Combustion Engines (ICEs). Fuel cell vehicles, and indeed the cells themselves, have been in existence for many years. The first fuel cell was constructed by the Welsh physicist Sir William Grove in 1839, and the first vehicles were developed in the late 1950s and early 1960s [4, 5]. The development work carried out in the mid-20th century was driven by the need for a highly efficient and lightweight power source for space applications; in this area, of course, the economic cost is not an important factor. It is only relatively recently, however, that the technical advances necessary for the application of hydrogen fuel cells for domestic transportation have put us within reach of a hydrogen-fuelled future.[1]

The idea of the widespread use of hydrogen as a transportation fuel first began to be taken seriously in the 1970s, following the 1973 oil crisis, and research and development work has continued ever since; albeit with varying levels of support [6, 7]. In 1974, the International Association for Hydrogen Energy (IAHE) was founded, and the International Journal of Hydrogen Energy (IJHE) was established the following year [8].

Later, in the 1990s, the automotive industry began in earnest to develop hydrogen fuel cell technology, with an initial view to producing zero emission cars [5]. This move undoubtedly helped bring the concept of a hydrogen energy-based transportation system closer to a practical reality. By the end of 2003, according to a report by the Breakthrough Technologies Institute in Washington, at least 16 of the world's leading car manufacturers were actively developing fuel cell technology and fuel cell vehicles [9].[2] Bus demonstration projects are taking placing

---

[1] Hydrogen can also be used to fuel ICEs, but their efficiency is lower than that of fuel cells and so this just increases the problem of the storage of low density hydrogen. The focus is therefore predominantly on fuel cells.

[2] The report summarises active fuel cell vehicle development programmes by BMW, Daihatsu, DaimlerChrysler, Fiat, Ford, General Motors, Honda, Hyundai, Mazda, Mitsubishi, Nissan, Peugeot Citroën, Renault, Suzuki, Toyota and Volkswagen AG. Solomon and Banerjee [7] also

worldwide, due to the commonly held view that hydrogen-powered buses provide an excellent route to fuel cell vehicle commercialisation. A vast array of both light and heavy duty vehicles powered by hydrogen fuel cells, including bicycles, scooters, golfcarts, forklift trucks, tractors and locomotives, have also been developed, together with boats, submarines and aeroplanes.

Since the 1970s the development of hydrogen fuel cell technology has been supported by governments and international organisations that have implemented various initiatives and networks to aid the transition. In 1977 the International Energy Agency (IEA) began its Hydrogen Program [11], which currently has representatives from 21 countries, together with the European Commission (EC). In 2002, a coalition of national hydrogen associations formed the Partnership for Advancing the Transition to Hydrogen (PATH), and the following year the International Partnership for the Hydrogen Economy (IPHE) was established in order to "accelerate the transition to a hydrogen economy".

As a result of the current level of research and development activity in the area, the amount of investment that has already been committed by the public sector and industry, and the increasing concerns surrounding climate change and energy security, the likelihood of a successful transition to a hydrogen-based energy network has never been greater;[3] however, there are a number of barriers that could either prevent the transition or at a minimum make it considerably more arduous.

1.2 Technological Barriers

The overall task of the hydrogen energy transition is immense, requiring large-scale changes to our energy supply networks and the transportation technology that we currently use. Although the barriers are not all technological,[4] solving the main technical problems would help pave the way. According to a 2004 report by the US National Research Council (NRC) and National Academy of Engineering

(Footnote 2 continued)
identify a number of companies with active hydrogen programmes. A recent report by the United States Council for Automotive Research (USCAR) LLC [10], a joint initiative between Chrysler, Ford and General Motors, strongly supports the continued funding of hydrogen fuel cell research, due to both the recent progress made in the required technology and the greater potential shown by hydrogen fuel cell vehicles compared to the alternative of non-hydrogen electric vehicle technology.

[3] It should be noted that although the idea of a hydrogen economy is gaining increasing levels of support it is not without its critics [12, 13]; however, a discussion of the arguments is beyond the scope of this introduction.

[4] For example, according to Robertson and Beard [14] of DaimlerChrysler, "the scientific, engineering, social, consumer acceptance, economic, and policy hurdles that must be overcome are unprecedented".

(NAE), there are four technological and economic challenges that must be addressed to make the widespread use of hydrogen-fuelled transportation a reality [15]. The first challenge is the development of cost-effective, durable, safe, and environmentally desirable fuel cell and hydrogen storage systems. Within this, onboard storage is described as a "major stumbling block" that is critical to the future of the use of hydrogen fuel cells in transportation technology. The other three challenges are (i) to develop the necessary infrastructure to deliver hydrogen to light vehicle users, (ii) to reduce the cost of hydrogen production from renewable energy sources and (iii) to capture and store the carbon dioxide that is a by-product of the production of hydrogen through coal gasification. Similarly, the US DOE has identified the three greatest challenges as improving hydrogen storage, reducing the cost of hydrogen production and delivery, and making fuel cells less expensive and more durable [16]. Although these are only two examples, both of which are from the US, many other studies and reports on the hydrogen energy transition also recognise onboard hydrogen storage as one of the major technological barriers [17].

Aside from the issues that relate to a sustainable hydrogen energy infrastructure (hydrogen production and delivery, and carbon capture and storage), it is clear that significant improvements in fuel cell performance and hydrogen storage technology would help the transition enormously. Let us first take a brief look at hydrogen fuel cells, the development of which has progressed significantly in recent years. There are a number of types of cell available: Alkaline, Molten Carbonate, Phosphoric Acid, Proton Exchange Membrane and Solid Oxide [4]. Of these, it is primarily the *Proton Exchange Membrane* (PEM) fuel cell that is being considered for transportation applications. This type of fuel cell, also known as the *Polymer Electrolyte Membrane* fuel cell, operates at relatively low temperatures and is now approaching the cost and performance levels necessary for widespread commercialisation. However, the cost and durability of these cells remains a problem. The high cost is due to the fuel cell membrane materials and the use of precious metal catalysts. Although the catalyst materials and membrane manufacture are still relatively expensive, it is thought that significant reductions can be achieved through higher production volumes, as well as other potential routes, such as catalyst reclamation and recycling [18]. The durability, meaning resistance or otherwise to performance degradation during use, is a more complex and challenging issue. The degradation is caused by many of the practical aspects of fuel cell usage, including load cycling, start/stop cycling, cold starts and low (sub-freezing) temperature operation, as well as fuel starvation and the presence of gaseous impurities in the fuel stream. The negative effects of such operational activities include chemical and physical degradation of the membrane, the dissolution and sintering of the platinum catalyst, and the corrosion of the carbon (catalyst) support. Many of the aspects of fuel cell durability and degradation are currently under investigation [19]. A better understanding of the fundamental mechanisms behind the deterioration of the membrane, the catalyst and the catalyst support should help in the overall improvement of fuel cell performance.

The other major part of the challenge is the storage of hydrogen. The driving range of a fuel cell vehicle, of a given tank-to-wheel efficiency, is ultimately limited by the amount of hydrogen stored onboard. In the case of petroleum ICE cars, a typical range is approximately 500 km. The approach taken by the automotive industry is to assume that the widespread introduction of hydrogen fuel cell cars will rely on their ability to match or exceed the performance of their petroleum ICE-driven counterparts. This requires the onboard storage of between 4 and 8 kg of hydrogen, which can be seen from the following real-life examples. Firstly, the 2007 Chevrolet Equinox Fuel Cell, a 5 door Sports Utility Vehicle (SUV), has a range of 320 km using 4.2 kg of hydrogen [5]. Secondly, the Chevrolet Sequel is a concept vehicle capable of the onboard storage of 8 kg of hydrogen, which gives it a range of over 480 km. Thirdly, the Honda FCX Clarity, which was launched in mid-2008, carries 3.9 kg of hydrogen and has a range of 240 miles (386 km).[5] The latter two examples may give the impression that the problem has nearly been solved but in the next section we shall see that this is not the case.

1.3 Hydrogen Storage Technology

Hydrogen stores a large amount of chemical energy per unit mass: 142 MJ kg^{-1} compared to 47 MJ kg^{-1} for a liquid hydrocarbon. The problem is that, unlike the liquid hydrocarbons, at atmospheric pressure hydrogen exists as a low density gas. At room temperature and atmospheric pressure, for example, 1 kg of hydrogen occupies a volume of 11 m^3 [20].

A number of methods can be used to achieve practical densities for the purpose of storage [21]. The hydrogen can be cooled to cryogenic temperatures and stored as a liquid, or compressed to high pressures and stored as a gas at near ambient temperature. Alternatively, it can be stored in non-reversible chemical form in hydrogen-rich liquids and solids, or in molecular or atomic form absorbed or adsorbed in solid state materials. The storage of hydrogen using materials can be broken down into further categories but at this stage we will treat it as one.

Firstly, *liquid hydrogen storage* requires very low storage temperatures. The boiling point of hydrogen is 20.3 K (−252.8°C), and its critical temperature, the temperature above which it can only exist as a gas, is 32.9 K (−240.2°C). During the required liquefaction process a significant amount of energy is consumed. According to von Helmolt and Eberle [5], this is 30% of the stored chemical energy, compared to 15% for compressed hydrogen gas at 70 MPa. There are also significant losses during both cooling and storage, due to a number of so-called *boil-off* mechanisms [22]. During storage, for example, heat leaks into the system due to thermal conduction through the system pipes and cables result in increased evaporation of the fluid and a subsequent rise in the vessel pressure. At around

[5] http://automobiles.honda.com/fcx-clarity/, accessed 5th September 2009.

1 MPa, a valve must be opened to vent some of the hydrogen. The actual losses depend on a number of factors, including the tank size and shape, but a figure of between 0.3 and 0.5% per day was used recently by de Wit and Faaij [23] in their well-to-wheels analysis of hydrogen storage technologies. However, this figure is for a 50 m^3 storage container; for a small onboard storage unit, the losses are significantly greater than this, and so cause a practical problem.

Compressed hydrogen gas storage can also achieve useful storage capacities. This is illustrated by its use in the vehicles mentioned in the previous section. Current state of the art storage tanks can hold hydrogen at pressures of up to 70 MPa, which allows, for example, the storage of 6 kg of hydrogen in a 260 L tank. However, this is rather bulky: as a consequence, the Chevrolet Sequel, with its impressive driving range of 480 km, was essentially designed around its hydrogen storage units (three 70 MPa pressure vessels). In compressed gas storage, the vessels themselves add significant weight to the system, although low density materials are used to construct the high pressure tanks. The tanks developed by Quantum Technologies Inc (USA), for example, consist of three layers: a polymer tank liner acting as a hydrogen permeation barrier, a second carbon composite shell that provides the mechanical strength, and an outer protection layer offering impact resistance. In the longer term, there are fundamental limits to the densities that can be achieved using compressed gas storage, which is problematic. Gray [24] recently calculated that, without even considering the tank construction material, pressures of 255 MPa would be required to meet the 2015 US DOE targets as they stood at the time. At pressures of 70 MPa there are already public acceptance issues and so, leaving aside the problems of developing the necessary tank construction materials, hydrogen storage at pressures this high is unlikely to be widely accepted and commercialised. There is also considerable energetic cost in the compression of the hydrogen gas, a problem that increases with increasing storage pressure. Although the 15% consumed during compression to 70 MPa is favourable compared to the losses of hydrogen liquefaction, the consumption required for even higher pressures would become prohibitive.

Another option, non-reversible *chemical hydrogen storage* in hydrogen-rich liquids and solids, could potentially achieve high energy storage densities; however, the non-reversible nature of these hydrogen stores means that a convenient method for off-board regeneration is required. Although it is currently an active area of research [25], the practical problems with off-board regeneration have previously led the automotive industry to conclude that it is not a viable hydrogen storage method [5]. Examples of the materials and liquids in this category include boron-based compounds, such as sodium tetrahydroborate ($NaBH_4$) and ammonia borane (NH_3BH_3) [26], and organic liquids, such as cyclohexane, decalin and carbazole [25, 27]. The potential of this solution can be seen in the high achievable gravimetric and volumetric storage densities. For example, ammonia borane stores 19.6 wt% and 0.145 kg L^{-1}, although the organic liquids do not unfortunately provide such impressive figures. However, the considerable practical difficulties with the use of this storage method will have to be overcome for it to become viable, with the development of suitable catalysts being a priority.

And so we come to the fourth option: the storage of hydrogen in solid state materials. In *solid state hydrogen storage* [28], a material that can reversibly absorb or adsorb hydrogen in atomic (H) or molecular (H_2) form is used to compress hydrogen (chemically or physically) to high storage densities. The advantage of this option is the density to which the hydrogen atoms can be compressed. Archetypal metal hydrides such as $LaNi_5H_x$ and PdH_x, for example, can store hydrogen at an atomic density far greater than that seen in its liquid form, without the need for cryogenic temperatures and the associated liquefaction. The problem with these two examples, and many other hydride-forming elements and compounds, is that the mass of the host lattice results in a storage unit that is simply too heavy for practical transportation applications. However, hydrides such as these are not the only solid state option. In the next section we will expand on the various classes of potential storage materials that are currently being considered.

1.4 Solid State Storage

There are various types of potential hydrogen storage materials. Examples include microporous materials, interstitial metal hydrides and complex hydrides. Each of these will be covered in more depth, along with other possible storage materials, in the next chapter; however, in this section we will present a brief overview.

The different types of material can be categorised in a number of ways. For example, by operating temperature: *microporous materials* tend to have high storage capacities at lower temperatures near the boiling point of liquid nitrogen, *interstitial hydrides* can absorb and desorb hydrogen effectively at temperatures just above or around ambient, whereas *complex hydrides* tend to require elevated temperatures of 100–300°C (373–573 K) or even higher. The different materials can also be categorised by the mechanism through which they adsorb or absorb hydrogen. Microporous materials adsorb molecular hydrogen through physisorption, interstitial hydrides absorb atomic hydrogen as an interstitial that is free to diffuse rapidly through the metal lattice, and complex hydrides contain ionically or covalently bound atomic hydrogen as part of their structure, which then decomposes upon desorption.[6]

Microporous materials have pore dimensions of less than 2 nm and extraordinarily large internal surface areas. These materials store hydrogen in molecular form at low temperatures. Within the micropores the hydrogen is effectively compressed to a high density, resulting in an increase in the volumetric density of the hydrogen in comparison with free gas at the same temperature and pressure.

[6] Eberle et al. [21] categorise adsorbents as a physical storage method alongside compressed gas and liquid storage, whereas interstitial and complex hydrides are considered chemical methods alongside non-reversible storage compounds.

This enrichment of the hydrogen density close to the pore walls is called *adsorption* and the aim is to find or develop a material that can adsorb large amounts of molecular hydrogen at low pressures and near ambient temperatures. Materials that fall into this category include activated carbons, zeolites, metal-organic frameworks and microporous organic polymers [29].

Metal hydrides store atomic hydrogen in the bulk of the material. In the case of interstitial metal hydrides, the molecular hydrogen in the gas is split into atomic hydrogen on the surface of the material and then diffuses into its bulk. Vast numbers of different metallic compounds exist that will absorb hydrogen in this manner. In most cases, however, the absorption does not occur at an appropriate temperature and hydrogen pressure for practical storage purposes and, generally speaking, the low mass of the hydrogen is overwhelmed by the mass of the host metal. However, there are a number of interstitial hydrides that show very favourable hydrogen absorption and desorption characteristics, in terms of the kinetics, operating temperatures and pressures, and gaseous impurity resistance. In this case, the challenge is to find a hydrogen-absorbing material that retains these favourable characteristics but is light enough for the mass of the stored hydrogen to still be significant for use in a hydrogen store. Materials that fall into the interstitial hydride category include the two examples given previously (Pd and $LaNi_5$), intermetallic compounds, such as TiFe and $TiCr_{1.8}$, and solid solution alloys based on vanadium [30].

Like the interstitial hydrides, complex hydrides also store atomic hydrogen, but in this case the hydrogen forms part of the structure of the host material. When the hydrogen is desorbed the structure decomposes into two or more decomposition products, one of which is molecular hydrogen. This reaction in some cases can be reversed by the presence of a catalyst; the addition of a Ti catalyst to sodium alanate ($NaAlH_4$), for example, renders the hydrogenation reaction reversible. For other systems, such as lithium imide and amide (Li_2NH and $LiNH_2$), a catalyst is not required, although the hydrogenation process can be very slow. Along with the alanates and the nitrides, another example from this group is lithium borohydride ($LiBH_4$). The strong interest in these complex hydrides is a result of their high potential storage capacities (18.36 wt% for $LiBH_4$); however, in each case there are currently limitations to their practical use [31], which can include slow kinetics and high sensitivity to air and moisture.

1.5 The Hydrogen Storage Properties of Materials

A number of different chemical and physical properties are important in assessing the suitability of a material for use in hydrogen storage,[7] but the most important

[7] Examples include its thermal conduction properties, toxicity and the pyrophoricity of the material upon hydrogen activation.

are those related to its gas phase hydrogen sorption behaviour. These include the amount of gaseous hydrogen taken up and released by the material, the rate at which this occurs, the temperature and hydrogen pressure required, and the reversibility of the process. The key characteristics can be derived, to a certain extent, from the list of US DOE requirements to which we referred in the opening section of this chapter: the reversible storage capacity (whether gravimetric or volumetric), the long term cycling stability, the hydrogen absorption and desorption kinetics, and the upper and lower operating pressures, although some important characterisation parameters, such as the enthalpy of hydride formation and decomposition or adsorption, are not referred to directly.

The storage capacity can be defined in a number of different ways. The *gravimetric capacity* is the ability of the material to store hydrogen relative to its mass. This can be expressed as a weight percentage (wt%), kg H_2 kg^{-1}, or mol kg^{-1}. The *volumetric capacity* is the amount of hydrogen stored per unit volume of material. The units for this include kg m^{-3} and mol m^{-3}. Both of these can also be expressed as an energy storage density. See, for example, the definition of the US DOE targets for the gravimetric and volumetric energy densities, which are expressed as kWh kg^{-1} and kWh L^{-1}. These two properties will define, respectively, the weight and the bulk of the storage tank for a given mass of stored hydrogen.

The *reversible storage capacity* of a material is normally lower than its maximum hydrogen storage capacity, although it is the latter figure that is often reported in the literature. The reversible capacity is the amount of hydrogen absorbed and desorbed between the lower and upper operating pressures of the hydrogen store [30]. The closeness in agreement between the total capacity and the reversible capacity will depend on the uptake behaviour of the material. In some cases they may be effectively equal, but if a material takes up a relatively large amount of hydrogen at low pressures there may be a considerable discrepancy. The US DOE targets suggest that the upper and lower pressures considered should be 3 and 100 atm (0.3 and 10.1 MPa); any change in this pressure range is likely to change the reversible capacity of a given material to a certain extent.

Materials can lose their reversible capacity as they are charged and discharged with hydrogen. The ability of a material to retain its reversible storage capacity is known as its *long term cycling stability*, or *cyclic stability*. The capacity loss can be due to a number of factors including the effects of impurities in the hydrogen gas and the partial breakdown of the host material, and is analogous to the loss in the capacity of rechargeable batteries over a number of charge and discharge cycles.

The sorption and desorption *kinetics* of a material define how quickly it can take up and release hydrogen. This is important because it will determine how rapidly the hydrogen storage tank can be refuelled and whether the store will release hydrogen at a rate sufficient to supply the fuel cell during use. This can be defined in a number of ways, but can practically be considered as the period the material requires to be recharged or emptied between the lower and upper operating pressures mentioned above. In the case of hydrides the kinetics are strongly dependent on temperature.

In addition to these considerations, there are a number of other relevant practical properties. These include the *gaseous impurity resistance* of a material, which will affect its cyclic stability. The impurities present in hydrogen gas, depending on its source, can include carbon monoxide, hydrogen sulphide and ammonia.[8] Any of these could affect the subsequent hydrogen sorption properties of a material by poisoning the surface, occupying and hence blocking adsorption sites, or by breaking down the host structure. Another important property is the ease of activation. *Activation* can be defined generally as the process by which the material is prepared for reversible hydrogen sorption. In the case of microporous materials this is the process of degassing the material to remove environmental adsorbates or remnants of the synthesis process from its pore structure and external surfaces. For hydrides it often involves the repeated hydrogenation and dehydrogenation of the material. In other cases the activation may essentially be part of the synthesis process itself.

The *thermodynamic properties* of a material will determine its operating pressure and temperature ranges. For hydrides, the main property of interest is the enthalpy of hydride formation and decomposition, and for microporous materials it is the enthalpy of hydrogen adsorption. Both of these are typically expressed in kJ mol^{-1} and determine the amount of hydrogen that can be stored by the material at a particular temperature and pressure under equilibrium conditions. These are thermodynamic parameters that must be considered in conjunction with the kinetic properties. It is possible, for example, for a hydrogen-absorbing material to have a suitable enthalpy of hydride formation and decomposition, but possess inappropriately slow sorption kinetics.

Each of these properties, and their importance with regard to each material type, will be discussed in more depth in Chap. 3.

1.6 Hydrogen Sorption Measurement

The properties described above can each be determined using gas sorption measurement techniques. Although testing a material in a prototype storage unit is probably the most effective method for determining how well it performs as a hydrogen store, this is not practical when developing, or searching for, new

---

[8] As an example, the US FreedomCAR Fuel Cell Technology Team have specified, for the purpose of testing fuel cell performance, a composition of hydrogen (> 99.9% purity), with 10 ppb H_2S, 0.1 ppm CO, 5 ppm CO_2 and 1 ppm NH_3 [19]. The FreedomCAR Hydrogen Storage Roadmap specifies 99.99% purity with the following impurity levels: 5 ppm H_2O, 2 ppm Hydrocarbons, 5 ppm O_2, 100 ppm He/N_2/Ar combined, 1 ppm CO_2, 0.2 ppm CO, 0.004 ppm S (total), 0.01 ppm Formaldehyde, 0.2 ppm Formic Acid, 0.1 ppm NH_3 and 0.05 ppm Halogenates [1].

materials. Therefore, laboratory gas sorption measurement techniques are routinely used to determine the hydrogen sorption properties of potential storage materials. Knowledge of the engineering properties of a material is also required for storage unit design, and so at every stage of the process, from materials discovery to storage unit development, this type of measurement is crucial.

Gas sorption measurements can be broadly separated into three categories: gravimetric, volumetric and temperature-programmed techniques. *Gravimetric techniques* determine the amount of hydrogen sorbed or desorbed by a sample by measuring the change in its mass. Generally speaking, the mass change is determined using a microbalance housed in a pressure and vacuum-rated vessel. *Volumetric techniques* determine the amount of hydrogen sorbed or desorbed by a sample by measuring the change in pressure in a fixed, known volume: absorption or adsorption results in a decrease in the observed pressure, whereas desorption results in an increase. These two techniques are used to measure isotherms, which are plots of the hydrogen uptake versus pressure, or vice versa, at a fixed temperature. *Temperature-programmed techniques*, on the other hand, measure the amount of hydrogen desorbed by a material as a function of temperature. All three of these can also, to a certain extent, be used to determine the kinetics of the sorption process as well as to quantify the reversible equilibrium hydrogen content of a material. These methods, and some of their variants, are described in more detail in Chap. 4.

Hydrogen sorption data determined using the gas sorption measurement methods introduced above can be complemented by the use of a range of other techniques. In the case of hydrides, for example, hydrogen can be absorbed and desorbed electrochemically, neutron scattering can be used to determine the presence and position of hydrogen within the crystal structure, and both X-ray and neutron diffraction can be used to determine hydride lattice expansion as a function of hydrogen loading. Thermal analysis and calorimetry can be used to measure the thermodynamic properties that can also be calculated from isotherm data. For microporous materials, gas adsorption measurements performed with adsorptives such as nitrogen and carbon dioxide can be used to determine effective surface areas, pore volumes and pore size distributions. These properties correlate to a certain extent with hydrogen uptake and can therefore be compared with measured hydrogen capacities. These and other complementary characterisation techniques will be covered in Chap. 5.

One motivation for the use of the complementary techniques mentioned above, aside from scientific interest and the further insight they offer, is the determination of the accuracy of a hydrogen sorption measurement. Good agreement and consistency between data measured using different techniques is likely to indicate that the hydrogen sorption measurements are reasonably accurate, providing the results can be correlated sufficiently with the observed uptake. Although high measurement accuracy is always important, it is particularly prominent in the field of hydrogen storage material research due in part to the controversy surrounding the hydrogen sorption properties of carbon nanotubes and nanofibres, which we shall describe briefly below.

Although the storage of hydrogen in microporous materials is not a new idea [32–34], recent interest was undoubtedly triggered[9] by the work of Dillon et al. [35], in which the authors reported potential room temperature hydrogen storage capacities of up to 10 wt% by carbon nanotubes. An estimated capacity of between 5 and 10 wt% was reported, based on temperature-programmed desorption measurements. Although these initial measurements appear to have been erroneous, a worldwide research effort was launched as a result. The later, and even more impressive, claim made by Chambers et al. [36], as a result of erroneous volumetric measurements, of up to 67 wt% room temperature storage in carbon nanofibres also fuelled the effort. A number of reports were subsequently published with a large variation in the claimed storage capacity of these materials and the topic attracted a great deal of debate. Despite some initial scepticism, carbon nanotubes and nanofibres were considered for a while to be a possible solution to the hydrogen storage problem [6], although it now seems clear that this is not the case [37]. Inadequate characterisation of the host materials played a significant role in this story, but inaccuracy in the measurement of the hydrogen sorption behaviour was also partly to blame.

As a result of the importance of measurement accuracy, and the ease with which mistakes can be made, Chap. 6 examines the possible sources of error that can contribute to inaccuracy in these types of measurement. Although the volumetric and gravimetric techniques are, in principle, straightforward a number of corrections must be applied to the measured data and, particularly at elevated hydrogen pressures, which are crucial for storage applications, significant errors can be introduced if they are not applied with care. We begin Chap. 6 by considering the areas that are common to most measurements to a certain extent. These include the various possible equations of state that can be used to describe the compressibility of hydrogen at elevated pressures and potential problems with gas purity. Both of these are relevant to volumetric and gravimetric measurements. In the latter part of Chap. 6 we cover the issues that are relevant to each particular technique. For example, the buoyancy effect corrections that are required in gravimetric measurement, and the error sources specific to the volumetric technique, such as the accumulation of errors through the measurement from each successive gas dosing step.

In Chap. 7 we conclude with a discussion of some aspects of measurement accuracy determination, such as the interlaboratory study of hydrogen sorption measurement and the availability of reference materials for testing instrumentation. We also propose some measurement guidelines, based on the discussion in

---

[9] In the intervening period there have been significant breakthroughs, including the exciting developments in the synthesis of metal-organic frameworks. It is therefore likely that these breakthroughs would have led to the re-emergence of the idea of adsorptive hydrogen storage in the absence of the work of Dillon et al. [35]; however, this early work certainly opened up the prospect of new candidate materials in this category and made adsorptive hydrogen storage a hot topic.

Chap. 6, and discuss the need for further research into the accuracy of hydrogen sorption measurements.

1.7 Terminology

In the penultimate section of this introductory chapter, we will define some of the terminology used in this book, partly by way of introducing the concepts but also for reasons of clarity.

Firstly, with regard to microporous materials, the term *micropore* has been defined in IUPAC (International Union of Pure and Applied Chemistry) guidelines to mean a pore with a dimension of less than 2 nm [38]. The *pore dimension*, or *pore size*, refers to the smallest pore dimension or the width of the pore. In the case of cylindrical pores, the pore size is therefore represented by the diameter and not its radius. The terms 'nanopore' or 'nanoporous material' are avoided here, because they have a less precise definition[10] and do not follow the current IUPAC conventions.

Hydrogen *adsorption* is the enrichment of the hydrogen density on or near the surface of a material whether in atomic or molecular form; hydrogen *absorption* is the incorporation of atomic hydrogen within the crystallographic structure or bulk of the material. Following the IUPAC guidelines on adsorption measurement [38], the non-adsorbed, free (hydrogen) gas is known as the *adsorptive*, the adsorbed gas as the *adsorbate*, and the adsorbing material as the *adsorbent*. *Physisorption* is the result of the weak interaction of molecular hydrogen with the surface of the adsorbent via van der Waals forces, whereas *chemisorption* involves the formation of chemical bonds. *Sorption* will be used as a term encompassing the different variations of adsorption and absorption, and *desorption* is the reverse process of any of the above sorption mechanisms.

The measured hydrogen sorption property in a hydrogen adsorption experiment is the *excess adsorption*. This is the difference between the amount of hydrogen that would be present in the volume of the adsorbate layer or region, in the absence of adsorption, and the total amount of adsorbate in the layer, the *total* or *absolute adsorption*. In an absorption experiment the amount of hydrogen uptake measured is the total uptake. If this is expressed as a weight percentage, the mass of the hydrogen in the hydride is included in the overall mass of the material [30].

In this book, the term *hydrogen storage material* refers to a solid state hydrogen store that is capable of reversible hydrogen storage. *Reversible storage* refers to the capability of a material to be reversibly charged with gaseous hydrogen in the

[10] 'Nanopore' typically refers to a pore size < 100 nm, which covers the IUPAC micro-, meso- and macroporous ranges. In hydrogen storage it is almost exclusively microporous materials that are of interest, although it is worth noting that the IUPAC definitions are approximate and materials will quite often have a mix of pore sizes; nevertheless, microporous materials will have pore dimensions that are predominantly in the micropore region.

fuel tank, not just re-hydrogenated off-board. The term *chemical hydride* is reserved for the non-reversible storage materials, such as ammonia borane and sodium tetrahydroborate, that require an off-board re-hydrogenation or regeneration process. *Complex hydride* is used as an umbrella term for any of the non-interstitial hydrides currently being considered as reversible hydrogen storage materials, including alanates, nitrides, and hydroborates [31].[11]

The *volumetric* sorption measurement technique refers to any technique in which the measurement of the volume, pressure and temperature of the gas present in a system leads to the calculation of the adsorbed or absorbed quantity. Strictly speaking, the measurement of the sorbed quantity through the measurement of a change in pressure is a *manometric* measurement; however, the term *volumetric* has become widely used to describe instruments that in actual fact perform manometry and so we will retain it. The *Sieverts Method* or *Sieverts' Technique* is the terminology commonly found in the hydride field for the same manometric measurements. True volumetric measurements, in which a change in the volume is measured at a fixed pressure, can be and are used, but these are not as common as the manometric equivalent. *Gravimetric sorption measurement* refers to the technique of measuring the amount of gas adsorbed or absorbed by measuring the change in the weight of the sample as a function of pressure, at a constant temperature, but not to refer to standard *Thermogravimetric Analysis* (TGA) in which samples are thermally decomposed into a flowing stream of inert carrier gas.

Finally, with regard to measurement accuracy, the *accuracy* of a measurement is the closeness of the agreement between the measured quantity and its true value. The *repeatability* is defined as the closeness of the agreement between the results of successive measurements carried out under the same conditions, and the *reproducibility* is the closeness of the agreement between the results of the measurement of the same quantity under changed measurement conditions. As the true value of a quantity can never really be known, under this definition, accuracy is qualitative and can be 'high' or 'low' but cannot have a specific value. The quantitative equivalent is the *uncertainty* of the measurement [39].[12] The measurement conditions referred to here can include the procedure, observer, instrument, experimental conditions and location.

---

[11] Under these definitions there is a blurred boundary between some of the chemical and complex hydrides, but we will retain the former for compounds in which hydrogen release is highly irreversible, and the latter for compounds that at a very minimum offer the promise of reversible storage.

[12] Note that in the most recent international vocabulary of metrology [40] some of the terminology has changed slightly. For example, the *measurement accuracy* is now defined as being the "closeness of the agreement between a measured quantity value and a true quantity value of a measurand", rather than just the measured value and true value. However, the underlying principles of the terms accuracy, repeatability, reproducibility and uncertainty remain the same.

1.8 Summary

In this chapter we have introduced the idea of the hydrogen energy transition and have demonstrated the high level of international support for its practical implementation. A number of technological barriers have been identified by many of the organisations, authors and analysts who have examined the potential of this idea, and these must therefore be overcome before a hydrogen-based energy network can become a reality. Of these barriers, the efficient onboard storage of hydrogen in fuel cell vehicles is a major concern. We have seen that one of the most promising solutions to this problem is the reversible storage of hydrogen in solid state materials. The potential materials for this application span a range of material types from high surface area, microporous materials that can store hydrogen at low temperatures, to high temperature, air and moisture sensitive complex hydrides. Gas sorption measurement techniques can be used to characterise the hydrogen storage properties of each of these materials, and we have introduced both the properties and the techniques in this chapter. The accuracy of this kind of measurement is a major issue because of the experimental errors that have substantially affected some of the data published in the literature to date. We have therefore briefly summarised the background to this issue, and have introduced some of the complementary techniques that can be used. We have also covered some definitions of common terminology that will be used throughout the book.

References

1. United States Department of Energy (2005) FreedomCAR and fuel partnership: hydrogen storage technologies roadmap. United States Department of Energy, Washington
2. Sperling D, Cannon JS (eds) (2004) The hydrogen energy transition: moving toward the post petroleum age in transportation. Elsevier Academic Press, Oxford
3. Rose R (2003) Fuel cells and hydrogen: the path forward. Breakthrough Technologies Institute, Washington
4. Perry ML, Fuller TF (2002) A historical perspective of fuel cell technology in the 20th century. J Electrochem Soc 149(7):S59–S67
5. von Helmolt R, Eberle U (2007) Fuel cell vehicles: status 2007. J Power Sources 165:833–843
6. Dunn S (2002) Hydrogen futures: toward a sustainable energy system. Int J Hydrogen Energy 27:235–264
7. Solomon BD, Banerjee A (2006) A global survey of hydrogen energy research, development and policy. Energy Policy 34:781–792
8. Ohta T, Van Vorst WD, Winter C-J (2008) IAHE establishes hydrogen energy trust to assist implementation. Int J Hydrogen Energy 33(18):4713–4714
9. Breakthrough Technologies Institute (2003) Fuel cell vehicle world survey 2003. Breakthrough Technologies Institute, Washington
10. United States Council for Automotive Research LLC (2009) Hydrogen research for transportation: the USCAR perspective. United States Council for Automotive Research, Southfield

11. Elam CC, Gregoire Padró CE, Sandrock G, Luzzi A, Lindblad P, Fjermestad Hagen E (2003) Realizing the hydrogen future: the International Energy Agency's efforts to advance hydrogen energy technologies. Int J Hydrogen Energy 28:601–607
12. Shinnar R (2003) The hydrogen economy, fuel cells, and electric cars. Technol Soc 25:455–476
13. Kreith F, West R (2004) Fallacies of a hydrogen economy: a critical analysis of hydrogen production and utilization. J Energy Resour Technol 126:249–257
14. Robertson BI, Beard LK (2004) Lessons learned in the deployment of alternative fueled vehicles. In: Sperling D, Cannon JS (eds) The hydrogen energy transition: moving toward the post petroleum age in transportation. Elsevier Academic Press, Oxford
15. National Research Council and National Academy of Engineering (2004) The hydrogen economy: opportunities, costs, barriers and R&D needs. National Academy Press, Washington
16. Chalk S, Inouye L (2004) The President's U.S. hydrogen initiative. In: Sperling D, Cannon JS (eds) The hydrogen energy transition: moving toward the post petroleum age in transportation. Elsevier Academic Press, Oxford
17. McDowall W, Eames M (2006) Forecasts, scenarios, visions, backcasts and roadmaps to the hydrogen economy: a review of the hydrogen futures literature. Energy Policy 34:1236–1250
18. von Dokkum J, Dasinger A (2008) Meeting the challenges in the transport sector. J Power Sources 181:378–381
19. Borup R, Meyers J, Pivovar B, Kim YS, Mukundan R, Garland N, Myers D, Wilson M, Garzon F, Wood D, Zelenay P, More K, Stroh K, Zawodzinski T, Boncella J, McGrath JE, Inaba M, Miyatake K, Hori M, Ota K, Ogumi Z, Miyata S, Nishikata A, Siroma Z, Uchimoto Y, Yasuda K, Kimijima K, Iwashita N (2007) Scientific aspects of polymer electrolyte fuel cell durability and degradation. Chem Rev 107:3904–3951
20. Züttel A (2003) Materials for hydrogen storage. Mater Today 6(9):24–33
21. Eberle U, Felderhoff M, Schüth F (2009) Chemical and physical solutions for hydrogen storage. Angew Chem Int Ed 48:6608–6630
22. Sherif SA, Zeytinoglu N, Veziroğlu TN (1997) Liquid hydrogen: potential, problems, and a proposed research program. Int J Hydrogen Energy 22(7):683–688
23. de Wit MP, Faaij APC (2007) Impact of hydrogen onboard storage technologies on the performance of hydrogen fuelled vehicles: a techno-economic well-to-wheel assessment. Int J Hydrogen Energy 32:4859–4870
24. Gray EM (2007) Hydrogen storage—status and prospects. Adv Appl Ceram 106(1–2):25–28
25. Biniwale RB, Rayalu S, Devotta S, Ichikawa M (2008) Chemical hydrides: a solution to high capacity hydrogen storage and supply. Int J Hydrogen Energy 33:360–365
26. Wang P, Kang XD (2008) Hydrogen-rich boron-containing materials for hydrogen storage. Dalton Trans 5400–5413
27. Crabtree RH (2008) Hydrogen storage in liquid organic heterocycles. Energy Environ Sci 1:134–138
28. Walker G (ed) (2008) Solid-state hydrogen storage: materials and chemistry. Woodhead Publishing, Cambridge
29. Thomas KM (2007) Hydrogen adsorption and storage on porous materials. Catal Today 120:389–398
30. Sandrock G (1999) A panoramic overview of hydrogen storage alloys from a gas reaction point of view. J Alloy Compd 293–295:877–888
31. Orimo S, Nakamori Y, Eliseo JR, Züttel A, Jensen CM (2007) Complex hydrides for hydrogen storage. Chem Rev 107:4111–4132
32. Carpetis C, Peschka W (1980) A study on hydrogen storage by use of cryoadsorbents. Int J Hydrogen Energy 5:539–554
33. Agarwal RK, Noh JS, Schwarz JA, Davini P (1987) Effect of surface acidity of activated carbon on hydrogen storage. Carbon 25(2):219–226
34. Chahine R, Bose TK (1994) Low-pressure adsorption storage of hydrogen. Int J Hydrogen Energy 19(2):161–164

35. Dillon AC, Jones KM, Bekkedahl TA, Kiang CH, Bethune DS, Heben MJ (1997) Storage of hydrogen in single-walled carbon nanotubes. Nature 386:377–379
36. Chambers A, Park C, Baker RTK, Rodriguez NM (1998) Hydrogen storage in graphite nanofibers. J Phys Chem B 102(22):4253–4256
37. Becher M, Haluska M, Hirscher M, Quintel A, Skakalova V, Dettlaff-Weglikovska U, Chen X, Hulman M, Choi Y, Roth S, Meregalli V, Parrinello M, Ströbel R, Jörissen L, Kappes MM, Fink J, Züttel A, Stepanek I, Bernier P (2003) Hydrogen storage in carbon nanotubes. C R Phys 4:1055–1062
38. Sing KSW, Everett DH, Haul RAW, Moscou L, Pierotti RA, Rouquérol J, Siemieniewska T (1985) Reporting physisorption data for gas/solid systems with special reference to the determination of surface area and porosity. Pure Appl Chem 57(4):603–619
39. International Organisation for Standardisation (1993) International vocabulary of basic and general terms in metrology. ISO, Geneva
40. Joint Committee for Guides in Metrology (2008) International vocabulary of metrology—basic and general concepts and terms (VIM). JCGM 200:2008

Chapter 2
Potential Storage Materials

A wide range of materials are currently being considered as potential reversible hydrogen storage media. In this chapter we will expand on the material types introduced in Sect. 1.4 and discuss some of their storage properties. We begin with microporous adsorbents, which physisorb molecular hydrogen at low temperatures. Secondly, we look at the interstitial hydrides, which are reactive metals that reversibly absorb dissociated atomic hydrogen into their bulk as an interstitial. We then cover the complex hydrides that bind atomic hydrogen either covalently or ionically and release it via solid state decomposition, and finally consider some alternative storage materials that do not fit readily into the other categories. A summary of some basic hydrogen storage properties of the material types covered in this chapter is given in Table 2.1.

2.1 Microporous Materials

Microporous materials adsorb molecular hydrogen. The narrow width of their micropores leads to an increase in the density of hydrogen in the pore network compared to the gas phase. This occurs primarily at relatively low, sub-ambient, temperatures. According to the current IUPAC classification scheme [1], pores can be divided by size into three categories: *micropores*, of dimensions below 2 nm, *mesopores*, between 2 and 50 nm, and *macropores*, which are greater than 50 nm. The porous materials being considered for physisorbed molecular hydrogen storage are predominantly microporous. They are also often in the subcategory of *ultramicroporous*, in which the pore dimensions are less than 0.7 nm; close to the size of a single hydrogen molecule. At these length scales the adsorption potentials of the opposing pore walls overlap, which significantly increases the density of the adsorbed hydrogen, compared to the gas phase, at any given temperature and pressure.

D. P. Broom, *Hydrogen Storage Materials*, Green Energy and Technology,
DOI: 10.1007/978-0-85729-221-6_2,

The use of these materials as gas storage media has been considered for a number of years. For example, various reports can be found in the literature that describe the use of microporous carbons, zeolites and other adsorbents to enhance the storage capacity of natural gas (methane) storage tanks [2] and the use of microporous carbons and zeolites for hydrogen storage has been discussed for the past 30 years[1] [5–7]. Recently, however, the level of interest in the use of these materials for hydrogen storage has increased dramatically. The proposal of carbon nanotubes as a possible storage medium first attracted significant attention in the late 1990s and the exciting progress made in the synthesis of new porous hybrid materials [8] has further invigorated the field. Progress continues with the synthesis of microporous organic polymers, as well as new hybrid material classes, and new candidate materials are likely to emerge as the search, as well as the interest in these materials for a range of other applications, continues. In this section we cover the main groups of adsorbents currently being considered as storage media.

Before beginning, however, we will briefly discuss the hydrogen uptake behaviour of microporous materials because it is broadly similar for all the materials considered in this section. According to the conventional classification of adsorption by IUPAC [1, 9], isotherms can be of six general types (I–VI). Hydrogen adsorption by microporous materials invariably corresponds to Type I, which is concave to the pressure axis and saturates at a finite limit. Note that the identification of supercritical hydrogen adsorption as Type I refers to the absolute hydrogen uptake rather than the experimentally determined excess uptake (see Sect. 3.1.1.3). The excess adsorption typically peaks at an elevated pressure and then decreases as the hydrogen gas density increases yet further. This behaviour is typical for high pressure adsorption at supercritical temperatures [10]. The amount of adsorption that occurs at any given pressure decreases with increasing temperature, although the form of the absolute adsorption isotherm will generally remain the same. As an example, Fig. 2.1 shows some hydrogen adsorption data for Na-X zeolite, determined gravimetrically up to a pressure of 2.0 MPa in the temperature range 87–237 K. It can be seen that, as expected for pure physisorption, the adsorption and desorption isotherms show good reversibility, with no observable hysteresis.

The kinetics of hydrogen adsorption by microporous materials are very rapid, which is practically advantageous for hydrogen storage (see Sect. 3.3), but low temperatures are required to achieve significant capacities at useful storage pressures. In practice, low temperature storage requires cooling, which inevitably adds to the weight of a storage unit and is therefore a significant disadvantage. The hydrogen adsorption capacity of microporous materials at ambient temperature is currently too low for practical use [12]. However, the hope is to enhance the interaction of hydrogen with the materials to allow this possibility in the future.

[1] Hydrogen adsorption by porous materials has, however, been studied since the early twentieth century [3, 4].

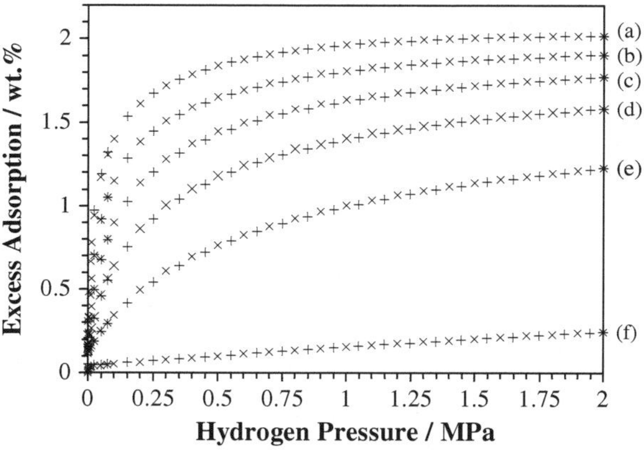

Fig. 2.1 Hydrogen adsorption (+) and desorption (×) isotherms for Na-X zeolite at a number of sub-ambient temperatures. The isotherms were measured gravimetrically at the following temperatures: (a) 87 K, (b) 97 K, (c) 107 K, (d) 117 K, (e) 137 K and (f) 237 K. Reproduced from Broom et al. [11] with permission from Hiden Isochema Ltd

Another drawback is their volumetric storage capacity, an issue that has occasionally been sidelined in favour of gravimetric considerations. In order to achieve high volumetric capacities, adsorbed hydrogen must be stored at relatively high densities within the pores and the microporous framework should also occupy a low proportion of the geometric volume of the material.[2]

2.1.1 Carbons

Several types of microporous carbons have attracted interest for hydrogen storage, including activated carbons, carbon nanotubes and nanofibres and, more recently, microporous templated carbons [13]. Carbon is very attractive technically as a host because of its low molar mass. It is also chemically stable and can be synthesised in a number of different forms. From a practical point of view, porous carbons are already commercially produced in large quantities for a broad range of applications and are relatively inexpensive.

2.1.1.1 Activated Carbon

Activated carbon is a porous form of carbon that can be synthesised via both chemical or physical activation methods. Depending on both the raw material and the activation method and conditions, the resultant carbon will have a specific, although not necessarily clearly defined, pore structure. Activated carbons can be predominantly macro, meso or microporous, but the latter are the main focus for hydrogen storage. Activated carbons tend to have slit-shaped pores [14, 15], but

---

[2] The definition of the geometric volume is the volume occupied by the sample including both closed and open pores (see the definition of geometric density in Sect. 6.2.1).

also exhibit a relatively wide pore size distribution. This is in contrast to the crystalline adsorbents, such as the zeolites or Metal-Organic Framework (MOF) materials covered in the following sections, which have a well defined pore size and pore geometry. There have been many studies of hydrogen adsorption by activated carbons and, according to Yürüm et al. [13], these show that gravimetric storage capacities can reach 5.5 wt% at 77 K. However, the pore structure of carbons can take many different forms, depending on the chosen activation method and the raw material used for the synthesis, and so a range of capacities have been reported in the literature. Carbons are also challenging to characterise accurately, as demonstrated by the variability in the results of a recent interlaboratory study [16], which will be discussed in more detail in the closing chapter of this book.

In addition to the experimental characterisation difficulties, modelling work on these materials is significantly limited by the lack of knowledge regarding their microstructure, which is fundamentally different to crystalline microporous adsorbents. The latter can be characterised crystallographically using powder diffraction (Sect. 5.3) but activated carbons are X-ray and neutron amorphous. The structure itself therefore has to be modelled, or an idealised slit-pore model used. However, estimated maximum theoretical capacities appear to agree reasonably well with those determined experimentally.

Jordá-Beneyto et al. [17] advocate the use of activated carbon monoliths for hydrogen storage because microporous activated carbon in this form has a greater bulk density than the same carbon in powder form. Equivalent monoliths formed from MOFs have not yet been reported in the literature, at the time of writing, and it is possible that activated carbon monoliths have significant advantages in terms of volumetric storage capacity. An experimental comparison would be worthwhile, however, because it may show that the higher gravimetric capacities reported for the best performing MOFs (see Sect. 2.1.3) are counterbalanced by lower practical volumetric capacities in real storage devices due to lower bulk material density.

2.1.1.2 Carbon Nanotubes and Other Carbon Nanostructures

The hydrogen sorption properties of carbon nanostructures, such as nanotubes and nanofibres, have been investigated extensively in recent years. As described briefly in Chap. 1, there has been considerable controversy over the storage properties of some of these materials since the publication of the first report of the potential for room temperature storage of hydrogen by carbon nanotubes [18]. Carbon nanotubes are cylindrical nanostructures formed from rolls of graphene. They can have diameters of 0.7 nm up to several nanometres and form single and multi-walled tubes that form close-packed bundles [19]. In 1997, Dillon et al. [18] reported potential room temperature storage capacities of 5–10 wt%, a figure derived from a rather optimistic extrapolation of thermal desorption data measured on a sample consisting of an estimated 0.1–0.2 wt% of nanotubes (the remainder of the sample was uncharacterised soot). The high temperature desorption peak claimed by Dillon et al. [18] to indicate the ambient temperature storage capabilities of the nanotube sample was later shown by Hirscher et al. [20] to be due to metal nanoparticles deposited during an ultrasonic purification process [19] (see Sect. 6.2). Although modelling work has predicted the possibility of relatively high storage capacities in nanotube structures the reported values are no greater than those for activated carbons and it therefore appears that they offer no significant advantage over other forms of carbon, which can be considerably easier to synthesise in large quantities.

Carbon nanofibres consist of graphene layers, stacked together in various orientations with respect to the axis of the fibre, including parallel and perpendicular, as well as intermediately angled (so-called *herringbone*) configurations. In 1998, Chambers et al. [21] reported incredibly high capacities for carbon nanofibres of up to 67 wt%. Questionable experimental procedures in this work that produced anomalous hydrogen storage capacities for well known and understood hydrides, together with the unphysical nature of the claimed capacity, which corresponds to a H/C atomic ratio of around 24, indicated that the measured capacities were significantly overestimated. Subsequent studies of both nanotubes and nanofibres produced a wide range of values, none of which substantiated these initial claims, and an interesting controversy developed. Tabulated values can be found in a number of articles, showing the spread of the data found in the literature at the time [13, 22, 23]. Although the issue has not yet been resolved completely, it seems fairly safe to conclude that both nanotubes and nanofibres are unlikely to be the hydrogen storage panacea that some authors have suggested. In addition to nanotubes and nanofibres, fullerenes and carbon nanohorns have also been investigated, either experimentally or theoretically, for their hydrogen storage properties [24, 25].

2.1.1.3 Templated Carbons

Templated carbons are a form of microporous or mesoporous carbon typically synthesised by introducing a carbon precursor, such as sucrose or acetonitrile, into the pores of an inorganic template. Carbonisation and the subsequent removal of the template results in a pore structure that is relatively well defined compared to their activated carbon counterparts [26, 27]. There have been a number of studies on these materials and they can exhibit impressive hydrogen adsorption properties, with the largest capacity reported to date being 6.9 wt% for a zeolite-templated carbon at 77 K and 20 bar [12, 26]. The uptake was Type I, with an estimated saturation capacity of 8.33 wt%. Unlike activated carbons, these materials show evidence of microstructural ordering through the appearance of Bragg peaks in their X-ray powder diffraction patterns. Another type of templating uses carbide precursors and can produce microporous carbons with very well defined pores sizes [28, 29]. According to Gogotsi et al. [29] the hydrogen uptake of these so-called *Carbide-Derived Carbons* (CDCs) can reach 4.7 wt% at 60 bar and 77 K. Although this is not particularly high, the CDCs offer the opportunity to further tune and optimise the pore size and continued investigation of the interaction of hydrogen with these materials would undoubtedly be valuable.

2.1.2 Zeolites

Zeolites are microporous aluminosilicates formed from AlO_4 and SiO_4 tetrahedra. They have a range of practical applications that exploit their ion exchange, molecular sieving and catalytic properties. The term *zeolite* is often used to describe compounds with similar structures that are formed by elements other than Al and Si, including P, Ga, Ge, B and Be, but these materials are also known as *zeotypes*. These materials form a variety of different crystallographic structures.[3] Their ordered crystalline nature gives them uniform cavities and channels with dimensions in the microporous regime. The framework structures are relatively rigid and, as with all microporous solids, they have high specific surface areas and large pore volumes. Their properties can be tuned by altering the Si/Al ratio of the framework, or the equivalent stoichiometry in the case of zeotypes, with limits in the range $0.5 < \mathrm{Si/Al} < \infty$. The anionic nature of zeolite frameworks leads to the presence of cations within their structure. Exchange of these cations can also significantly alter their properties.

Although zeolites are highly porous, with high surface areas, the hydrogen storage capacities reported in the literature have, to date, been fairly low. In their study of low pressure (< 0.1 MPa) hydrogen uptake by a range of porous solids, Nijkamp et al. [31] concluded that zeolites are less likely to be effective as hydrogen storage materials, in comparison to porous carbons, due to their limited pore volume. Vitillo et al. [32] tabulated a number of the experimentally determined hydrogen uptakes, with the highest reported uptake at 77 K or above being 1.81 wt%, as reported by Langmi et al. [33]. This figure was found for Na-Y zeolite at a pressure of 1.5 MPa. According to Anderson [34], the highest reported uptake until recently was 2.55 wt% for Na-X at 77 K and 40 bar (4.0 MPa) [35].

The results of the molecular mechanics simulations performed by Vitillo et al. [32] predict that the maximum storage capacities for a range of zeolites are in the region of 2.65 and 2.86 wt%, whereas van den Berg et al. [36] found that, theoretically, the hydrogen content could reach (4.8 ± 0.5) wt% for sodalite (SOD) structures.[4] In the study by Vitillo et al. [32] a lower figure of 1.92 wt% was determined for SOD. In later work, van den Berg et al. [37] found lower values for the saturation uptake using Grand Canonical Monte Carlo (GCMC) simulations, although the higher figures are supported by Song and No [38], who found an

---

[3] 176 different structure types are listed by Baelocher et al. [30].

[4] Van den Berg et al. [36] and Vitillo et al. [32] both calculated the maximum hydrogen uptake using molecular mechanics simulations but used different convergence criteria to define when the hydrogen capacity had reached saturation. In addition, the latter study included a correction for the zero point motion of hydrogen, whereas the former did not.

uptake of 4.45 wt% for Mg-X. However, even these higher capacities are low in comparison to the challenging US DOE storage system targets.

The temperature dependence of hydrogen adsorption by Na-X zeolite is illustrated in Fig. 2.1. These isotherms were measured gravimetrically up to a pressure of 2 MPa in the temperature range 87–237 K. At room temperature and moderate pressure, hydrogen uptake by zeolites is low. Hydrogen storage at higher temperatures has been proposed through the mechanism of *encapsulation* [39–41], a process by which, in the case of zeolites A, X and Y, hydrogen molecules enter the sodalite cages. However, this does not appear to be a mechanism capable of delivering practical storage capacities.[5]

Enthalpies of adsorption, a measure of the strength of the hydrogen-adsorbent interaction (see Sect. 3.2.1), for zeolites can have a range of values. Garrone et al. [42] report $-\Delta H^0$ (standard adsorption enthalpy) values in the range 3.5–18 kJ mol^{-1}, with the largest of these values being found for (Mg, Na)-Y zeolite [43]. These values were determined using infrared spectroscopy (see Sect. 5.4.3).

Zeolites have a number of significant practical advantages over other microporous adsorbents. They possess high thermal stability in comparison to metal-organic frameworks and organic polymers, for example. Thorough degassing of the framework can therefore be performed at 350°C (623 K) with no decomposition. Their crystalline nature allows easy characterisation of the host material and they have a well defined and understood pore size in comparison with activated carbon. Current industrial production of synthetic zeolites also demonstrates the feasibility of their synthesis in large quantities, although this is also the case for carbons. However, it seems unlikely that zeolites will be useful as practical hydrogen storage media for onboard applications. Felderhoff et al. [44] support this view and argue that they do not show the required capacities for practical storage applications. Zeolites are, however, certainly valuable as model systems for the further investigation of the interaction of hydrogen with microporous materials [45], and so their continued study is both likely and worthwhile. Anderson [34] covered hydrogen storage in zeolites in detail and this review is recommended to interested readers.

2.1.3 Metal-Organic Frameworks

Metal-Organic Frameworks (MOFs) are crystalline inorganic-organic hybrid solids consisting of metal ions or clusters linked by organic bridges [8, 46, 47]. The archetypal MOF is $Zn_4O(bdc)$, where bdc = 1,4-benzenedicarboxylate, which

[5] Hydrogen encapsulation has been investigated by a number of authors [39–41] by loading zeolites at elevated temperatures under a hydrogen atmosphere, then cooling the sample to ambient and performing TPD up to temperatures of 673 K to desorb the encapsulated hydrogen. However, storage capacities were found to be low; for example, 0.6 wt% for Na-X at hydrogenation pressures of 13.3 kpsi (91.7 MPa) [39].

is commonly known as either MOF-5 or IRMOF-1.[6] This material consists of zinc oxide clusters joined by benzene linkers. The metallic clusters are known generically as Secondary Building Units (SBUs) and in MOF-5 the structure forms a highly porous cubic network. A number of reviews specifically addressing the storage of hydrogen using MOFs can be found in the literature and these are recommended to interested readers [12, 48–52]. Hundreds of different MOFs have been studied for their hydrogen adsorption properties. Murray et al. [52], for example, list hydrogen uptake data[7] for 177 different MOFs and the current level of interest in these materials is such that since this recent review the amount of data in the literature is likely to have grown considerably.

Hydrogen storage by metal-organic frameworks was first reported by Rosi et al. [53] in a study including MOF-5, IRMOF-6 and IRMOF-8. Their initial claimed capacity for MOF-5 of 4.5 wt% at 77 K and 0.07 MPa met with some scepticism [54], and was subsequently reduced in a later report [55]. Nevertheless, following their discovery there have been many more studies on hydrogen storage by other framework materials and the erroneously high capacity claimed for low pressure adsorption has since been exceeded at higher pressures. Tabulated data compiled by Collins and Zhou [49] and Thomas [12] indicate maximum hydrogen uptakes at 77 K of between approximately 1.0 and 7.5 wt% for a range of MOFs. The highest value of 7.5 wt% was found for MOF-177 [Zn_4O(btb), where btb = 1,3,5-benzenetribenzoate], as reported by Yaghi et al. [55, 56]. This uptake was found at a measurement pressure of 7.0 MPa; the uptake at 0.1 MPa was 1.25 wt%. The highest reported value at room temperature (298 K) is 1.4 wt% reported for a pressure of 9.0 MPa, for Mn(btt), where btt = 1,3,5-benzenetristetrazolate, by Dincă et al. [57]. This MOF also adsorbs 6.9 wt% at 77 K and a hydrogen pressure of 9.0 MPa.

Aside from the large number of possible combinations of different SBUs and organic linkers, and hence possible pore geometries, the presence within the frameworks' pores of *exposed metal sites* is another reason for the interest in MOFs. To either increase the operating temperature of a microporous hydrogen store, or increase the maximum storage capacity at near ambient temperature, an increase in the strength of the hydrogen-surface interaction is required. Molecular hydrogen is known to form so-called *Kubas complexes* with nearly all transition metals [58] and the possibility of enhancing the hydrogen-surface interaction through a similar mechanism offers the promise of developing MOFs with much improved storage properties. This aspect of hydrogen adsorption in MOFs was the

[6] Metal-organic frameworks tend to be known by the initials assigned to them by the researchers responsible for the original synthesis. These initials do not follow any particular pattern but tend to refer to either the material type or the researchers' institution. Examples include MOF (Metal-Organic Framework), MIL (Materials of Institute Lavoisier), IRMOF (IsoReticular Metal-Organic Framework) and UMCM (University of Michigan Crystalline Material).

[7] Note that the majority of the data reports the hydrogen uptake at atmospheric pressure, which gives limited information with regard to the reversible capacity for storage purposes (Sect. 3.1.1), and therefore further studies at elevated pressures are required.

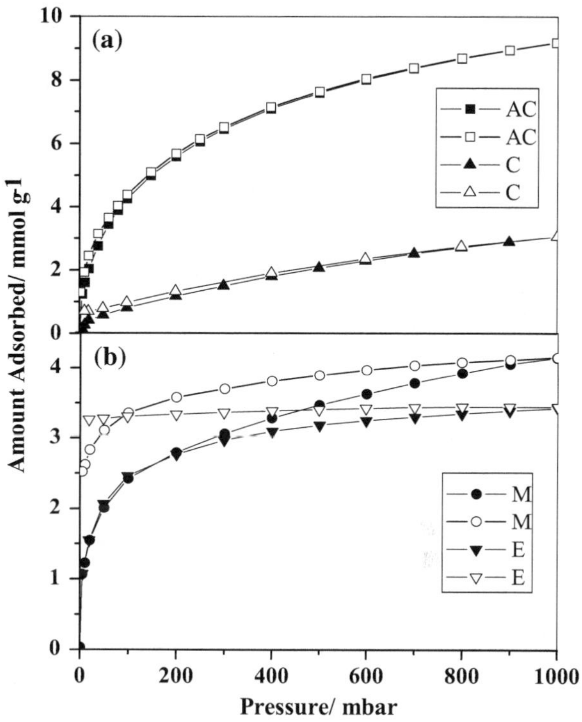

Fig. 2.2 Hydrogen adsorption and desorption isotherms measured at 77 K for an activated carbon (AC), and three metal-organic frameworks (C, M and E). Two of the MOFs (M and E) exhibit hysteretic hydrogen adsorption behaviour, whereas the activated carbon and MOF C exhibit conventional reversible Type I adsorption behaviour. From Zhao et al. [60]. Reprinted with permission from AAAS

focus of the review by Dincă and Long [51], and was also addressed recently by Hoang and Antonelli [59].

Another interesting feature is *framework flexibility*. In 2004, Zhao et al. [60] reported two MOFs that showed hysteretic hydrogen adsorption, due to structural flexibility, so that for a given temperature the materials desorbed hydrogen at a significantly higher pressure than that of adsorption. The hysteresis is shown in Fig. 2.2, which includes isotherm data illustrating the hysteresis in comparison with an activated carbon and a MOF exhibiting conventional reversible Type I behaviour. The hysteresis is unusual for molecular hydrogen physisorption, which is normally fully reversible at any given temperature. Thomas [12] gives a number of examples of MOF materials that show flexibility during the adsorption and desorption of guest molecules, either with or without the breaking of chemical bonds in their structure. The flexibility can lead to structural transformations that involve stretching, rotational, breathing and scissoring mechanisms. The review on framework flexibility by Fletcher et al. [61] is recommended to interested readers.

Many studies have focused on the enthalpy of hydrogen adsorption for MOFs, since this is a measure of the strength of interaction between hydrogen and the surface or pore structure (see Sect. 3.2.1). Reported values of the isosteric enthalpy of adsorption are in the range from 3.8 kJ mol^{-1} for MOF-5/IRMOF-1 [62] to 12.3 kJ mol^{-1} for a mixed zinc/copper metal-organic framework, M′MOF 1 [63].[8]

MOFs hold greater promise as potential storage media than zeolites, as the reported gravimetric capacities are significantly higher and the unique features, such as structural flexibility and exposed metal sites, give them more potential for future development. However, these materials tend to be less robust than zeolites and microporous carbons, because they exhibit lower thermal stability. Nevertheless, their commercialisation is already underway, principally by BASF who market a series of framework materials under the tradename Basolite™, and so the practical application and use of these materials on an industrial scale is clearly feasible.

2.1.4 Organic Polymers

There are three main classes of microporous organic polymer that have emerged recently as potential candidates for adsorptive hydrogen storage: *Polymers of Intrinsic Microporosity* (PIMs), *Hypercrosslinked Polymers* (HCPs) and *Covalent Organic Frameworks* (COFs) [64]. The latter are crystalline organic analogues of MOFs, while both PIMs and HCPs are non-crystalline (X-ray and neutron amorphous) materials that have a disordered structure that more closely resembles activated carbon than crystalline materials such as zeolites, MOFs and COFs.

PIMs are rigid and contorted macromolecules that consist of fused-ring components. They form a microporous network as they are unable to pack space efficiently and, as a consequence, have high BET surface areas in the range 500–1100 m^2 g^{-1} [65, 66]. HCPs derive a similarly high degree of microporosity, and hence high BET surface areas, from a high density of crosslinks, the covalent chemical bonds that occur between macromolecules in polymeric materials [67, 68]. COFs, meanwhile, are crystalline networks formed exclusively from the light elements H, B, C, O and Si, that are linked by strong covalent bonds (B–O, C–O, C–C, B–C, and Si–C) [69, 70].

Hydrogen capacities of up to 2.7 wt%, at 1.0 MPa and 77 K, have been reported for a triptycene-based polymer (trip-PIM) [66]. The reported capacities for HCPs are higher, reaching 3.68 wt% at 1.5 MPa and 77 K for a polymer based on BCMBP [4,4′-bis(chloromethyl)-1,1′-biphenyl] [68]. Although the uptakes of these amorphous organic polymers are not outstanding, they are attractive as potential storage media due to the light elements from which they are formed, and further development and the synthesis of new materials may yet produce a high storage capacity polymer. One disadvantage in comparison to other microporous media is their relatively low thermal stability, which means that care must be taken not to induce thermal decomposition during the degassing process (see Sect. 6.4.1).

---

[8] M′MOF 1 is $Zn_3(bdc)_3[Cu(pyen)]$, where $pyenH_2$ = 5-methyl-4-oxo-1,4-dihydro-pyridine-3-carbaldehyde.

Compared to the relatively conservative capacities reported so far for PIMs and HCPs, the predicted and measured gravimetric capacities of COFs are impressive. Furukawa and Yaghi [70] reported experimental uptakes of 72.4 mg g^{-1} and 70.5 mg g^{-1} for COF-102 and COF-103, respectively. These measurements were made gravimetrically at 77 K and at pressures of up to around 9 MPa. COF-102 and COF-103 are prepared by self-condensation reactions of *tetra*(4-dihydroxyborylphenyl)methane (TBPM) and its silane analogue (TBPS) [71]. Simulated total uptakes reported by Han et al. [69] for COF-105 and COF-108 both exceed 18 wt% at 10.0 MPa and 77 K, with values of 18.3 and 18.9 wt%, respectively. However, these two materials have large free volumes and the volumetric capacity is lower than that of COF-102, which shows a lower simulated and experimental gravimetric capacity.

With regard to the enthalpy of adsorption for hydrogen on organic polymers, Spoto et al. [72] determined a value of approximately 4 kJ mol^{-1} H_2 for a HCP (cross-linked poly(styrene-*co*-divinylbenzene) polymer) using infrared spectroscopy (see Sect. 5.4.3). Wood et al. [68], however, determined higher values in the range 6–7.5 kJ mol^{-1} H_2 for the isosteric enthalpies of adsorption from hydrogen isotherms measured at 77 and 87 K. These higher figures were in general agreement with those obtained from accompanying simulations. Measured isosteric enthalpies of adsorption for COFs have been found to be in the range 4–7 kJ mol^{-1} H_2 [70]. The materials that exhibit the largest gravimetric uptake possess the lowest enthalpy of adsorption, perhaps reflecting the fact that larger pore volumes and pore dimensions lead to higher overall hydrogen uptakes, but a concomitant decrease in adsorption potential due to the increased pore dimension.

In addition to PIMs, HCPs and COFs, the hydrogen adsorption behaviour of a number of other organic polymers have been reported in the literature, including dipeptide-based materials [73] and 3,3′,4,4′-tetra(trimethylsilylethynyl)biphenyl, a type of organic zeolite [74], although the uptakes of these materials are relatively modest. Early results indicating high hydrogen uptake by HCL-treated conducting polymers, polyaniline and polypyrrole, were not successfully reproduced by other investigators [75]. Another new group of materials reported by Rose et al. [76] are *Element Organic Frameworks* (EOFs). These authors reported hydrogen uptake in two of these materials, poly(1,4-phenylene)silane (EOF-1) and poly(4,4′-biphenylene)silane (EOF-2), of 0.94 and 1.21 wt%, respectively, at 77 K and 0.1 MPa. Although, to date, none of these compounds have been proven as a practical storage material, it is likely that significant further progress will be made as synthetic chemists continue to discover interesting new microporous media. Further study will also undoubtedly improve our understanding of the interaction of hydrogen with organic microporous materials.

2.2 Interstitial Hydrides

These are formed from metallic elements or compounds that react with gaseous hydrogen to produce binary, or higher, hydrides. Molecular hydrogen dissociates into atomic hydrogen on the surface of the host material and enters the bulk via diffusion between interstitial sites in the host lattice. Most elemental metals will absorb hydrogen in this manner under certain conditions of temperature and hydrogen pressure; however, binary hydrides (MH_x, where M is a metallic element and x is the hydride stoichiometry) are generally either too unstable or too stable for use as practical storage materials. The former means that the pressures at which the host reversibly absorbs and releases hydrogen are too high at practical storage temperatures, and the latter means the pressures are too low. In terms of formation or decomposition enthalpies (see Sect. 3.2.2), ΔH is too high in the former case (either small negative numbers or positive) and too low in the latter (a relatively large negative value). However, when two or more metallic elements are combined, particularly one that forms a stable hydride and one that does not, the resultant alloy or intermetallic tends to form a hydride of intermediate stability.

The study of metallic hydrides began nearly 150 years ago with the discovery of the hydrogen-absorbing properties of palladium by Thomas Graham [77]. Work beginning in the 1960s resulted in the later commercialisation of Nickel-Metal Hydride (Ni-MH) batteries in which the negative electrode material forms an intermetallic hydride. Although this has been the most commercially successful application of interstitial hydrides, metal hydride technology is also exploited in a number of other application areas, including gas separation and purification, temperature sensing, thermal compression, and refrigeration or cryocooling [78–80]. For mobile storage applications, the gravimetric capacity of many interstitial hydrides is relatively low, but some of these compounds show remarkable and practical hydrogen absorption and desorption characteristics, and they are likely to play an important role in a future hydrogen economy. A thorough account of the physics of interstitial hydrides, along with many of their basic properties, can be found in the monograph by Fukai [81] and this is recommended to interested readers. Further extensive coverage can be found in the volumes edited by Alefeld and Völkl [82, 83], Schlapbach [84, 85] and Wipf [86].

In this section, we look at a number of the different types of interstitial hydride, and cover intermetallic compounds, solid solution alloys, modified binary hydrides, and amorphous and nanostructured hydrides, including mechanically milled materials, amorphous alloys produced using other methods, and quasicrystals.

2.2.1 Intermetallic Compounds

The host materials in this group are ordered stoichiometric compounds typically formed from two metallic components, A and B. The A and B components tend to form hydrides AH_x and BH_y, with enthalpies of formation ΔH_A and ΔH_B, representing a stable and an unstable hydride, respectively. As mentioned above, the resultant intermetallic hydride $A_mB_nH_z$, where m and n are integers and z is a real number, will then tend to have an enthalpy of formation, ΔH_{AB}, where $\Delta H_A < \Delta H_{AB} < \Delta H_B$. Varying the ratio n/m then shifts the value of ΔH_{AB} in either direction. The components A and B can generally be fully or partially substituted by other elements of relatively similar size or chemistry. Hydrogen-absorbing intermetallics form a number of different groups, which can be distinguished by their stoichiometries, including AB_5, A_2B_7, AB_3, AB_2, AB and A_2B compounds [54, 87]. Early reports of the reversible hydriding properties of intermetallic compounds include those by Libowitz et al. [88], published in 1958, on $ZrNiH_3$ and the early work of Reilly and Wiswall [89, 90] in the late 1960s on the hydrogenation of Mg_2Ni and Mg_2Cu. These reports pre-date the discovery of the reversible hydrogenation of $LaNi_5$ around 1970 [91] and the subsequent increase in the amount of metal hydride research that ultimately led to the commercialisation of Ni-MH battery technology, and the research into metal hydride-based hydrogen storage that continues today.

A vast number of A and B element combinations are now known to form reversible hydrides, although relatively few do so at useful temperatures and pressures. Calculated ΔH values for 1265 different A_5BH_x, A_2BH_x, ABH_x, AB_2H_x and AB_5H_x intermetallic hydrides, along with measured values for 63 binary hydrides and 135 ternary hydrides, were tabulated by Griessen and Riesterer [92]. Similarly, Buschow et al. [93] provide tabulated data and references on a large number of compounds, including studies using many of the complementary techniques we introduce in Chap. 5. In this section we will cover some of the combinations and stoichiometries that are of practical interest for hydrogen storage. Note that different stoichiometries can often be found within the binary phase diagram of any given pair of constituents but, because of the dramatic effect that the ratio of elements has on the hydriding properties, it is likely that only one of these has a formation enthalpy that gives the material hydrogen absorption and desorption capabilities in a practical temperature and pressure regime. As an example, let us consider La and Ni, which combine to produce the well known hydride-forming compound $LaNi_5$. La forms a very stable hydride, LaH_2, with $\Delta H = -104$ kJ mol^{-1} H, while Ni forms a very unstable hydride ($\Delta H = -3$ kJ mol^{-1} H) [92]. The calculated values of Griessen and Riesterer [92] for the La–Ni–H system are −13.3, −23.0, −40.3, −60.6 and −82.5 kJ mol^{-1} H for $LaNi_5$, $LaNi_2$, LaNi, La_2Ni and La_5Ni, respectively. Although these values do not match the experimental values particularly well,[9] the dependence on the ratio of the two constituent metal elements can be seen clearly and matches the trend observed experimentally: the lower the Ni content of the host intermetallic, the more stable the associated hydride compound.

Some intermetallics, principally Mg-based compounds such as Mg_2Ni and Mg_2Cu, form stoichiometric hydride complexes upon hydrogenation and can therefore be categorised as complex transition metal hydrides rather than

[9] The calculated values were obtained using the semi-empirical band structure model of Griessen and Driessen [94]. The discrepancy between these values and experiment most likely originates from the implicit assumption in this model that each hydrogen atom sits in the same environment, surrounded by an average number of A and B atoms, rather than on crystallographically distinct interstitial sites.

interstitial intermetallic hydrides. Furthermore, there are interesting cases, such as the intermetallic compound $LaMg_2Ni$ that forms a mixed hydride ($LaMg_2NiH_7$) consisting of NiH_4 tetrahedra and 'interstitial' H^- ions, which blur the boundary between the interstitial intermetallic hydrides and their complex counterparts, which are covered in Sect. 2.3 [95]. To add to the already slightly fuzzy picture, many of the intermetallics can be under or over-stoichiometric. The distinction between the intermetallics and solid solution alloys (Sect. 2.2.2) still exists, however, because the A and B components of an intermetallic tend to occupy different lattice sites rather than randomly occupying all sites throughout the host lattice. In each of the following sections we will briefly summarise the main groups of interstitial hydride-forming intermetallics that are of interest for hydrogen storage, and we therefore cover AB_5, AB_2 and AB compounds. We will cover some of the A_2B compounds in Sect. 2.3.4.

2.2.1.1 AB_5 Compounds

The archetypal AB_5 intermetallic is $LaNi_5$. This compound readily forms a hydride under fairly moderate hydrogen pressures and ambient temperatures. It has an enthalpy of formation of -15.7 kJ mol^{-1} H and an enthalpy of decomposition of -15.1 kJ mol^{-1} H [96]. Hydrogen capacity exceeds $LaNi_5H_6$, giving a reversible gravimetric capacity in the region of 1.25 wt% [87]. $LaNi_5$ and some other binary AB_5 compounds, such as $CaNi_5$, can be subject to significant disproportionation and therefore lose their reversible capacity during hydrogen cycling (see Sect. 3.1.2). However, these compounds can be modified via partial substitution to reduce the disproportionation. The most effective substituent for this purpose for $LaNi_5$ is Sn, with a composition of $LaNi_{5-x}Sn_x$, where $x \approx 0.2$ [97]. For economic reasons, mischmetal (Mm), a naturally occurring mixture of rare earths, can be substituted for the La as the A component. As well as the $LaNi_5$, $LaNi_{5-x}Sn_x$ $MmNi_5$ and $CaNi_5$ intermetallics, hydride-forming AB_5 compounds can be composed of many other elemental combinations, with the A elements tending to be either one or more of the lanthanides or Ca, as in the compounds mentioned above, or Y and Zr, and the B elements any of a range of elements including Co, Al, Mn, Fe, Cu, Sn, Si and Ti, as either full or partial constituents [87, 98]. The hydrogen capacities, plateau pressures and the enthalpies of hydride formation of 36 binary, ternary and quaternary AB_5 compounds were tabulated by Ivey and Northwood [99] and 477 records for AB_5 compounds are currently listed in the Sandia National Laboratories (US) Metal Hydride Properties database[10] [100].

It can be seen clearly that the gravimetric storage capacity of these materials is substantially lower than the current US DOE target for mobile hydrogen storage applications. However, the AB_5-based intermetallics show some remarkable cycling properties including excellent resistance to gaseous impurity contamination, good long term cycling stability and a high volumetric storage density, and

[10] http://hydpark.ca.sandia.gov/DBFrame.html, accessed 2nd January 2010.

are therefore a prime example of a practically effective reversible hydrogen storage material.

2.2.1.2 AB_2 Compounds

AB_2 compounds can be formed from the combination of many different elements. According to Sandrock [87], the A elements are typically from group 4 (Ti, Zr, Hf) or the lanthanoids (La, Ce, Pr, and so forth), whereas the B element can be a transition or non-transition metal, with a preference for V, Cr, Mn and Fe. Feng et al. [101] summarise the elements of interest for electrochemical applications of AB_2 compounds as A = Mg, Zr and Ti, and B = V, Cr, Mn and Ni, although many more can be used as partial substituents. The Sandia database[10] currently includes 625 records for AB_2 compounds.

These intermetallics crystallise in either the hexagonal C14 or cubic C15 Laves phase structure. Like the AB_5 compounds, they can show a range of hydriding properties depending on the elemental composition. This allows the hydriding properties of a given AB_2 material to be tuned via partial elemental substitution. A number of examples of multicomponent AB_2 compounds can be found in Young et al. [102], a recent report on $Ti_xZr_{1-x}(VNiCrMnCoAl)_2$, a C14 Laves phase AB_2 proposed for use as a battery electrode material. In addition to partially substituted compounds, sub- or superstoichiometric compositions can also be formed. This modification also affects the hydriding properties of a material [79]. Bowman and Fultz [79] gave a good example of the effects of non-stoichiometry using the Zr–Mn–H system. Substoichiometric $ZrMn_{2-x}$, where x is positive, has a lower plateau pressure and a significantly lower reversible hydrogen storage capacity than $ZrMn_2$, whereas superstoichiometric $ZrMn_{2+x}$ has a higher plateau pressure and only a slightly smaller reversible capacity than the stoichiometric compound [79, 103].

AB_2 intermetallics have been used practically as reversible hydrogen storage materials. In the 1980s, a fleet of Daimler vans and automobiles were operated using AB_2-based hydrogen stores containing the non-stoichiometric compound $Ti_{0.98}Zr_{0.02}Cr_{0.05}V_{0.43}Fe_{0.09}Mn_{1.5}$ [104]. This intermetallic exhibits very fast kinetics and good long term cycling stability (see Sect. 3.1.2) [79]. On the other hand, high material costs and a gravimetric storage capacity of 1.8 wt% do not, in view of the current US DOE targets, make it particularly practical for widespread use. However, despite this it is another example, along with the AB_5-based compounds, of a conventional interstitial metallic hydride that works well in real applications by satisfying many of the required performance criteria.

2.2.1.3 AB Compounds

In comparison to the number of different AB_2 and AB_5 compositions reported in the literature, the number of AB compounds of interest for hydrogen storage is fairly limited. The earliest study of an intermetallic hydride, by Libowitz et al. [88], featured an ABH_x intermetallic hydride, $ZrNiH_3$, and many other metallic AB compounds form hydrides; 179 records are currently listed in the Sandia database.[10] However, it is only TiFe that is of practical interest for storage applications. The hydrogen-absorbing properties of this compound were originally discovered by Reilly and Wiswall around 1970 [105].[11] It achieves a total gravimetric capacity of 1.86 wt% at $H/M = 0.975$, with a reversible capacity of 1.5 wt% [87]. There have been many studies on the hydrogen absorption properties of TiFe and it has reversible hydrogen storage characteristics in a useable range. It has an ordered Body-Centered Cubic (BCC) structure, and shows two distinct plateaus in its hydrogen absorption isotherm. Partial substitution can again be used to modify the hydrogen absorption behaviour. Examples include the partial substitution of Fe with Mn and Ni, which lower the pressure of the first ambient temperature isotherm plateau, and therefore stabilise the hydride with respect to pure TiFe. TiFe and $TiFe_{0.85}Mn_{0.15}$ show good long term cycling stability, they are low cost, and tend not to be pyrophoric [87], unlike the AB_5 compounds. Unfortunately their activation is relatively difficult and they are significantly more sensitive to gaseous impurities than AB_5 intermetallics. As with most of the intermetallic hydrogen storage compounds, their gravimetric capacities are low in comparison to the current US DOE targets, but they are nonetheless favourable to many other interstitial hydrides.

2.2.2 Solid Solution Alloys

Solid solution alloys formed by dissolving one or more hydrogen-absorbing metallic elements in another also show interesting hydrogen storage properties. Unlike the intermetallics described above, these materials do not necessarily have stoichiometric or near-stoichiometric compositions. They can be formed from a number of host solvents, including Pd, Ti, Zr and V; thermodynamic data for many solid solution alloy hydride compositions are given by Fukai [81]. From a hydrogen storage point of view, Pd-based alloys suffer from low gravimetric capacities and the high cost of palladium, and Ti and Zr-based solid solutions tend to be too stable. However, vanadium-based alloys have been found to possess favourable absorption properties, and distinct advantages in terms of gravimetric capacities over some intermetallics. Although pure vanadium is prohibitively expensive, the use of low cost ferrovanadium has shown promise and so Fe-containing V-based solid solution alloys make feasible hydrogen storage materials [87].

---

[11] Results had been presented at symposia 4 to 5 years prior to the report published in 1974 [105].

Tabulated data presented by Sakintuna et al. [106] indicate that typical gravimetric hydrogen capacities can approach 4 wt% for Ti–V–Fe ($Ti_{43.5}V_{49.0}Fe_{7.5}$) and Ti–V–Cr–Mn alloys, although the latter requires elevated temperatures in the range 520–745 K. The pressure-composition isotherms for these materials show two plateaus, of which only the higher pressure plateau region can be exploited for hydrogen storage applications. Therefore, in a practical pressure range, the reversible capacity is significantly lower than the maximum capacity quoted above at around 2.5 wt%. In their study of Ti–V–Fe alloys, Nomura and Akiba [107] defined the "Available Hydrogen Quantity" (AHQ) as being the difference between the hydrogen content at 100 kPa on the absorption and desorption isotherms at 253 and 573 K, respectively. They found the AHQ to be 2.4 wt% for $Ti_{43.5}V_{49.0}Fe_{7.5}$. Cho et al. [108], meanwhile, reported a reversible capacity of 2.3 wt% for $Ti_{0.32}Cr_{0.43}V_{0.25}$; they also found that the alloy exhibits good cyclic stability, maintaining a reversible capacity of approximately 2 wt% over the course of 1000 hydriding and dehydriding cycles.

Many different compositional variations of the solid solution alloys can be found in the literature [108–112]. The so-called Laves phase-related BCC solid solution alloys are related materials, which are Ti–V–Mn, Ti–V–Cr and Ti–V–Cr–Mn compounds that contain both Laves and BCC phases. The hydrogen storage properties of these materials were reviewed by Akiba and Okada [113].

2.2.3 *Modified Binary Hydrides*

MgH_2 is the binary hydride that has attracted by far the most attention as a potential storage material; however, there are two other binary hydrides that we should also mention, namely AlH_3 and PdH_x. The former is of interest due to its high gravimetric storage capacity of 10.1 wt% [114–116]. However, it is effectively non-reversible within a realistic hydrogen pressure range for a practical storage unit and therefore requires off-board regeneration. This process is economically and energetically costly, and non-reversible hydrides are beyond the scope of our discussion.[12] Palladium, meanwhile, is impractical as a storage material because of its relatively high operating temperature and low gravimetric capacity, as well as its high cost. However, research into nanoscale palladium [117–121] is being actively pursued due to the interesting differences seen in its hydrogen absorption behaviour compared to the bulk material. The hydrogen solubility of the hydride phase and equilibrium hydrogen pressures are reduced for small particles. The plateau region also slopes and narrows, and the critical temperature, T_{crit}, is significantly reduced in Pd clusters compared to the bulk

[12] AlH_3 is an example of a kinetically stabilised hydrogen storage material [114, 116], which are hydrides that have high equilibrium hydrogen pressures at ambient temperature but do not desorb appreciable amounts of hydrogen at this temperature due to kinetic limitations.

[120, 121]. Concomitantly, there is also a reduction in the enthalpies of hydride phase formation and decomposition. Similar effects are also seen for nanocrystalline samples [117]. From a practical point of view, the cost of Pd would seem to be prohibitive for widespread use, although it is possible that nanoscale Pd could find use as a catalyst in practical storage materials. Nevertheless, Pd hydride can be regarded as a model system for the investigation of size effects [118–120] and therefore continued study of this system is likely to provide further valuable insight into the effects of particle size reduction in other interstitial metal hydride systems.

Returning to magnesium, MgH_2 is an attractive potential storage material because it has a gravimetric capacity of 7.66 wt%. However, in addition to its high thermodynamic stability ($\Delta H \approx -75$ kJ mol^{-1} H_2), the kinetics of hydride formation and decomposition of the bulk material are too slow for practical purposes [122–124]. The kinetics have, however, been improved significantly using mechanical milling both with and without catalytic additives. The positive effects of ball-milling pure MgH_2 have been known since the late 1990s [123, 125], although the temperatures required for desorption are still too high for practical applications. The reviews of Huot et al. [123] and Zaluska et al. [125] both show the improvement in the kinetics at 573 K, for example. However, alternative methods of producing nanoscale magnesium have also been reported (see Gross et al. [126] and references therein), and include an electrochemical synthesis method reported recently by Aguey-Zinsou and Ares-Fernández [127], which results in an average particle size of approximately 5 nm.

With regard to the catalytic enhancement of the hydrogen absorption and desorption properties of MgH_2, the most successful catalyst found to date appears to be Nb_2O_5. The hydrogen desorption rates obtained using various oxide additives are shown in Fig. 2.3 [128]. The reasons for the enhanced absorption and desorption rates are not yet understood and it is possible that the additives do not act catalytically but instead induce further MgH_2 particle or grain size reduction during the milling process [129]. The results of a number of different additives,

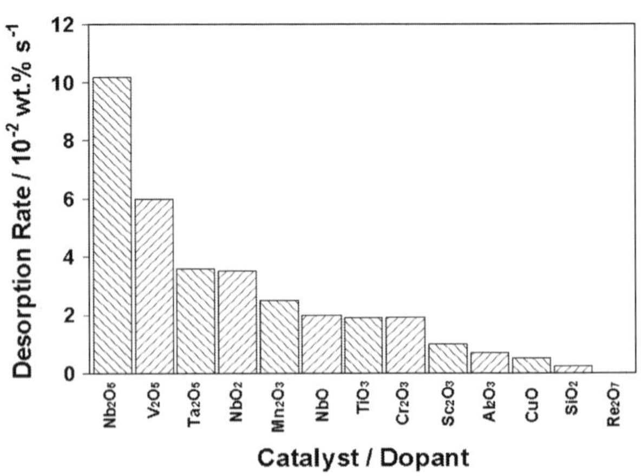

Fig. 2.3 The catalytic effect of various transition metal oxides on the hydrogen desorption rate for MgH_2 [128]. Reaction rates were determined between 20 and 80% of the maximum hydrogen storage capacity in each case. Reprinted with permission from Barkhordian et al. [128]. Copyright 2006 American Chemical Society

including AB_5 and AB_2 intermetallics and elemental metals such as Ni and Ti, were tabulated by Sakintuna et al. [106]. In addition to the kinetic enhancement, there is some theoretical evidence that a reduced particle size can alter the thermodynamics of this system. Ab initio Hartree-Fock and density functional theory calculations, reported by Wagemans et al. [130], show that the stability of Mg hydride is reduced for very small cluster sizes below approximately 1.3 nm. Although this dimension is significantly smaller than the grain size achieved by mechanical milling, whether with or without an additive, these calculations demonstrate that there may be some potential for nanoscale magnesium hydride to provide hydrogen storage capabilities in a practical temperature range, although further work in this area is required. The use of magnesium hydride for hydrogen storage was recently reviewed by Grant [131].

2.2.4 Amorphous and Nanostructured Alloys

In this section we will look primarily at three types of amorphous and nanostructured alloy. Firstly, we will cover compounds that are synthesised or processed through the use of mechanical milling. This can have different effects on a material, depending on a number of factors. One possible effect is amorphisation, which is the loss of crystallographic order by an originally crystalline material. There are, however, a number of other ways to synthesise or process similarly amorphous alloys and so, secondly, we will discuss amorphous materials that can be produced using alternative routes. Thirdly, we will look at an interesting class of materials known as quasicrystals.

2.2.4.1 Mechanical Milling and Alloying

A significant area of research into new hydrogen storage materials involves the use of mechanical milling, as introduced in the section above for the modification of MgH_2, for which it is well suited. The use of the technique originates from the development of Mechanical Alloying (MA), a powder processing method developed in the 1960s to produce oxide dispersion strengthened alloys [123, 132]. In materials processing there are a number of different types of milling with specific terminology used in each case [132]. For our purposes, the approaches applied to hydrogen storage material synthesis or modification can be broadly separated into three areas. Firstly, the synthesis of nanostructured or amorphous hydrogen-absorbing alloys using either elemental metals or the combination of elemental metals with crystalline intermetallics or alloys. Secondly, the nanostructuring or amorphisation of existing crystalline alloys or intermetallics through milling and, thirdly, the doping of an existing crystalline alloy or intermetallic with a catalytic additive. The latter is essentially the same as the approach used widely for magnesium hydride.

In each case mentioned above, the materials can benefit from the effects of the mechanical milling process, which includes the introduction of nanoscale structural features. Such nanostructuring can lead to size effects that are known to dramatically alter the behaviour of materials compared to bulk samples [133, 134] and the approach has a broad range of applications in materials science. Materials that have undergone milling can possess a smaller grain size, amorphous and disordered structural features, metastable and high pressure phases, an increased proportion of grain boundaries, a higher density of defects and an increased surface area [132]. All of these features can affect the hydrogen absorption properties of the material in comparison with unmilled, single phase, homogeneous crystalline samples [119, 120, 135]. If hydrogen diffusion occurs more rapidly through grain boundary regions, a smaller grain size will enhance the rate at which hydrogen is absorbed or desorbed. Also, amorphous phases of some compounds can show favourable absorption characteristics [136]. In addition, increased defect densities can enhance hydrogen diffusion rates. The activation process, which in many metallic absorbers involves the introduction of high dislocation densities into the host structure, can also be accelerated by milling the material. All of these factors can therefore potentially contribute to the improvement of the hydrogen storage properties of a material through milling. Although the process does not always have the desired effect, there is significant experimental evidence for the improvement of the hydrogen storage properties of materials through mechanical milling and alloying [123, 125, 137].

The different approaches mentioned above have been applied to a range of interstitial hydrides and hydride-forming hydrogen storage materials. Catalytic additives, such as Pd and Ni, have been milled with TiFe to improve the difficult activation process for this compound [138, 139]. Amorphous and crystalline hydrogen storage compounds that have been synthesised from the milling of elemental metals include $Ni_{1-x}Zr_x$ amorphous alloys [140], nanocrystalline $LaNi_5$ [141], amorphous and crystalline TiFe [142, 143], Ti–V–Mn alloys [144], and nanocrystalline Mg_2AlNi_2 [145], although many more examples can be found in the literature. The modification of crystalline alloys and intermetallics by milling, meanwhile, has included work on $LaNi_5$ [141, 146, 147], other AB_5-based compounds [148, 149], Ti–Cr [150] and Ti–Cr–V [151, 152] alloys, and the Mg–Ni alloys [137, 153, 154] that can form complex transition metal hydrides (see Sect. 2.3.4). A further approach to the use of mechanical alloying is reactive milling in which the materials are processed in a hydrogen atmosphere [147, 151]. As intimated above, the milling process can also be damaging to the hydrogen sorption properties, as well as beneficial, with hydrogen absorption sometimes prevented entirely as a result of milling [144], or favourable isotherm behaviour being lost along with a significant amount of storage capacity [147, 150, 152]. However, increases in capacity as a result of milling have also been reported [149]. For interested readers, a recent monograph by Varin et al. [155] focuses primarily on the synthesis of hydrides for hydrogen storage through the use of mechanical milling.

2.2.4.2 Amorphous Alloys

In addition to the use of milling methods, amorphous alloys can also be produced using alternative synthesis routes. The methods include the rapid quenching of a melt, thermal evaporation, sputtering, electrodeposition and ion implantation [156, 157]. Many amorphous materials produced using these techniques can absorb significant amounts of hydrogen. Those of interest for hydrogen storage were summarised by Bowman [156] as being of the form $A_{1-y}B_y$ where A is an early transition metal or rare earth and B is a late transition metal, together with $Pd_{1-y}Si_y$. Examples of the former include $Ti_{1-y}Ni_y$, where $0.40 \leq y \leq 0.67$, $Zr_{1-y}Ni_y$, where $0.30 \leq y \leq 0.90$, $Zr_{1-y}Fe_y$, where y is 0.24 and 0.25, and $Y_{1-y}Fe_y$, where $0.25 \leq y \leq 0.55$. These compounds have maximum observed hydrogen-to-metal atom ratios (*H*/*M*) of 1.25, 1.4, 2.46 and 1.8, respectively. Unlike the crystalline intermetallic hydrides, these amorphous alloys tend to exhibit no plateau in their hydrogen absorption and desorption isotherms. This would seem to be a disadvantage in storage applications, in which the relatively large uptakes over small pressure changes in the plateau region are exploited (see Sect. 3.1.1). However, it has been argued that because this indicates the absence of hydride phase formation, the decrepitation that is associated with this process does not occur, which may lead to greater long term cycling stability. However, it is worth noting that the metastable state of these amorphous materials means that they can undergo recrystallisation if subjected to too high a temperature. For storage purposes, it would seem that these amorphous alloys would primarily be of interest if they absorb significantly more hydrogen than their crystalline counterparts, but this is only the case for some of these materials, such as $Ti_{1-y}Cu_yH_x$ and $Pd_{1-y}Si_yH_x$, although there appears to be some discrepancies in the literature. See the review of Eliaz and Eliezer [157] for further discussion of these points. Amorphous alloy systems are discussed in more detail by Fukai [81], with a focus on their thermodynamics, structure and the distributions of site energies in these materials, which differ significantly from ordered hydrides. It is worth noting that hydrogen absorption can itself induce amorphisation in some originally crystalline materials [81], although this phenomenon does not necessarily have an application in the field of hydrogen storage.

2.2.4.3 Quasicrystals

Quasicrystals are an interesting class of materials that show long range order but no translational symmetry. Since the first report in 1984 [158], over one hundred different quasicrystals have been synthesised [159, 160] and they have recently been reported to occur in nature [160]. Several quasicrystals have been investigated for their hydrogen storage properties, including alloys based on Ti–Zr–Ni [161, 162], Ti–Hf–Ni [162], and Mg–Al–Zn [163]. The promise of enhanced hydrogen storage capacities in these materials derives from the large number of potential interstitial sites that exist in their structure. Takasaki and Kelton [162] found that a Ti-based quasicrystal of composition $Ti_{61}Zr_{22}Ni_{17}$ had a hydrogen storage capacity of 2.8 wt%, following electrochemical hydrogenation. The capacity of the $Mg_{44}Al_{15}Zn_{41}$ material studied by Bystrzycki et al. [163], meanwhile, was only 0.6 wt% at 100 bar (10.0 MPa) and temperatures of 573 and 673 K. Furthermore, the material was found to decompose into a $MgZn_2$ phase and MgH_2 following hydrogenation. Although this latter result does not seem particularly promising, further work is undoubtedly necessary to determine whether other quasicrystalline compounds could serve as effective hydrogen storage materials.

2.3 Complex Hydrides

In the case of complex hydrides, atomic hydrogen is bound either ionically or covalently into the bulk of the storage material. It is then released via the decomposition of the host into two or more components. These materials are generally formed from alkali or alkaline earth metals and $[AlH_4]^-$, $[NH_2]^-$ and $[BH_4]^-$ anionic hydrides. As with many of the materials covered in this chapter, complex hydrides were first synthesised many years ago but it is only through recent work that their potential use as practical reversible hydrogen storage materials has been realised. This has occurred, most notably, in the case of sodium alanate ($NaAlH_4$), due to the discovery of the catalytic enhancement by Ti-doping of the hydrogenation/dehydrogenation process by Bogdanović and Schwickardi [164] , but also by the discovery of the reversible hydrogen sorption properties of the Li–N–H system by Chen et al. [165] . The term *complex hydride* has become an umbrella term encompassing the alanates, nitrides and borohydrides that are currently being considered for hydrogen storage [166, 167]. In this section, we cover each of these in turn and also look at the A_2BH_x compounds that are classed as complex transition metal hydrides.

2.3.1 Alanates

The prototype hydrogen storage material in this group is sodium alanate, $NaAlH_4$. Its structure consists of sodium atoms surrounded by $[AlH_4]^-$ tetrahedra. During the dehydrogenation process this phase decomposes into an intermediate Na_3AlH_6 phase with an associated release of gaseous hydrogen,

$$3NaAlH_4 \rightarrow Na_3AlH_6 + 2Al + 3H_2. \tag{2.1}$$

A second reaction step then results in further hydrogen evolution,

$$Na_3AlH_6 \rightarrow 3NaH + Al + \frac{3}{2}H_2. \quad (2.2)$$

Reaction (2.1) occurs between approximately 210 and 220°C (483 and 493 K), and reaction (2.2) around 250°C (523 K). The dehydrogenation of NaH does not occur until approximately 425°C (698 K) and does not, therefore, play a practical role in the reversible hydrogen storage process. The two stage reaction results in an isotherm with two plateaus. At 210°C (483 K), plateau pressures of 15.4 and 2.1 MPa were determined by Dymova et al. [168]. The irreversibility of this reaction, the instability of the hydride and the slow desorption kinetics meant that this material was not considered a particularly promising storage material. The breakthrough came in the mid 1990s when Bogdanović and Schwickardi [164] discovered the remarkable effect that the addition of $TiCl_3$ has on the reversibility and the kinetics of the hydrogenation process. Jensen et al. [169, 170] subsequently improved on Bogdanović and Schwickardi's wet chemistry doping method and showed that further improvements could be made through mechanical mixing of the alanate and the dopant compounds. Since this early work, many other effective dopants have been identified, including $ScCl_3$, $CeCl_3$ and $PrCl_3$ [168, 171].

Ti-doped $NaAlH_4$ readily desorbs hydrogen at temperatures in the region of 120°C (393 K) and can be rehydrogenated at 170°C (443 K) in 15 MPa of hydrogen [172]. Although the role of the Ti catalyst is not yet fully understood [173], this material has already been used in practical hydrogen storage units, and so it has proved to a certain extent that complex hydrides can be practically applied to storage applications. Applied research into the use of sodium alanate is now at a reasonably advanced stage; a recent study, for example, investigated the safety of a sodium alanate store by experimentally simulating a tank failure and the subsequent expulsion of sodium alanate powder [174]. The expelled dust cloud did not spontaneously ignite, but the presence of an external ignition source resulted in a flame of reacting powder. Ignition also occurred when water was sprayed onto the expelled dust cloud. This result is relatively positive because spontaneous ignition may be expected but was not observed.

Other alanates that are being considered for hydrogen storage include $LiAlH_4$, $KAlH_4$, $Mg(AlH_4)_2$ and $Ca(AlH_4)_2$, which have gravimetric hydrogen capacities of 10.54, 5.71, 9.27 and 7.84 wt%, respectively. A number of mixed alanates have also been reported in the literature, including Na_2LiAlH_6, K_2NaAlH_6, K_2LiAlH_6 and $LiMg(AlH_4)_3$ [115, 175] and recent work has investigated various mixed alanate combinations, including Mg–Li–Al–H [176], Mg–Ca–Al–H, Li–Ca–Al–H and Na–Ca–Al–H [177], and Mg–Na–Al–H, Mg–K–Al–H and Ca–K–Al–H [178], although work on these systems is still at an early stage. Continued research in the area will hopefully reveal new alanates, new alanate phases or mixed alanate combinations, as well as new catalysts, particularly if significant progress can be made in understanding the role of the Ti dopant. Hydrogen storage using $NaAlH_4$, $LiAlH_4$, $KAlH_4$ and $Mg(AlH_4)_2$ was reviewed by Jensen et al. [168], with a focus on Ti-doped sodium alanate.

2.3.2 Nitrides, Amides and Imides

The potential of the Li–N–H system as a potential storage material was first reported by Chen et al. [165]. The system forms three stoichiometric ternary compounds: *lithium imide* (Li_2NH), *lithium amide* ($LiNH_2$) and *lithium nitride hydride* (Li_4NH). The latter, however, is not involved in the proposed reversible hydrogen storage process. This begins with lithium nitride (Li_3N), which is hydrogenated to form a combination of lithium imide and lithium hydride (LiH). Further hydrogenation results in the formation of lithium amide and further lithium hydride, and the complete process is represented by the following reaction,

$$Li_3N + 2H_2 \rightarrow Li_2NH + LiH + H_2 \leftrightarrow LiNH_2 + 2LiH. \qquad (2.3)$$

Dehydrogenation of the imide requires high vacuum and temperatures above 600 K [165]. These are unsuitable conditions for reversible hydrogen storage, but the reaction between the imide and the amide is reversible under more moderate conditions of both temperature and pressure. The mixed hydride/imide on the right-hand side of reaction (2.3) gives a high theoretical hydrogen capacity of 10.4 wt%, but the reversible capacity between the imide and amide is approximately 6.5 wt% (one H_2 molecule released from $LiNH_2$ + 2LiH).[13]

A number of other similar materials have been studied as hydrogen storage media, including the ternary compounds $Mg(NH_2)_2$, $RbNH_2$, $CsNH_2$ and Ca–N–H, and the quarternary and higher systems Li–Ca–N–H, Li–Al–N–H, Na–Mg–N–H, Na–Ca–N–H, Mg–Ca–N–H and Li–Mg–Ca–N–H [167]. It can be seen that these materials have the potential to provide high gravimetric storage capacities and are therefore of great interest. However, the Li–N–H system suffers from a number of drawbacks, including the high hydrogenation and dehydrogenation temperatures, and air or moisture sensitivity. Another is the evolution of ammonia during the dehydrogenation reaction [180]. As well as contributing to the degradation of samples during long term cycling, the evolution of ammonia is also an issue because it is a very effective poison for PEM fuel cell membranes [181]. The storage of hydrogen using imides and amides, including mixed systems, was reviewed recently by Gregory [182].

2.3.3 Borohydrides

Borohydrides have the highest gravimetric hydrogen storage capacities of any of the complex hydrides. $LiBH_4$ contains 18.5 wt% hydrogen, and releases it through one of the following two reactions [183],

[13] A theoretical capacity of 11.5 wt% is occasionally quoted but this value, depending on the definition of hydrogen capacity, is an error from the original Chen et al. [165] paper that has since appeared in other reports [179].

$$LiBH_4 \rightarrow Li + B + 2H_2 \quad (2.4)$$

or,

$$LiBH_4 \rightarrow LiH + B + \frac{3}{2}H_2. \quad (2.5)$$

However, the decomposition temperature is too high for practical purposes. According to Orimo et al. [167], $LiBH_4$ releases three of its four hydrogen atoms upon melting at 280°C (553 K), with an enthalpy of decomposition of −88.7 kJ mol^{-1} H_2. LiH is very stable and its dehydrogenation occurs only above a temperature of 727°C (1000 K). Nevertheless, the dehydrogenation reaction of $LiBH_4$ is a reversible process, although rehydrogenation requires elevated pressures and temperatures of 35.0 MPa at 600°C (873 K) and 20.0 MPa at 690°C (963 K). As for most of the complex hydrides, the hydrogenation mechanism of LiH and B is not yet understood, but it is thought to result either from the intermediate reaction of LiH with B, to form a compound that is subsequently filled with hydrogen, or from the reaction of B and H to form diborane, which then spontaneously reacts with LiH to give the full borohydride [167].

A number of other alkali metal and alkaline earth metal borohydrides have high gravimetric and volumetric hydrogen storage capacities. For example, the theoretical gravimetric capacities of $NaBH_4$, KBH_4, and $Mg(BH_4)_2$ are 10.6, 7.4 and 14.8 wt%, with volumetric capacities of 113.1, 87.1 and 146.5 kg m^{-3}, respectively. However, these are all theoretical values and cannot be reversibly achieved in practice at practical temperatures; borohydrides are also moisture sensitive [184] and Eberle et al. [173] suggest that the possible evolution of volatile boranes, even at trace levels, would be problematic due to storage capacity loss and fuel cell damage. However, as for all of the complex hydrides, their high theoretical capacities mean that further work on these materials would be valuable. Nakamori and Orimo [184] recently reviewed hydrogen storage using borohydrides, including those suitable for both reversible and non-reversible chemical hydrogen storage. Walker [185], meanwhile, discussed the destabilisation of complex hydrides via mixing with other compounds, focusing on the use of this strategy for $LiBH_4$ with various additives including hydrides (MgH_2), magnesium salts (MgF_2, MgS_2 and $MgSe_2$), elemental metals (Al), alloys, oxides (TiO_2) and carbon. This offers an interesting route for the modification of borohydrides, as well as other complex hydrides, for hydrogen storage applications, and is likely to be the subject of continued research.

2.3.4 Complex Transition Metal Hydrides

A number of complex transition metal hydrides have been known for some time to reversibly desorb and absorb hydrogen. Examples of these materials include Mg_2FeH_6, Mg_2NiH_4 and Mg_2CoH_5 [95]. Some of the host metallic constituents form stable intermetallics, such as Mg_2Ni and Mg_2Cu, but most do not. The complex transition metal hydrides include compounds with impressive storage capabilities, including a *H*/*M* ratio of 4.5 in the case of $BaReH_9$ and a volumetric capacity of approximately 150 g L^{-1}, twice that of liquid hydrogen, as is the case for Mg_2FeH_6 [95].

Mg_2NiH_4 is the material in this group that has probably received the most attention as a candidate storage material, due in part to its gravimetric hydrogen capacity of 3.6 wt%. Its synthesis was first reported by Reilly and Wiswall [90] and it was long considered to be an interstitial hydride, but its structure is now known to consist of tetrahedral NiH_4 complexes [186–188]. Complex transition metal hydrides generally suffer from the same problem as the other complex hydrides currently being considered for hydrogen storage, namely the high absorption and desorption temperatures. Mg_2NiH_4 has an enthalpy of formation of −32.3 kJ mol^{-1} H and requires temperatures above 520 K for hydrogen desorption, and higher temperatures still for absorption. Progress has, however, been made on the destabilisation of this material through mechanical milling (Sect. 2.2.4) [186]. A decrease of the hydrogen desorption temperature in mechanically milled Mg_2Ni was attributed by Orimo and Fujii [137, 186] to the presence of a large proportion of grain boundaries following the milling process. However, the hydrogen storage capacity was concomitantly decreased to 1.6 wt% in the milled material.

Although these materials have not recently received as much attention as the alanates, amides and imides, and borohydrides, they are prominent examples of another type of non-interstitial hydride. It seems likely that there are many more, as yet undiscovered, multinary complex hydride compounds that could potentially serve as effective hydrogen storage materials. Therefore, although they do not necessarily exhibit gravimetric capacities in excess of the US DOE targets, continued work into new complex hydride compounds, such as the complex transition metal hydrides, may yet prove fruitful.

2.4 Other Materials

In the last three sections we have covered a range of potential reversible storage materials grouped into the porous adsorbents, interstitial hydrides and complex hydrides. In this section we look at materials that do not fit readily into these categories, which include the clathrates and ionic liquids. We shall also look at materials that show enhanced hydrogen storage through the mechanism of hydrogen spillover. Although the latter are principally porous adsorbents, the proposed storage mechanism is fundamentally different to molecular physisorption, as covered in Sect. 2.1; hence its separate treatment.

2.4.1 Clathrates

Clathrate hydrates are a novel proposed solution to the hydrogen storage problem. They are inclusion compounds formed from hydrogen-bonded networks of water molecules and have long been known due to their occurrence in natural gas and oil pipelines [189]. The guests can include species such as methane, nitrogen and argon, but the discovery of molecular hydrogen encapsulation by Mao et al. [190] has led to the study of clathrates for solid state hydrogen storage [191]. Although pure hydrogen clathrate hydrates are stable only at high pressures or low temperatures, recent work has shown that they can be stabilised under near-ambient conditions using a promoter such as Tetrahydrofuran (THF) [191, 192]. There are three clathrate hydrate structures, namely sI, sII and sH, and the latter two have been studied for their hydrogen storage properties.

Some controversy has surrounded the capacity of sII after an initial report by Lee et al. [193] of a potential hydrogen storage capacity of 4 wt% and the ability to tune the clathrates by varying the THF concentration. These results have not been independently verified, and it seems that the consensus is a practical maximum storage capacity closer to 1 wt% [194]. Recently, hydrogen storage in the sH phase has been proposed, with estimated capacities in the region of 1.4 wt% [195–197]. The stabilisation of sH has been achieved using alternative promoters, including Methyl *tert*-Butyl Ether (MTBE) and 1,1-Dimethylcyclohexane (DMCH) [196, 197]. The reduced capacities of the stabilised clathrates, and hence the source of some of the controversy, originates from the occupancy of a proportion of the cages in the clathrate structures by the promoter molecules rather than hydrogen. In the case of sII, it was argued that reducing the quantity of THF used to stabilise the clathrate allowed hydrogen to occupy some of the larger cages occupied by THF at higher concentrations, thus allowing significantly higher hydrogen storage capacities to be achieved [193]. However, the occupation of these larger cavities by hydrogen was subsequently disputed by other authors [194]. In addition to hydrogen storage in clathrate hydrates, the possibility of hydrogen storage in the organic clathrate hydroquinone (1,4-benzenediol) has also been reported recently [198–200]. As for many of the materials covered in this chapter, hydrogen storage using both clathrate hydrates and alternative organic clathrates is the subject of ongoing research.

One significant disadvantage of clathrates is the slow rate of clathrate formation. In recent work, Cooper et al. have demonstrated that clathrates stabilised in the pores of emulsion templated polymers[14] [201] and formed in hydrophilic water-swellable polymer network[15]-based hydrogels [202] show potential for hydrogen storage applications. Although the reported capacities are low, such supported clathrate formation appears to greatly enhance the kinetics of the process and is therefore an interesting area for further research.

---

[14] Ultralow density, emulsion-templated polymerized High Internal Phase Emulsion (polyHIPE) material.

[15] Lightly crosslinked poly(acrylic acid) sodium salt (PSA).

2.4.2 Ionic Liquids

Ionic liquids have been attracting significant amounts of attention due to their use in catalysis, as so-called green solvents and in a vast array of other applications [203, 204]. They are defined as materials composed of cations and anions which melt at or below 100°C (373 K) [203], and their negligible vapour pressures mean that they are environmentally-friendly in comparison to the volatile organic solvents that they can potentially replace in many industrial and chemical processes. Stracke et al. [205] recently reported the potential of imidazolium ionic liquids for hydrogen storage, with a volumetric hydrogen capacity of up to 30 g L^{-1}. However, the dehydrogenation temperature of the Pd/C-catalysed material[16] was in the region 230 to 300°C (503 to 573 K) and the hydrogenation time was approximately 100 hours. Although this performance does not appear particularly encouraging, it is worth bearing in mind that, according to Plechkova and Seddon [203], there are over one million simple ionic liquids, and many more binary and ternary systems. Therefore, further work is certainly required if a meaningful conclusion on the suitability of this interesting class of materials for hydrogen storage applications is to be drawn.

2.4.3 The Use of Hydrogen Spillover

Extensive research on the exploitation of spillover for hydrogen storage has been carried out in recent years [206, 207]. *Spillover* is a mechanism by which molecular hydrogen dissociates on catalytically active particles and subsequently migrates to the surface of a solid state support that would not otherwise adsorb or absorb the atomic hydrogen under the same conditions [208]. For this to occur, a porous material must be doped with suitably catalytic nanoparticles to facilitate the dissociation process. It is a well documented, yet poorly understood, phenomenon in heterogeneous catalysis [208] and has been proposed as a way of enhancing the storage capacity of porous materials at near-ambient temperature. The proposed methods for producing high capacity storage materials include physical mixing of a supported catalyst with a secondary (receptor) material, the use of carbon bridges between a supported catalyst and the receptor, and the direct doping of the receptor with the catalyst [206]. The first two of these use so-called *secondary spillover*. In the case of the carbon bridge-building approach, the carbon bridges are formed from a carbon precursor, such as glucose, which is subsequently carbonised by heating. This forms a carbon bridge between the supported catalyst, such as Pd on carbon, and a secondary receptor, such as an activated carbon. At 298 K, this method was found to enhance the storage capacity of an activated carbon (AX-21) by a factor of 2.9 [209]. The carbon bridge-building approach has also been

[16] 1-alkyl(aryl)-3-methylimidazolium *N*-bis(trifluoromethanesulfonyl) salt.

applied to MOFs using a carbon precursor that does not require carbonisation temperatures high enough to decompose the metal-organic framework support, although simple physical mixing of a supported catalyst has also been used. The doping of various metallic catalysts, such as Ni, Pd, Pt and Ru, has been performed via chemical, ultrasonic and plasma-assisted methods on a range of carbon supports [206]. The capacities claimed for room temperature hydrogen storage using the spillover mechanism show improvement over undoped materials; however, it remains to be seen whether significant enhancement in the reversible storage density of hydrogen can be achieved using this approach.

2.4.4 *Organic and Inorganic Nanotubes*

We will close this section by mentioning some nanostructured materials closely related to the carbon nanotubes covered in Sect. 2.1.1.2. As we saw in that section, carbon nanotubes have received a great deal of attention as potential storage materials, but many other *organic* and *inorganic nanotube materials* also exist. The review by Rao and Nath [210] provides a good overview of inorganic nanotubes and their synthesis, including chalcogenide, oxide and nitride nanotube materials. A number of these, in particular those consisting of lighter elements, have been investigated for their hydrogen storage properties. Three types were covered by Seayad and Antonelli [211]: boron nitride (BN) [212, 213], titanium sulfide (TiS_2) [214] and molybdenum sulfide (MoS_2) nanotubes [215], with reported capacities reaching 4.2 wt% for collapsed BN nanotubes at 10 MPa and ambient temperature [216]. In addition, other materials in nanotube form have since been investigated either experimentally or theoretically for their hydrogen storage properties, including titanium oxide (TiO_2) [211, 217], tungsten carbide (WC) [218], silicon [219] and silicon carbide (SC) [220], and the hydrogen sorption properties of some of these materials appear very interesting. For example, the ability of some multiwalled nanotubes to intercalate other species [210, 211] provides a possible proposed mechanism that could allow significant amounts of hydrogen to be stored at practical temperatures [217]. However, the suitability of these nanostructured materials for a large scale application, such as automotive transportation, remains open to question.

2.5 Summary

In this chapter we have presented an overview of the various types of materials that are currently being considered as potential reversible hydrogen storage media. We began with microporous materials, which include carbons, zeolites, metal-organic frameworks and microporous organic polymers, before moving on to the interstitial hydrides. Microporous materials store molecular hydrogen adsorbed in

Table 2.1 A comparison of the basic hydrogen storage properties of a number of reversible hydrogen storage material types, and other physical and chemical storage methods. Note that absolute comparisons between the values given in this table, and others published widely elsewhere, are difficult because of the use of different definitions of storage capacity, particularly between different storage material types, and the significant effect of the hydrogen pressure and the temperature on quoted capacity values. In addition, isotherm shape can greatly affect the reversible capacity in comparison with the total capacity. As a result, these values should be treated only as an approximate guide and care should be exercised in attempts to draw a firm conclusion from direct comparisons. Further discussion of the different capacity definitions and other related considerations are given in the following chapter

Category / storage method	Material / storage medium	Gravimetric capacity[a] (wt%) (P/MPa)	Volumetric capacity[a] ($g\ L^{-1}$ or $kg\ m^{-3}$)	Operating temperature region	References
Pure H_2	Compressed (70 MPa)	100(70) 4.8, with tank	39.0[b] 23.0, with tank	ambient	[44]
	Liquid	100 5.11, with tank	70.8	20 K	[221]
Chemical hydrides (hydrocarbons)	Cyclohexane (C_6H_{12})	7.2[c]	28.0	508–623 K[d]	[221]
	cis-decalin	7.3[e]	32.7	483 K[f]	[221]
	trans-decalin	7.3[e]	31.7	483 K[f]	[221]
Chemical hydrides (other)	$NaBH_4$	10.57	113.1	778 K	[167]
	$Mg(NH_3)_6Cl_2$	9.2	110	> 400 K	[222]
Binary hydrides	AlH_3	10.1	148	< 373 K	[116]
Microporous adsorbents	Zeolite-templated carbon	6.9(2.0)	–	77 K	[26]
	Na-X zeolite	2.55(4.0)	–	77 K	[35]
	MOF-5	5.25(4.85)	31.0	77 K	[56]
	MOF-177	7.52(6.85)	32.1	77 K	[56]
	trip-PIM	2.7(1.0)	–	77 K	[66]
	HCP (BCMBP-based)	3.68(1.5)	–	77 K	[68]
	COF-102	7.24(4.0)	29 (excess)[g] 48 (absolute)[g]	77 K	[70]
	EOF-2	1.21(0.1)	–	77 K	[76]

(continued)

Table 2.1 (continued)

Category / storage method	Material / storage medium	Gravimetric capacity[a] (wt%) (P/MPa)	Volumetric capacity[a] ($g\ L^{-1}$ or $kg\ m^{-3}$)	Operating temperature region	References
Interstitial hydrides	TiFe	1.86/*1.5*	100.4/*83.7*	near ambient	[87, 223]
	$Ti_{0.98}Zr_{0.02}Cr_{0.05}V_{0.43}Fe_{0.09}Mn_{1.5}$	1.9/*1.3*	*63.6*	near ambient	[87]
	$LaNi_5$	1.49/*1.28*	*87.0*	near ambient	[87]
	$LaNi_{4.8}Sn_{0.2}$	1.4/*1.24*	*85.4*	near ambient	[87]
	$(V_{0.9}Ti_{0.1})_{0.95}Fe_{0.05}$	1.95/*1.8*	*82.0*	near ambient	[87]
	Nb_2O_5-doped MgH_2	≈ 7	112.1 [h]	≈ 573 K	[128, 223]
Complex hydrides	$NaAlH_4$	7.47/*5.6*	–	373–473 K	[167]
	$LiNH_2$.+ 2LiH	10.4/*6.5*	–	≈ 573 K	[224]
	$LiBH_4$	18.36/*5–10*[i]/ *13.5*[j]	122.5	453–923 K	[167]
	Mg_2NiH_4	3.6/*3.3*	98.8/*87.0*	> 528 K	[87, 223]
Clathrates	THF-stabilised H_2 clathrate	≈ *1*	–	270–280 K	[191]
RTILs	Imidazolium RTIL	–	30	503–573 K	[205]
Nanotubes	"Collapsed" BN	4.2(10.0)	–	ambient	[216]
	TiO_2	≈ 4(0.6)	–	77 K	[217]

[a] Reversible capacity shown in *italics*, where appropriate
[b] Density at 70.0 MPa and 300 K calculated using NIST REFPROP database [225]
[c] Hydrogen released during dehydrogenation to benzene
[d] Temperature required for dehydrogenation [221]
[e] Hydrogen released during dehydrogenation to naphthalene
[f] Temperature required for dehydrogenation [226]
[g] Estimated from adsorption isotherm at approximately 8.5 MPa
[h] Calculated from the value for MgH_2 given by Wiswall [223]
[i] Observed rehydrogenation
[j] Observed first dehydrogenation

their pores at relatively low temperatures and, in certain cases, can achieve gravimetric storage capacities on a materials basis in excess of the current 2015 US DOE system storage target. The interstitial hydrides absorb atomic hydrogen into the bulk of a metallic host material and, although they do not possess particularly high gravimetric capacities, they exhibit impressive volumetric hydrogen storage capabilities and some favourable hydrogen storage properties. The third type of storage material we have considered are the complex hydrides. This group of compounds, like the interstitial hydrides, store atomic hydrogen in their bulk but, in this case, bond the hydrogen in complexes. Upon hydrogen desorption, their host structure decomposes to one or more additional decomposition products. We concluded the chapter by considering some materials that do not fit readily into the other three main categories, including clathrates, ionic liquids and inorganic nanotubes. Some of the basic hydrogen storage properties of materials covered in this chapter are summarised in Table 2.1.

References

1. Sing KSW, Everett DH, Haul RAW, Moscou L, Pierotti RA, Rouquérol J, Siemieniewska T (1985) Reporting physisorption data for gas/solid systems with special reference to the determination of surface area and porosity. Pure Appl Chem 57(4):603–619
2. Menon VC, Komarneni S (1998) Porous adsorbents for vehicular natural gas storage: a review. J Porous Mater 5:43–58
3. McBain JW (1909) The mechanism of the adsorption ("sorption") of hydrogen by carbon. Philos Mag Ser 6 18(108):916–935
4. Frolich PK, White A (1930) Adsorption of methane and hydrogen on charcoal at high pressure. Ind Eng Chem 22(10):1058–1060
5. Carpetis C, Peschka W (1980) A study on hydrogen storage by use of cryoadsorbents. Int J Hydrogen Energy 5:539–554
6. Agarwal RK, Noh JS, Schwarz JA, Davini P (1987) Effect of surface acidity of activated carbon on hydrogen storage. Carbon 25(2):219–226
7. Chahine R, Bose TK (1994) Low-pressure adsorption storage of hydrogen. Int J Hydrogen Energy 19(2):161–164
8. Férey G (2008) Hybrid porous solids: past, present, future. Chem Soc Rev 37:191–214
9. Rouquerol F, Rouquerol J, Sing K (1999) Adsorption by powders and porous solids: principles, methodology and applications. Academic Press, London
10. Menon PG (1968) Adsorption at high pressures. Chem Rev 68(3):277–294
11. Broom DP, Walton A, Book D, Benham MJ (2007) The accurate determination of the temperature dependence of hydrogen uptake by Na-X zeolite. Presented at the 15th International Zeolite Conference, Beijing, China, 12–17 August 2007
12. Thomas KM (2009) Adsorption and desorption of hydrogen on metal-organic framework materials for storage applications: comparison with other nanoporous materials. Dalton Trans 1487–1505
13. Yürüm Y, Taralp A, Veziroglu TN (2009) Storage of hydrogen in nanostructured carbon materials. Int J Hydrogen Energy 34:3784–3798
14. Rzepka M, Lamp P, de la Casa-Lillo MA (1998) Physisorption of hydrogen on microporous carbon and carbon nanotubes. J Phys Chem B 102:10894–10898
15. Jordá-Beneyto M, Suárez-García F, Lozano-Castelló D, Cazorla-Amorós D, Linares-Solano A (2007) Hydrogen storage on chemically activated carbons and carbon nanomaterials at high pressures. Carbon 45:293–303

16. Zlotea C, Moretto P, Steriotis T (2009) A Round Robin characterisation of the hydrogen sorption properties of a carbon based material. Int J Hydrogen Energy 34(7):3044–3057
17. Jordá-Beneyto M, Lozano-Castelló D, Suárez-García F, Cazorla-Amorós D, Linares-Solano Á (2008) Advanced activated carbon monoliths and activated carbons for hydrogen storage. Microporous Mesoporous Mater 112:235–242
18. Dillon AC, Jones KM, Bekkedahl TA, Kiang CH, Bethune DS, Heben MJ (1997) Storage of hydrogen in single-walled carbon nanotubes. Nature 386:377–379
19. Züttel A, Orimo S (2002) Hydrogen in nanostructured, carbon-related, and metallic materials. MRS Bull 27(9):705–711
20. Hirscher M, Becher M, Haluska M, Dettlaff-Weglikowska U, Quintel A, Duesberg GS, Choi Y-M, Downes P, Hulman M, Roth S, Stepanek I, Bernier P (2001) Hydrogen storage in sonicated carbon materials. Appl Phys A 72:129–132
21. Chambers A, Park C, Baker RTK, Rodriguez NM (1998) Hydrogen storage in graphite nanofibers. J Phys Chem B 102(22):4253–4256
22. Lamari Darkrim F, Malbrunot P, Tartaglia GP (2002) Review of hydrogen storage by adsorption in carbon nanotubes. Int J Hydrogen Energy 27:193–202
23. Blackman JM, Patrick JW, Snape CE (2006) An accurate volumetric differential pressure method for the determination of hydrogen storage capacity at high pressures in carbon materials. Carbon 44:918–927
24. Pubysheva OV, Farajian AA, Yakobson BI (2008) Fullerene nanocage capacity for hydrogen storage. Nano Lett 8(3):767–774
25. Xu W-C, Takahashi K, Matsuo Y, Hattori Y, Kumagai M, Ishiyama S, Kaneko K, Iijima S (2007) Investigation of hydrogen storage capacity of various carbon materials. Int J Hydrogen Energy 32:2504–2512
26. Yang Z, Xia Y, Mokaya R (2007) Enhanced hydrogen storage capacity of high surface area zeolite-like carbon materials. J Am Chem Soc 129:1673–1679
27. Hu Q, Lu Y, Meisner GP (2008) Preparation of nanoporous carbon particles and their cryogenic hydrogen storage capacities. J Phys Chem C 112:1516–1523
28. Gogotsi Y, Dash RK, Yushin G, Yildirim T, Laudisio G, Fischer JE (2005) Tailoring of nanoscale porosity in carbide-derived carbons for hydrogen storage. J Am Chem Soc 127:16006–16007
29. Gogotsi Y, Portet C, Osswald S, Simmons JM, Yildirim T, Laudisio G, Fischer JE (2009) Importance of pore size in high-pressure hydrogen storage by porous carbons. Int J Hydrogen Energy 34:6314–6319
30. Baerlocher Ch, Yoshikawa T, McCusker LB, Olson DH (2007) Atlas of zeolite framework types, 6th edn. Elsevier, Amsterdam
31. Nijkamp MG, Raaymakers JEMJ, van Dillen AJ, de Jong KP (2001) Hydrogen storage using physisorption - materials demands. Appl Phys A 72:619–623
32. Vitillo JG, Ricchiardi G, Spoto G, Zecchina A (2005) Theoretical maximal storage of hydrogen in zeolitic frameworks. Phys Chem Chem Phys 7:3948–3954
33. Langmi HW, Walton A, Al-Mamouri MM, Johnson SR, Book D, Speight JD, Edwards PP, Gameson I, Anderson PA, Harris IR (2003) Hydrogen adsorption in zeolites A, X, Y and RHO. J Alloy Compd 356–357:710–715
34. Anderson PA (2008) Storage of hydrogen in zeolites. In: Walker G (ed) Solid-state hydrogen storage: materials and chemistry. Woodhead Publishing, Cambridge
35. Du X, Wu E (2006) Physisorption of hydrogen in A, X and ZSM-5 types of zeolites at moderately high pressures. Chin J Chem Phys 19(5):457–462
36. van den Berg AWC, Bromley ST, Jansen JC (2005) Thermodynamic limits on hydrogen storage in sodalite framework materials: a molecular mechanics investigation. Microporous Mesoporous Mater 78:63–71
37. van den Berg AWC, Bromley ST, Wojdel JC, Jansen JC (2006) Adsorption isotherms of H_2 in microporous materials with SOD structure: a grand canonical Monte Carlo study. Microporous Mesoporous Mater 87:235–242

38. Song MK, No KT (2007) Molecular simulation of hydrogen adsorption in organic zeolite. Catal Today 120:374–382
39. Fraenkel D, Shabtai J (1977) Encapsulation of hydrogen in molecular sieve zeolites. J Am Chem Soc 99:7074–7076
40. Efstathiou AM, Suib SL, Bennett CO (1990) Encapsulation of molecular hydrogen in zeolites at 1 atm. J Catal 123:456–462
41. Weitkamp J, Fritz M, Ernst S (1995) Zeolites as media for hydrogen storage. Int J Hydrogen Energy 20(12):967–970
42. Garrone E, Bonelli B, Otero Areán C (2008) Enthalpy-entropy correlation for hydrogen adsorption on zeolites. Chem Phys Lett 456:68–70
43. Otero Areán C, Turnes Palomino G, Llop Carayol MR (2007) Variable temperature FT-IR studies on hydrogen adsorption on the zeolite (Mg, Na)-Y. Appl Surf Sci 253:5701–5704
44. Felderhoff M, Weidenthaler C, von Helmolt R, Eberle U (2007) Hydrogen storage: the remaining scientific and technological challenges. Phys Chem Chem Phys 9:2643–2653
45. van den Berg AWC, Otero Areán C (2008) Materials for hydrogen storage: current research trends and perspectives. Chem Commun 668-681
46. Rosseinsky MJ (2004) Recent developments in metal-organic framework chemistry: design, discovery, permanent porosity and flexibility. Microporous Mesoporous Mater 73:15–30
47. Rowsell JLC, Yaghi OM (2004) Metal-organic frameworks: a new class of porous materials. Microporous Mesoporous Mater 73:3–14
48. Rowsell JLC, Yaghi OM (2005) Strategies for hydrogen storage in metal-organic frameworks. Angew Chem Int Ed 44:4670–4679
49. Collins DJ, Zhou HC (2007) Hydrogen storage in metal-organic frameworks. J Mater Chem 17:3154–3160
50. Zhao D, Yuan D, Zhou H-C (2008) The current status of hydrogen storage in metal-organic frameworks. Energy Environ Sci 1:222–235
51. Dincă M, Long JR (2008) Hydrogen storage in microporous metal-organic frameworks with exposed metal sites. Angew Chem Int Ed 47(36):6766–6779
52. Murray LJ, Dincă M, Long JR (2009) Hydrogen storage in metal-organic frameworks. Chem Soc Rev 38:1294–1314
53. Rosi NL, Eckert J, Eddaoudi M, Vodak DT, Kim J, O'Keefe M, Yaghi OM (2003) Hydrogen storage in microporous metal-organic frameworks. Science 300:1127–1129
54. Züttel A (2003) Materials for hydrogen storage. Mater Today 6(9):24–33
55. Rowsell JLC, Millward AR, Park KS, Yaghi OM (2004) Hydrogen sorption in functionalized metal-organic frameworks. J Am Chem Soc 126:5666–5667
56. Wong-Foy AG, Matzger AJ, Yaghi OM (2006) Exceptional H_2 saturation uptake in microporous metal-organic frameworks. J Am Chem Soc 128:3494–3495
57. Dincă M, Dailly A, Liu Y, Brown CM, Neumann DA, Long JR (2006) Hydrogen storage in a microporous metal-organic framework with exposed Mn^{2+} coordination sites. J Am Chem Soc 128:16876–16883
58. Kubas GJ (2007) Fundamentals of H_2 binding and reactivity on transition metals underlying hydrogenase function and H_2 production and storage. Chem Rev 107:4152–4205
59. Hoang TKA, Antonelli DM (2009) Exploiting the Kubas interaction in the design of hydrogen storage materials. Adv Mater 21:1787–1800
60. Zhao X, Xiao B, Fletcher AJ, Thomas KM, Bradshaw D, Rosseinsky MJ (2004) Hysteretic adsorption and desorption of hydrogen by nanoporous metal-organic frameworks. Science 306:1012–1015
61. Fletcher AJ, Thomas KM, Rosseinsky MJ (2005) Flexibility in metal-organic framework materials: impact on sorption properties. J Solid State Chem 178(8):2491–2510
62. Panella B, Hirscher M, Pütter H, Müller U (2006) Hydrogen adsorption in metal-organic frameworks: Cu-MOFs and Zn-MOFs compared. Adv Funct Mater 16:520–524
63. Chen B, Zhao X, Putkham A, Hong K, Lobkovsky EB, Hurtado EJ, Fletcher AJ, Thomas KM (2008) Surface interactions and quantum kinetic molecular sieving for H_2 and D_2 adsorption on a mixed metal-organic framework material. J Am Chem Soc 130:6411–6423

64. Makowski P, Thomas A, Kuhn P, Goettmann F (2009) Organic materials for hydrogen storage applications: from physisorption on organic solids to chemisorption in organic molecules. Energy Environ Sci 2:480–490
65. McKeown NB, Budd PM (2006) Polymers of intrinsic microporosity (PIMs): organic materials for membrane separations, heterogeneous catalysis and hydrogen storage. Chem Soc Rev 35:675–683
66. Budd PM, Butler A, Selbie J, Mahmood K, McKeown NB, Ghanem B, Msayib K, Book D, Walton A (2007) The potential of organic polymer-based hydrogen storage materials. Phys Chem Chem Phys 9:1802–1808
67. Tsyurupa MP, Davankov VA (2006) Porous structure of hypercrosslinked polystyrene: state-of-the-art mini-review. React Funct Polym 66(7):768–779
68. Wood CD, Tan B, Trewin A, Niu H, Bradshaw D, Rosseinsky MJ, Khimyak YZ, Campbell NL, Kirk R, Stöckel E, Cooper AI (2007) Hydrogen storage in microporous hypercrosslinked organic polymer networks. Chem Mater 19:2034–2048
69. Han SS, Furukawa H, Yaghi OM, Goddard WA (2008) Covalent organic frameworks as exceptional hydrogen storage materials. J Am Chem Soc 130:11580–11581
70. Furukawa H, Yaghi OM (2009) Storage of hydrogen, methane, and carbon dioxide in highly porous covalent organic frameworks for clean energy applications. J Am Chem Soc 131:8875–8883
71. El-Kaderi HM, Hunt JR, Mendoza-Cortés JL, Côté AP, Taylor RE, O'Keefe M, Yaghi OM (2007) Designed synthesis of 3D covalent organic frameworks. Science 316:268–272
72. Spoto G, Vitillo JG, Cocina D, Damin A, Bonino F, Zecchina A (2007) FTIR spectroscopy and thermodynamics of hydrogen adsorbed in a cross-linked polymer. Phys Chem Chem Phys 9:4992–4999
73. Comotti A, Bracco S, Distefano G, Sozzani P (2009) Methane, carbon dioxide and hydrogen storage in nanoporous dipeptide-based materials. Chem Commun 284-286
74. Msayib KJ, Book D, Budd PM, Chaukura N, Harris KDM, Helliwell M, Tedds S, Walton A, Warren JE, Xu M, McKeown NB (2009) Nitrogen and hydrogen adsorption by an organic microporous crystal. Angew Chem Int Ed 48:3273–3277
75. Panella B, Kossykh L, Dettlaff-Weglikowska U, Hirscher M, Zerbi G, Roth S (2005) Volumetric measurement of hydrogen storage in HCL-treated polyaniline and polypyrrole. Synth Met 151:208–210
76. Rose M, Böhlmann W, Sabo M, Kaskel S (2008) Element-organic frameworks with high permanent porosity. Chem Commun 2462-2464
77. Graham T (1866) On the absorption and dialytic separation of gases by colloid septa. Philos Trans R Soc Lond 156:399–439
78. Sandrock G, Suda S, Schlapbach L (1992) Applications. In: Schlapbach L (ed) Topics in applied physics vol. 67: hydrogen in intermetallic compounds II. Surface and dynamic properties, applications. Springer-Verlag, Berlin
79. Bowman RC Jr, Fultz B (2002) Metallic hydrides I: hydrogen storage and other gas-phase applications. MRS Bull 27(9):688–693
80. Sandrock G, Bowman RC Jr (2003) Gas-based hydride applications: recent progress and future needs. J Alloy Compd 356, 357:794–799
81. Fukai Y (2005) The metal-hydrogen system. Basic bulk properties, 2nd edn. Springer, Berlin
82. Alefeld G, Völkl J (eds) (1978) Topics in applied physics vol. 28: hydrogen in metals I. Basic properties. Springer-Verlag, Berlin
83. Alefeld G, Völkl J (eds) (1978) Topics in applied physics vol. 29: hydrogen in metals II. Application-oriented properties. Springer-Verlag, Berlin
84. Schlapbach L (ed) (1988) Topics in applied physics Vol. 63: hydrogen in intermetallic compounds I. Electronic, thermodynamic and crystallographic properties, preparation. Springer-Verlag, Berlin
85. Schlapbach L (ed) (1992) Topics in applied physics vol. 67: hydrogen in intermetallic compounds II. Surface and dynamic properties, applications. Springer-Verlag, Berlin

86. Wipf H (ed) (1997) Topics in applied physics vol. 73: hydrogen in metals III. Properties and applications. Springer-Verlag, Berlin
87. Sandrock G (1999) A panoramic overview of hydrogen storage alloys from a gas reaction point of view. J Alloy Compd 293–295:877–888
88. Libowitz GG, Hayes HF, Gibb TRP Jr (1958) The system zirconium-nickel and hydrogen. J Phys Chem 62(1):76–79
89. Reilly JJ, Wiswall RH (1967) The reaction of hydrogen with alloys of magnesium and copper. Inorg Chem 6(12):2220–2223
90. Reilly JJ, Wiswall RH (1968) The reaction of hydrogen with alloys of magnesium and nickel and the formation of Mg_2NiH_4. Inorg Chem 7(11):2254–2256
91. van Vucht JHN, Kuijpers FA, Bruning HCAM (1970) Reversible room-temperature absorption of large quantities of hydrogen by intermetallic compounds. Philips Res Rep 25(2):133–140
92. Griessen R, Riesterer T (1988) Heat of formation models. In: Schlapbach L (ed) Topics in applied physics vol. 67: hydrogen in intermetallic compounds I. Surface and dynamic properties, applications. Springer-Verlag, Berlin
93. Buschow KHJ, Bouten PCP, Miedema AR (1982) Hydrides formed from intermetallic compounds of two transition metals: a special class of ternary alloys. Rep Prog Phys 45:937–1039
94. Griessen R, Driessen A (1984) Heat of formation and band structure of binary and ternary metal hydrides. Phys Rev B 30(8):4372–4381
95. Yvon K (2003) Hydrogen in novel solid-state metal hydrides. Z Kristallogr 218:108–116
96. Luo S, Clewley JD, Flanagan TB, Bowman RC Jr, Wade LA (1998) Further studies of the isotherms of $LaNi_{5-x}Sn_x$-H for x = 0–0.5. J Alloy Compd 267:171–181
97. Bowman RC Jr, Luo CH, Ahn CC, Witham CK, Fultz B (1995) The effect of tin on the degradation of $LaNi_{5-y}Sn_y$ metal hydrides during thermal cycling. J Alloy Compd 217:185–192
98. Chandra D, Reilly JJ, Chellappa R (2006) Metal hydrides for vehicular applications: the state of the art. JOM 58(2):26–32
99. Ivey DG, Northwood DO (2003) Storing energy in metal hydrides: a review of the physical metallurgy. J Mater Sci 18:321–347
100. Sandrock G, Thomas G (2001) The IEA/DOE/SNL on-line hydride databases. Appl Phys A 72:153–155
101. Feng F, Geng M, Northwood DO (2001) Electrochemical behaviour of intermetallic-based metal hydrides used in Ni/metal hydride (MH) batteries: a review. Int J Hydrogen Energy 26:725–734
102. Young K, Fetcenko MA, Li F, Ouchi T (2008) Structural, thermodynamic, and electrochemical properties of $Ti_xZr_{1-x}(VNiCrMnCoAl)_2$ C14 Laves phase alloys. J Alloy Compd 464:238–247
103. Luo W, Clewley JD, Flanagan TB, Oates WA (1992) Thermodynamic characterization of the Zr-Mn-H system Part 1. Reaction of H_2 with single-phase $ZrMn_{2+x}$ C-14 Laves phase alloys. J Alloy Compd 185:321–338
104. Töpler J, Feucht K (1989) Results of a test fleet with metal hydride motor cars. Z Phys Chem NF 164:1451–1461
105. Reilly JJ, Wiswall RH (1974) Formation and properties of iron titanium hydride. Inorg Chem 13(1):218–222
106. Sakintuna B, Lamari-Darkrim F, Hirscher M (2007) Metal hydride materials for solid hydrogen storage: a review. Int J Hydrogen Energy 32:1121–1140
107. Nomura K, Akiba E (1995) H_2 Absorbing-desorbing characterization of the Ti-V-Fe alloy system. J Alloy Compd 231:513–517
108. Cho S-W, Shim G, Choi G-S, Park C-N, Yoo J-H, Choi J (2007) Hydrogen absorption-desorption properties of $Ti_{0.32}Cr_{0.43}V_{0.25}$ alloy. J Alloy Compd 430:136–141
109. Seo C-Y, Kim J-H, Lee PS, Lee J-Y (2003) Hydrogen storage properties of vanadium-based b.c.c solid solution metal hydrides. J Alloy Compd 348:252–257

110. Song XP, Pei P, Zhang PL, Chen GL (2008) The influence of alloy elements on the hydrogen storage properties in vanadium-based solid solution alloys. J Alloy Compd 455:392–397
111. Mazzolai G, Coluzzi B, Biscarini A, Mazzolai FM, Tuissi A, Agresti F, Lo Russo S, Maddalena A, Palade P, Principi G (2008) Hydrogen-storage capacities and H diffusion in bcc TiVCr alloys. J Alloy Compd 466:133–139
112. Wang J-Y (2009) Comparison of hydrogen storage properties of $Ti_{0.37}V_{0.38}Mn_{0.25}$ alloys prepared by mechanical alloying and vacuum arc melting. Int J Hydrogen Energy 34:3771–3777
113. Akiba E, Okada M (2002) Metallic hydrides III: body-centered-cubic solid-solution alloys. MRS Bull 27(9):699–703
114. Graetz J, Reilly JJ (2007) Kinetically stabilized hydrogen storage materials. Scr Mater 56:835–839
115. Hauback BC (2008) Structures of aluminium-based light weight hydrides. Z Kristallogr 223:636–648
116. Graetz J (2009) New approaches to hydrogen storage. Chem Soc Rev 38:73–82
117. Kuji T, Matsumura Y, Uchida H, Aizawa T (2002) Hydrogen absorption of nanocrystalline palladium. J Alloy Compd 330–332:718–722
118. Suleiman M, Jisrawi NM, Dankert O, Reetz MT, Bähtz C, Kirchheim R, Pundt A (2003) Phase transition and lattice expansion during hydrogen loading of nanometer sized palladium clusters. J Alloy Compd 356–357:644–648
119. Pundt A (2004) Hydrogen in nano-sized metals. Adv Eng Mater 6(1–2):11–21
120. Pundt A, Kirchheim R (2006) Hydrogen in metals: microstructural aspects. Annu Rev Mater Res 36:555–608
121. Yamauchi M, Kobayashi H, Kitagawa H (2009) Hydrogen storage mediated by Pd and Pt nanoparticles. ChemPhysChem 10:2566–2576
122. Stampfer JF Jr, Holley CE Jr, Suttle JF (1960) The magnesium-hydrogen system. J Am Chem Soc 82(14):3504–3508
123. Huot J, Liang G, Schulz R (2001) Mechanically alloyed metal hydride systems. Appl Phys A 72:187–195
124. Corey RL, Ivancic TM, Shane DT, Carl EA, Bowman RC Jr, Bellosta von Colbe JM, Dornheim M, Bormann R, Huot J, Zidan R, Stowe AC, Conradi MS (2008) Hydrogen motion in magnesium hydride by NMR. J Phys Chem C 112:19784–19790
125. Zaluska A, Zaluski L, Ström-Olsen JO (2001) Structure, catalysis, and atomic reactions on the nano-scale: a systematic approach to metal hydrides for hydrogen storage. Appl Phys A 72:157–165
126. Gross AF, Ahn CC, Van Atta SL, Liu P, Vajo JJ (2009) Fabrication and hydrogen sorption behaviour of nanoparticulate MgH_2 incorporated in a porous carbon host. Nanotechnology 20:204005
127. Aguey-Zinsou K-F, Ares-Fernández J-R (2008) Synthesis of colloidal magnesium: a near room temperature store for hydrogen. Chem Mater 20:376–378
128. Barkhordarian G, Klassen T, Bormann R (2006) Catalytic mechanism of transition-metal compounds on Mg hydrogen sorption reaction. J Phys Chem 110:11020–11024
129. Aguey-Zinsou K-F, Ares Fernandez JR, Klassen T, Bormann R (2007) Effect of Nb_2O_5 on MgH_2 properties during mechanical milling. Int J Hydrogen Energy 32:2400–2407
130. Wagemans RWP, van Lenthe JH, de Jongh PE, van Dillen AJ, de Jong KP (2005) Hydrogen storage in magnesium clusters: quantum chemical study. J Am Chem Soc 127:16675–16680
131. Grant D (2008) Magnesium hydride for hydrogen storage. In: Walker G (ed) Solid-state hydrogen storage: materials and chemistry. Woodhead Publishing, Cambridge
132. Suryanarayana C (2001) Mechanical alloying and milling. Prog Mater Sci 46:1–184
133. Suryanarayana C, Koch CC (2000) Nanocrystalline materials—current research and future directions. Hyperfine Interact 130:5–44
134. Suryanarayana C (2002) The structure and properties of nanocrystalline materials: issues and concerns. JOM 54(9):24–27

135. Bérubé V, Radtke G, Dresselhaus M, Chen G (2007) Size effects on the hydrogen storage properties of nanostructured metal hydrides: a review. Int J Energy Res 31:637–663
136. Maeland AJ, Tanner LE, Libowitz GG (1980) Hydrides of metallic glass alloys. J Less-Common Met 74:279–285
137. Orimo S, Fujii H (1998) Effects of nanometer-scale structure on hydriding properties of Mg-Ni alloys: a review. Intermetallics 6:185–192
138. Zaluski L, Zaluska A, Tessier P, Ström-Olsen JO, Schulz R (1995) Effects of relaxation on hydrogen absorption in Fe-Ti produced by ball-milling. J Alloy Compd 227:53–57
139. Bououdina M, Fruchart D, Jacquet S, Pontonnier L, Soubeyroux JL (1999) Effect of nickel alloying by using ball milling on the absorption properties of TiFe. Int J Hydrogen Energy 24:885–890
140. Harris JH, Curtin WA, Schultz L (1988) Hydrogen storage characteristics of mechanically alloyed amorphous metals. J Mater Res 3(5):872–883
141. Liang G, Huot J, Schulz R (2001) Hydrogen storage properties of the mechanically alloyed $LaNi_5$-based materials. J Alloy Compd 320:133–139
142. Hotta H, Abe M, Kuji T, Uchida H (2007) Synthesis of Ti-Fe alloys by mechanical alloying. J Alloy Compd 439:221–226
143. Abe M, Kuji T (2007) Hydrogen absorption of TiFe alloy synthesized by ball milling and post-annealing. J Alloy Compd 446–447:200–203
144. Huot J, Enoki H, Akiba E (2008) Synthesis, phase transformation, and hydrogen storage properties of ball-milled $TiV_{0.9}Mn_{1.1}$. J Alloy Compd 453:203–209
145. Parente A, Nale A, Catti M, Kopnin E, Caracino P (2008) Hydrogenation properties of Mg_2AlNi_2 and mechanical alloying in the Mg-Al-Ni system. J Alloy Compd 477(1–2): 420–424
146. Corré S, Bououdina M, Kuriyama N, Fruchart D, Adachi G (1999) Effects of mechanical grinding on the hydrogen storage and electrochemical properties of $LaNi_5$. J Alloy Compd 292:166–173
147. Fujii H, Munehiro S, Fujii K, Orimo S (2002) Effect of mechanical grinding under Ar and H_2 atmospheres on structural and hydriding properties in $LaNi_5$. J Alloy Compd 330–332:747–751
148. Ares JR, Cuevas F, Percheron-Guégan A (2004) Influence of thermal annealing on the hydrogenation properties of mechanically milled AB_5-type alloys. Mater Sci Eng B 108:76–80
149. Singh A, Singh BK, Davidson DJ, Srivastava ON (2004) Studies on improvement of hydrogen storage capacity of AB_5 type: $MmNi_{4.6}Fe_{0.4}$ alloy. Int J Hydrogen Energy 29:1151–1156
150. Takeichi N, Senoh H, Takeshita HT, Oishi T, Tanaka H, Kiyobayashi T, Kuriyama N (2004) Hydrogenation properties and structure of Ti-Cr alloy prepared by mechanical grinding. Mater Sci Eng B 108:100–104
151. Santos SF, Costa ALM, de Castro JFR, dos Santos DS, Botta WJ, Ishikawa TT (2004) Mechanical and reactive milling of a TiCrV BCC solid solution. J Metastable Nanocrystalline Mater 20–21:291–296
152. Singh BK, Shim G, Cho S-W (2007) Effects of mechanical milling on hydrogen storage properties of $Ti_{0.32}Cr_{0.43}V_{0.25}$ alloy. Int J Hydrogen Energy 32:4961–4965
153. Orimo S, Züttel A, Ikeda K, Saruki S, Fukunaga T, Fujii H, Schlapbach L (1999) Hydriding properties of the MgNi-based systems. J Alloy Compd 293–295:437–442
154. Terashita N, Takahashi M, Kobayashi K, Sasai T, Akiba E (1999) Synthesis and hydriding/dehydriding properties of amorphous $Mg_2Ni_{1.9}M_{0.1}$ alloys mechanically alloyed from $Mg_2Ni_{0.9}M_{0.1}$ (M = none, Ni, Ca, La, Y, Al, Si, Cu and Mn) and Ni powder. J Alloy Compd 293–295:541–545
155. Varin RA, Czujko T, Wronski ZS (2009) Nanomaterials for solid state hydrogen storage. Springer, New York
156. Bowman RC Jr (1988) Preparation and properties of amorphous hydrides. Mater Sci Forum 31:197–228

157. Eliaz N, Eliezer D (1999) An overview of hydrogen interaction with amorphous alloys. Adv Perform Mater 6:5–31
158. Shechtman D, Blech I, Gratias D, Cahn JW (1984) Metallic phase with long-range orientational order and no translational symmetry. Phys Rev Lett 53(20):1951–1953
159. Tsai AP (2008) Icosahedral clusters, icosaheral order and stability of quasicrystals - a view of metallurgy. Sci Technol Adv Mater 9:013008
160. Bindi L, Steinhardt PJ, Yao N, Lu PJ (2009) Natural quasicrystals. Science 324:1306–1309
161. Takasaki A, Kelton KF (2002) High-pressure hydrogen loading in $Ti_{45}Zr_{38}Ni_{17}$ amorphous and quasicrystal powders synthesized by mechanical alloying. J Alloy Compd 347:295–300
162. Takasaki A, Kelton KF (2006) Hydrogen storage in Ti-based quasicrystal powders produced by mechanical alloying. Int J Hydrogen Energy 31:183–190
163. Bystrzycki J, Polanski M, Malka IE, Komuda A (2009) Hydriding properties of Mg-Al-Zn quasicrystal powder produced by mechanical alloying. Z Kristallogr 224:105–108
164. Bogdanović B, Schwickardi M (1997) Ti-doped alkali metal aluminium hydrides as potential novel reversible hydrogen storage materials. J Alloy Compd 253–254:1–9
165. Chen P, Xiong Z, Luo J, Lin J, Tan KL (2002) Interaction of hydrogen with metal nitrides and imides. Nature 420:302–304
166. Schüth F, Bogdanović B, Felderhoff M (2004) Light metal hydrides and complex hydrides for hydrogen storage. Chem Commun 2249-2258
167. Orimo S, Nakamori Y, Eliseo JR, Züttel A, Jensen CM (2007) Complex hydrides for hydrogen storage. Chem Rev 107:4111–4132
168. Jensen C, Yang Y, Chou MY (2008) Alanates as hydrogen storage materials. In: Walker G (ed) Solid-state hydrogen storage: materials and chemistry. Woodhead Publishing, Cambridge
169. Jensen CM, Zidan R, Mariels N, Hee A, Hagen C (1999) Advanced titanium doping of sodium aluminium hydride: segue to a practical hydrogen storage material? Int J Hydrogen Energy 24:461–465
170. Zidan RA, Takara S, Hee AG, Jensen CM (1999) Hydrogen cycling behavior of zirconium and titanium-zirconium-doped sodium aluminium hydride. J Alloy Compd 285:119–122
171. Bogdanović B, Felderhoff M, Pommerin A, Schüth F, Spielkamp N (2006) Advanced hydrogen-storage materials based on Sc-, Ce-, and Pr-doped $NaAlH_4$. Adv Mater 18:1198–1201
172. Jensen CM, Gross KJ (2001) Development of catalytically enhanced sodium aluminium hydride as a hydrogen-storage material. Appl Phys A 72:213–219
173. Eberle U, Felderhoff M, Schüth F (2009) Chemical and physical solutions for hydrogen storage. Angew Chem Int Ed 48:6608–6630
174. Lohstroh W, Fichtner M, Breitung W (2009) Complex hydrides as solid storage materials: first safety tests. Int J Hydrogen Energy 34:5981–5985
175. Graetz J, Lee Y, Reilly JJ, Park S, Vogt T (2005) Structures and thermodynamics of the mixed alkali alanates. Phys Rev B 71:184115
176. Léon A, Zabara O, Sartori S, Eigen N, Dornheim M, Klassen T, Muller J, Hauback B, Fichtner M (2009) Investigation of (Mg, Al, Li, H)-based hydride and alanate mixtures produced by reactive ball milling. J Alloy Compd 476:425–428
177. Sartori S, Léon A, Zabara O, Muller J, Fichtner M, Hauback BC (2009) Studies of mixed hydrides based on Mg and Ca by reactive ball milling. J Alloy Compd 476:639–643
178. Sartori S, Qi X, Eigen N, Muller J, Klassen T, Dornheim M, Hauback BC (2009) A search for new Mg- and K-containing alanates for hydrogen storage. Int J Hydrogen Energy 34:4582–4586
179. Gregory DH (2008) Lithium nitrides, imides and amides as lightweight, reversible hydrogen stores. J Mater Chem 18:2321–2330
180. Hino S, Ichikawa T, Ogita N, Udagawa M, Fujii H (2005) Quantitative estimation of NH_3 partial pressure in H_2 desorbed from the Li-N-H system by Raman spectroscopy. Chem Commun 3038-3040

181. Uribe FA, Gottesfeld S, Zawodzinski TA Jr (2002) Effect of ammonia as potential fuel impurity on proton exchange membrane fuel cell performance. J Electrochem Soc 149(3):A293–A296
182. Gregory DH (2008) Imides and amides as hydrogen storage materials. In: Walker G (ed) Solid-state hydrogen storage: materials and chemistry. Woodhead Publishing, Cambridge
183. Züttel A, Rentsch S, Fischer P, Wenger P, Sudan Mauron P, Emmenegger C (2003) Hydrogen storage properties of $LiBH_4$. J Alloy Compd 356–357:515–520
184. Nakamori Y, Orimo S (2008) Borohydrides as hydrogen storage materials. In: Walker G (ed) Solid-state hydrogen storage: materials and chemistry. Woodhead Publishing, Cambridge
185. Walker G (2008) Multicomponent hydrogen storage systems. In: Walker G (ed) Solid-state hydrogen storage: materials and chemistry. Woodhead Publishing, Cambridge
186. Orimo S, Fujii H (2001) Materials science of Mg-Ni-based new hydrides. Appl Phys A 72:167–186
187. Blomqvist H, Rönnebro E, Noréus D, Kujii T (2002) Competing stabilisation mechanisms in Mg_2NiH_4. J Alloy Compd 330–332:268–270
188. Häussermann U, Blomqvist H, Noréus D (2002) Bonding and stability of the hydrogen storage material Mg_2NiH_4. Inorg Chem 41:3684–3692
189. Mao WL, Koh CA, Sloan ED (2007) Clathrate hydrates under pressure. Phys Today 60(10):42–47
190. Mao WL, Mao H, Goncharov AF, Struzhkin VV, Guo Q, Hu J, Shu J, Hemley RJ, Somayazulu M, Zhao Y (2002) Hydrogen clusters in clathrate hydrate. Science 297:2247–2249
191. Struzhkin VV, Militzer B, Mao WL, Mao HK, Hemley RJ (2007) Hydrogen storage in molecular clathrates. Chem Rev 107:4133–4151
192. Florusse LJ, Peters CJ, Schoonman J, Hester KC, Koh CA, Dec SF, Marsh KN, Sloan ED (2004) Stable low-pressure hydrogen clusters stored in a binary clathrate hydrate. Science 306:469–471
193. Lee H, Lee J, Kim DY, Park J, Seo Y-T, Zeng H, Moudrakovski IL, Ratcliffe CI, Ripmeester JA (2005) Tuning clathrate hydrates for hydrogen storage. Nature 434:743–746
194. Papadimitriou NI, Tsimpanogiannis IN, Papaioannou AT, Stubos AK (2008) Evaluation of the hydrogen-storage capacity of pure H_2 and binary H_2-THF hydrates with Monte Carlo simulations. J Phys Chem C 112:10294–10302
195. Papadimitriou NI, Tsimpanogiannis IN, Peters CJ, Papaioannou AT, Stubos AK (2008) Hydrogen storage in sH hydrates: a Monte Carlo study. J Phys Chem B 112:14206–14211
196. Duarte ARC, Shariati A, Rovetto LJ, Peters CJ (2008) Water cavities of sH clathrate hydrate stabilized by molecular hydrogen: phase equilibrium measurements. J Phys Chem B 112(7):1888–1889
197. Strobel TA, Koh CA, Sloan ED (2008) Water cavities of sH clathrate hydrate stabilized by molecular hydrogen. J Phys Chem B 112(7):1885–1887
198. Daschbach JL, Chang T-M, Corrales LR, Dang LX, McGrail P (2006) Molecular mechanisms of hydrogen-loaded β-hydroquinone clathrate. J Phys Chem B 110:17291–17295
199. Strobel TA, Kim Y, Andrews GS, Ferrell JR III, Koh CA, Herring AM, Sloan ED (2008) Chemical-clathrate hybrid hydrogen storage: storage in both guest and host. J Am Chem Soc 130:14975–14977
200. Yoon J-H, Lee Y-J, Park J, Kawamura T, Yamamoto Y, Komai T, Takeya S, Han SS, Lee J-W, Lee Y (2009) Hydrogen molecules trapped in interstitial host channels of α-hydroquinone. ChemPhysChem 10:352–355
201. Su F, Bray CL, Tan B, Cooper AI (2008) Rapid and reversible hydrogen storage in clathrate hydrates using emulsion-templated polymers. Adv Mater 20:2663–2666
202. Su F, Bray CL, Carter BO, Overend G, Cropper C, Iggo JA, Khimyak YZ, Fogg AM, Cooper AI (2009) Reversible hydrogen storage in hydrogel clathrate hydrates. Adv Mater 21:1–5

203. Plechkova NV, Seddon KR (2008) Applications of ionic liquids in the chemical industry. Chem Soc Rev 37:123–150
204. Olivier-Bourbigou H, Magna L, Morvan D (2010) Ionic liquids and catalysis: recent progress from knowledge to applications. Appl Catal A 373:1–56
205. Stracke MP, Ebeling G, Cataluña R, Dupont J (2007) Hydrogen-storage materials based on imidazolium ionic liquids. Energy Fuels 21:1695–1698
206. Wang L, Yang RT (2008) New sorbents for hydrogen storage by hydrogen spillover–a review. Energy Environ Sci 1:268–279
207. Cheng H, Chen L, Cooper AC, Sha X, Pez GP (2008) Hydrogen spillover in the context of hydrogen storage using solid-state materials. Energy Environ Sci 1:338–354
208. Conner WC, Falconer JL (1995) Spillover in heterogeneous catalysis. Chem Rev 95: 759–788
209. Lachawiec AJ, Qi G, Yang RT (2005) Hydrogen storage in nanostructured carbons by spillover: bridge-building enhancement. Langmuir 21:11418–11424
210. Rao CNR, Nath M (2003) Inorganic nanotubes. Dalton Trans 1-24
211. Seayad AM, Antonelli DM (2004) Recent advances in hydrogen storage in metal-containing inorganic nanostructures and related materials. Adv Mater 16(9–10):765–777
212. Ma R, Bando Y, Zhu H, Sato T, Xu C, Wu D (2002) Hydrogen uptake in boron nitride nanotubes at room temperature. J Am Chem Soc 124:7672–7673
213. Oku T, Kuno M, Narita I (2004) Hydrogen storage in boron nitride nanomaterials studied by TG/DTA and cluster calculation. J Phys Chem Solids 65:549–552
214. Chen J, Li S-L, Tao Z-L, Shen Y-T, Cui C-X (2003) Titanium disulfide nanotubes as hydrogen-storage materials. J Am Chem Soc 125:5284–5285
215. Chen J, Li SL, Tao ZL (2003) Novel hydrogen storage properties of MoS_2 nanotubes. J Alloy Compd 356–357:413–417
216. Tang C, Bando Y, Ding X, Qi S, Golberg D (2002) Catalyzed collapse and enhanced hydrogen storage of BN nanotubes. J Am Chem Soc 124(49):14550–14551
217. Bavykin DV, Lapkin AA, Plucinski PK, Friedrich JM, Walsh FC (2005) Reversible storage of molecular hydrogen by sorption into multilayered TiO_2 nanotubes. J Phys Chem B 109:19422–19427
218. Pan H, Feng YP, Lin J (2007) Hydrogen adsorption by tungsten carbide nanotube. Appl Phys Lett 90:223104
219. Lan J, Cheng D, Cao D, Wang W (2008) Silicon nanotube as a promising candidate for hydrogen storage: from the first principle calculations to Grand Canonical Monte Carlo simulations. J Phys Chem C 112:5598–5604
220. Mpourmpakis G, Froudakis GE, Lithoxoos GP, Samios J (2006) SiC nanotubes: a novel material for hydrogen storage. Nano Lett 6(8):1581–1583
221. Binewale RB, Rayalu S, Devotta S, Ichikawa M (2008) Chemical hydrides: a solution to high capacity hydrogen storage and supply. Int J Hydrogen Energy 33:360–365
222. Christensen CH, Johannessen T, Sørensen RZ, Nørskov JK (2006) Towards an ammonia-mediated hydrogen economy? Catal Today 111:140–144
223. Wiswall R (1978) Hydrogen storage in metals. In: Alefeld G, Völkl J (eds) Topics in applied physics vol. 29: hydrogen in metals II. Application-oriented properties. Springer-Verlag, Berlin
224. Bull DJ, Weidner E, Shabalin IL, Telling MTF, Jewell CM, Gregory DH, Ross DK (2010) Pressure-dependent deuterium reaction pathways in the Li-N-D system. Phys Chem Chem Phys 12:2089–2097
225. Lemmon EW, Huber ML, McLinden MO (2007) NIST standard reference database 23: reference fuid thermodynamic and transport properties-REFPROP, version 8.0, National Institute of Standards and Technology, Standard Reference Data Program, Gaithersburg
226. Hodoshima S, Arai H, Saito Y (2003) Liquid-film-type catalytic decalin dehydrogeno-aromatization for long-term storage and long-distance transportation of hydrogen. Int J Hydrogen Energy 28:197–204

Chapter 3
Hydrogen Sorption Properties of Materials

A number of factors should be considered in order to assess the suitability of a material for widespread use as a hydrogen store. The important physical and chemical properties include those related to safety, such as the pyrophoricity of the material and its toxicity, as well as those that affect the operation and hence the design of hydrogen storage systems, such as the thermal conductivity of the material in an activated state. Other practical considerations include the cost of the material and its raw constituents, and their natural abundance.[1] However, the properties of prime technological importance are those relating to hydrogen sorption, and so in this chapter we will introduce and discuss the most important of these.

In the first section we will look at practical storage properties, such as storage capacity and long term cycling stability. We shall then discuss the equilibrium thermodynamic properties that affect the temperature and pressure at which a hydrogen store can operate effectively, before covering the kinetics of the hydrogen sorption and desorption processes. We conclude the chapter by describing some of the models that can be used to fit experimental data measured under both equilibrium and dynamic conditions for both adsorbing and absorbing materials.

3.1 Practical Storage Properties

The practical storage properties covered in this section are the reversible storage capacity, including the excess and absolute adsorbed quantities for adsorbents, the long-term cycling stability, the resistance of a material to gaseous impurities in the

[1] The natural abundance is significant if a material is to be used in bulk quantities in the automotive industry [1], but less so for niche applications.

D. P. Broom, *Hydrogen Storage Materials*, Green Energy and Technology,
DOI: 10.1007/978-0-85729-221-6_3,

hydrogen fuel supply and the ease with which a material can be activated to allow the reversible storage of hydrogen.

3.1.1 Reversible Storage Capacity

The *reversible storage capacity* of a material is the quantity of hydrogen sorbed and desorbed between the lower and upper operating pressures of a hydrogen store. This figure is more technologically important than the total or maximum storage capacity. The difference between the two values will depend on the uptake behaviour of the material under consideration. The most important aspect is the isotherm shape, because this will determine the pressure range over which the majority of the reversible uptake will occur. However, it can also be affected by the rate of hydrogen uptake and release as it is possible that some of the stored hydrogen will not be released in a practical timeframe due to kinetic limitations,[2] which could also prevent full hydrogenation upon recharging. With regard to isotherm shape, we will consider two straightforward examples: an interstitial hydride with a single plateau in its isotherm and a microporous material showing typical Type I adsorption isotherm behaviour.

Let us first consider the interstitial hydride. The reversible storage capacity in this case will be determined by the width of the plateau in the Pressure–Composition Isotherm (PCI), because this is the main part of the isotherm exploited for storage purposes. If there is only a small amount of uptake in either single phase region, the reversible hydrogen storage capacity will be close to the maximum hydrogen content of the compound.[3] The reversible capacity is indicated in Fig. 3.1 using the hydrogen absorption and desorption PCIs for an intermetallic (AB_5) hydride. The PCIs shown were determined for $LaNi_5H_x$ at a temperature of 60°C (333 K), and the reversible capacity is based on charging and delivery pressures of 2 and 0.3 MPa, respectively.

The temperature of the sorption process will also, however, affect the reversible capacity. For any given hydride, including binary compounds, the absorption plateau pressure will increase with increasing temperature, and will generally follow the van 't Hoff relation, as discussed in more detail later,

$$\ln P = \frac{\Delta H}{RT} - \frac{\Delta S}{R} \tag{3.1}$$

[2] The materials representing the extreme of this case are the kinetically stabilised hydrides [2].

[3] There will always be some discrepancy because isotherms are not typically vertical in the single phase regions (on a pressure against composition plot) and, in addition, some residual hydrogen is usually trapped in the sample after dehydrogenation.

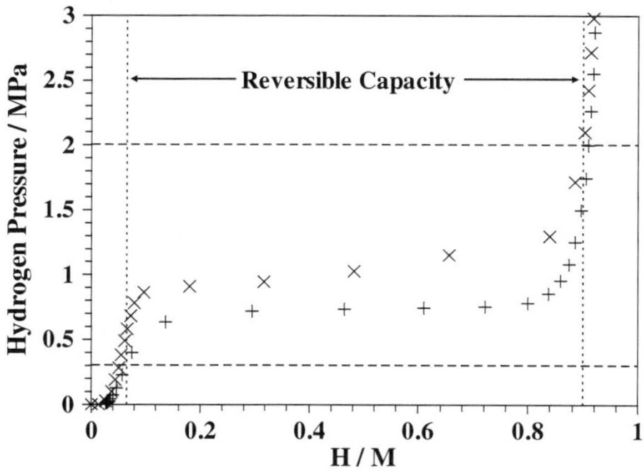

Fig. 3.1 The hydrogen absorption and desorption PCIs for $LaNi_5$ at 60°C (333 K), indicating the reversible capacity, in terms of the hydrogen-to-metal atom ratio. The capacity has been determined assuming a charging pressure of 2 MPa and a delivery pressure of 0.3 MPa, and the vertical dotted lines therefore correspond to the hydrogen content on the hydrogen absorption isotherm at 2 MPa and the content on desorption isotherm at 0.3 MPa. Data plotted from [3], with permission from Hiden Isochema Ltd

where P is the pressure, ΔH is the enthalpy of hydride formation or decomposition, ΔS is the entropy of formation or decomposition, R is the universal gas constant and T is the temperature.

As the temperature increases, the phase diagram of the hydride approaches a critical point above which there is no phase transition between the α and β phases that co-exist in the plateau region. The closer the temperature is to this critical point, the narrower the plateau region, and so therefore the lower the reversible capacity. This is illustrated in Fig. 3.2, which shows a series of isotherms measured at different temperatures for a Pd foil sample. The reversible capacity of a hydride showing this single plateau uptake behaviour therefore depends predominantly on the width of the plateau in the isotherm at the chosen operating temperature of the hydrogen store. Note that the upper temperature isotherm in Fig. 3.2 was measured at 300°C (573 K), which is just above the critical temperature of the Pd–H system, T_{crit} = 293°C (566 K) [4].[4]

Let us now consider a microporous adsorbent operating near liquid nitrogen temperature. In contrast to the shape of the hydride isotherm, a microporous material at low temperature will typically show Type I adsorption isotherm behaviour, as described in Sect. 2.1 and seen here in Fig. 3.3, which shows the hydrogen adsorption behaviour of Na–X zeolite at 87 K. As expected for Type I behaviour, the equilibrium uptake initially increases significantly as a function of pressure, but at higher pressure reaches a plateau, at which point the hydrogen

[4] The critical H/M ratio is x_{crit} = 0.29 and the critical pressure, P_{crit} = 2.015 MPa [4].

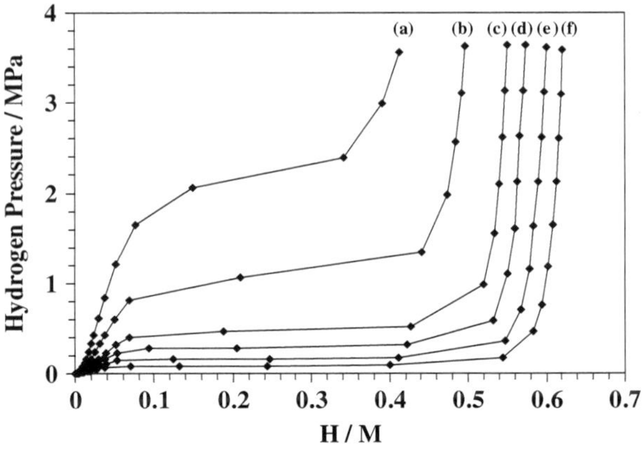

Fig. 3.2 Hydrogen absorption PCIs for a Pd foil sample measured at temperatures of (a) 300°C (573 K), (b) 250°C (523 K), (c) 200°C (473 K), (d) 175°C (448 K), (e) 150°C (423 K) and (f) 125°C (398 K). Unpublished data, reproduced with permission from Hiden Isochema Ltd

uptake saturates and will not increase further to any significant extent. In contrast to the interstitial hydride, there is no above ambient pressure plateau, over which large increases in capacity occur for relatively small pressures changes. The reversible capacity determined from the isotherm at a particular temperature therefore depends mainly on the choice of the lower storage pressure and, to a lesser extent, on the upper pressure. Figure 3.3 shows two reversible capacities, one for a lower operating pressure around ambient and one for a lower operating pressure of 0.304 MPa (3 atm), which is the lower figure identified in the current 2015 US DOE hydrogen storage targets. The reversible capacities determined using these pressures are considerably lower than the maximum storage capacity and clearly depend significantly on the operating pressure range (see Sect. 3.2.1). Although the choice of the upper and lower operating pressures is crucial, the temperature of operation also has a significant effect on the reversible storage capacity. The uptake at a chosen pressure generally decreases significantly with increasing temperature, as seen in the variable temperature hydrogen adsorption data for Na–X zeolite shown in Fig. 2.1. It is worth emphasising the larger variation in the reversible capacity of porous materials with temperature, compared to hydrides. Although there is some variation with temperature for the latter, providing the hydride phase can form, the plateau width will not vary significantly. For microporous materials, however, there is a rapid decrease in capacity with increasing temperature for any given set of upper and lower operating pressures. In reality it is likely that the practical design of a low temperature adsorptive hydrogen store would allow the temperature to be increased during desorption in order to deliver a greater amount of hydrogen than the reversible capacity indicated in Fig. 3.3. However, there is a compromise to be made between the amount of energy required to heat and cool the store and the amount of additional hydrogen that would be delivered by the store as a result of the temperature increase. The hydrogen content of the material at the lower operating pressure, or the hydrogen delivery pressure, would still dictate the maximum amount of

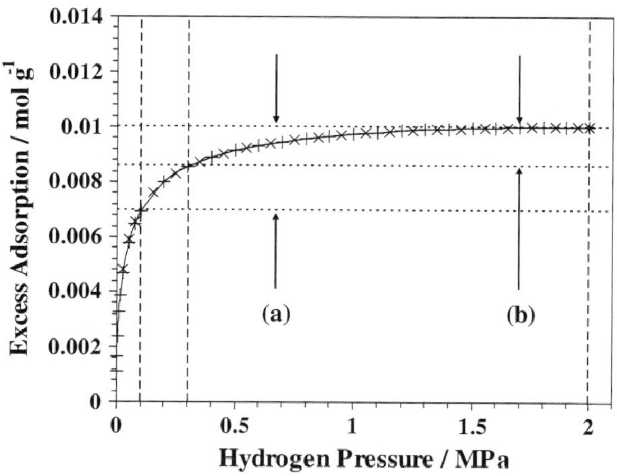

Fig. 3.3 Hydrogen adsorption (+) and desorption (×) isotherms for Na–X zeolite measured at 87 K, indicating (a) the reversible capacity for a lower storage pressure of 0.1 MPa and (b) the reversible capacity for a lower storage pressure of 3 atm (0.304 MPa). Data replotted from Broom et al. [5], with permission from Hiden Isochema Ltd

reversibly stored hydrogen in this case, but it would be determined from the isotherm at the upper operating temperature of the storage unit, rather than an isotherm at a single temperature as shown in Fig. 3.3.

In these examples, the hydride capacity is defined in terms of the hydrogen-to-metal atomic ratio and the adsorbent capacity in terms of the mass ratio of hydrogen to the host material (mol g^{-1}); without conversion these units do not express the practical storage capacity of a hydrogen storage material. Two practically useful definitions are the gravimetric and volumetric storage capacities. The definitions of both these quantities are straightforward, in principle, but in practice can be more difficult to define, and we will now take a look at each of these in turn.

3.1.1.1 Gravimetric Storage Capacity

The *gravimetric storage capacity* is the amount of hydrogen stored per unit mass of material. In the hydride case, this is a clearly defined quantity and is typically calculated from the ratio of the mass of hydrogen stored within the metal or compound to the mass of the host material including the absorbed hydrogen [6] so that the capacity in wt%, $C_{\mathrm{wt\%}}$, is given by,

$$C_{\mathrm{wt\%}} = \left(\frac{(H/M)M_H}{M_{\mathrm{Host}} + \left({}^{H}\!/_{M} \right) M_H} \times 100 \right) \% \tag{3.2}$$

where H/M is the hydrogen-to-metal or material host atom ratio, M_H is the molar mass of hydrogen, and M_{Host} is the molar mass of the host material or metal [7].

For adsorbents the analogous calculation cannot be performed because of the difficulty in defining the total adsorbed quantity of hydrogen, or the amount of hydrogen that can be attributed to the adsorbed phase,[5] as discussed further in Sect. 3.1.1.3. A typical unit for the measured quantity, the excess adsorption, is the number of moles of adsorbate per unit mass of adsorbent; for example, mol g^{-1}, mmol g^{-1}, or μmol g^{-1}.[6] As the molar mass of the adsorbate is known, this can be readily converted into a wt% uptake but this is not the same as the wt% uptake definition for hydride materials given above.

Nevertheless, a calculation equivalent to Eq. 3.2 can be performed, in which the excess adsorption is included in the denominator [9],

$$C_{\mathrm{wt\%}} = \left(\frac{n_a M_H}{n_{\mathrm{Host}} M_{\mathrm{Host}} + n_a M_H} \times 100 \right) \% \tag{3.3}$$

where n_a is the excess adsorption in moles and n_{Host} is the number of moles of host material in the sample; however, this does not account for the total amount of adsorbed hydrogen because it only uses the excess adsorbed quantity. To calculate a total gravimetric adsorption capacity, the total quantity of hydrogen in the adsorbed phase must be used, and this is where the ambiguity lies in adsorption because an assumption must be made in order to convert the excess to the total, or absolute, adsorbed quantity (see Sect. 3.1.1.3). n_a in Eq. 3.3 should therefore represent the absolute adsorption in order to define the total gravimetric adsorption capacity of an adsorbent.

3.1.1.2 Volumetric Storage Capacity

The *volumetric storage capacity* defines the amount of hydrogen stored per unit volume of the material. In the case of hydrides, this quantity is the amount of hydrogen stored per unit volume of the bulk metal.[7] Leaving aside the expansion of the host lattice due to hydrogen absorption, and any other real material phenomena, this is clearly defined, because every absorbed hydrogen atom enters the bulk of the material and is therefore contained within the boundaries of the crystal lattice. If we assume that the lattice (unit cell) volume does not change significantly with hydrogen absorption we can use the geometric density of the host compound, and the number of moles of absorbed hydrogen, for the calculation. In reality there will always be lattice expansion and so, strictly speaking, this should be accounted for in the calculation of the volumetric storage capacity.

In the case of adsorbents, the quantity is more problematic to define for equivalent reasons to the gravimetric case, as we do not know the volume of the

---

[5] The term *adsorbed phase* is used to describe the hydrogen that can be considered adsorbed and so therefore not in the bulk hydrogen gas phase.

[6] The unit recommended in the IUPAC adsorption measurement guidelines is mol g^{-1} [8].

[7] This does not, however, determine the actual *volumetric storage capacity* of a practical hydrogen store constructed using the material because this will depend on the bulk density of the storage bed (see Sect. 6.2.1).

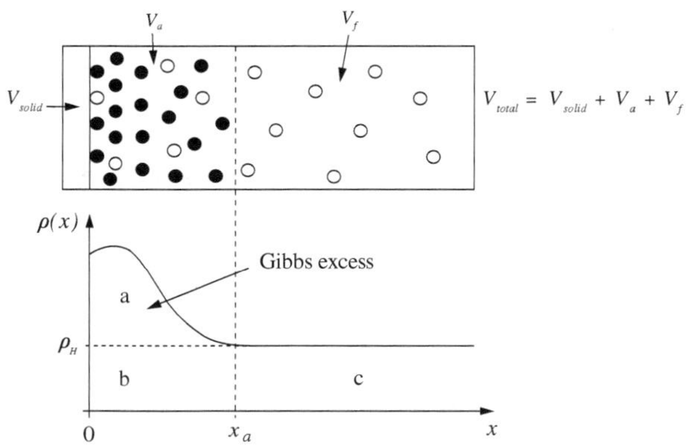

Fig. 3.4 A schematic diagram illustrating the concept of the Gibbs excess and total adsorbed quantities. The volumes of the solid, V_{solid}, the adsorbed layer, V_a, and the fluid (gas) phase, V_f, are shown. The adsorbed phase volume, V_a, is defined as the volume between the solid surface (at $x = 0$) and the boundary a distance, x_a, from the surface of the solid. The position of $x = 0$ is known as the Gibbs dividing surface. The total fluid (gas) in the system is shown in the lower plot as the region (a + b + c), the Gibbs excess is shown as region (a) and total adsorbed quantity is shown by region (a + b). In the upper part of the figure, the *white circles* indicate molecules that can be attributed to the region (b + c) and the *black circles* indicate molecules that can be attributed to the Gibbs excess (a). As the pressure of the bulk fluid increases, ρ_H increases and as the pressure decreases, ρ_H will decrease. At very low pressures, region (b) may become negligible compared to (a) and, in this case, $m_{\text{tot}} \approx m_{\text{excess}}$. At higher pressures the region (b) will become of increasing significance to the total adsorption and then $m_{\text{tot}} > m_{\text{excess}}$; if, at high pressure, the region (a) becomes insignificant compared to region (b), the surface excess will approach zero. At higher pressures still, the excess can become negative. Figure and caption adapted from Broom [7]

adsorbed phase. However, the assumption that the adsorbed phase occupies the entire volume of the pores, in what is termed below the *total pore volume approximation*, can be applied for the purpose of volumetric capacity calculation. The discussion of which assumption more closely represents the true adsorption capacity of a material is perhaps academic in the practical application of hydrogen storage materials because the most important point is to identify the material that operates most effectively in a practical storage unit. This will include other parameters such as the bulk (packing) density of the material (see Sect. 6.2.1) in the storage container. However, for the characterisation of hydrogen storage materials for storage applications, it is crucial that different materials are compared on a like-for-like basis and that a reported capacity is not increased unreasonably by an inappropriate choice of assumption.

3.1.1.3 Total and Excess Adsorption

In hydrogen adsorption measurement, the measured quantity is the *excess adsorption*. This is also known as the *Gibbs excess* or *Gibbsian surface excess*. This quantity and its relation to the total, or absolute, adsorbed quantity is illustrated in Fig. 3.4. The excess is the difference between the amount of gas phase hydrogen that would be present in the equivalent volume of the adsorbed phase, in the absence of adsorption, and the actual total amount in the phase. The total quantity is known as the *total* or *absolute adsorption*. The total adsorbed mass, m_{tot}, can be expressed as,

$$m_{\text{tot}} = \int_{V(A)} \rho(\mathbf{x})dV \tag{3.4}$$

where $\rho(\mathbf{x})$ is the gravimetric adsorbed hydrogen density at a point $\mathbf{x} = (x_1, x_2, x_3)$ and $V(A)$ is the volume of a set, A, of hydrogen molecules; in this case, all of those that are considered to be in the adsorbate phase [10]. At a sufficient, but unknown, distance from the surface of the adsorbent the hydrogen is assumed to approach an equilibrium state of constant bulk density, ρ_H,

$$\rho_H = \lim_{x \to \infty} \rho(\mathbf{x}) \tag{3.5}$$

where ρ_H can be described by an accurate equation of state for hydrogen (see Sect. 6.1.1). In Fig. 3.4, the unknown distance at which $\rho(\mathbf{x})$ approaches ρ_H is labelled x_a. The excess adsorbed mass, m_{excess}, can therefore be expressed as,

$$m_{\text{excess}} = \int_{V(A)} (\rho(\mathbf{x}) - \rho_H)dV. \tag{3.6}$$

This is the difference between the total adsorption, m_{tot}, or the total adsorbed mass, and the mass of the hydrogen that would be present in the volume of the adsorbed phase, $V_a = V(A)$, in the absence of hydrogen–surface interactions or, in other words, if there is no change in the average local hydrogen density due to the presence of the surface and therefore,

$$m_{\text{excess}} = m_{\text{tot}} - \rho_H V_a. \tag{3.7}$$

The conversion between the two quantities m_{excess} and m_{tot} requires knowledge of V_a, the volume of the adsorbed phase, which cannot be measured experimentally. The gravimetric and volumetric gas adsorption measurement methods described in the next chapter can only measure m_{excess} because they only determine the quantitative difference between the actual amount of hydrogen adsorbed and the amount of hydrogen that would be present in the absence of any interactions, which is, by definition, the excess. The conversion is therefore necessary to determine the total or absolute adsorption from experimental data.[8] This is less

[8] The reverse is also necessary in order to compare theoretical, or simulated, hydrogen adsorption capacities directly with experimental data, although this is less of a concern here because we are interested in the experimental determination of hydrogen adsorption capacity.

significant at lower pressures at which the bulk density of the hydrogen, ρ_H, is low, providing there is a significant amount of adsorption, because m_{excess} can be considered to be approximately equal to m_{tot} under these conditions. However, for hydrogen storage, which necessarily involves higher pressures, the conversion is important because ρ_H becomes more significant with increasing pressure and therefore $m_{\text{excess}} \neq m_{\text{tot}}$. The assumption that $m_{\text{excess}} \approx m_{\text{tot}}$ will generally hold if the adsorbate phase density is much greater than the bulk gas phase density. This is often the case at near and subcritical adsorptive temperatures, as the adsorbate is expected to condense to a liquid-like state on the surface, but is less likely to be the case with hydrogen adsorption at supercritical temperatures. Furthermore, V_a, and hence the phase boundary, is more clearly defined in subcritical adsorption because there is a gas and a liquid phase. At supercritical temperatures there is not necessarily such a clear, well-defined phase boundary [11].

At this point we can treat the problem using either the volumetric or the gravimetric measurement case and we will look at each of these in turn. Firstly, let us consider hydrogen adsorption measurement in a system with a known volume, V_{total}, given by,

$$V_{\text{total}} = V_{\text{solid}} + V_a + V_f \tag{3.8}$$

where V_{solid} is the solid adsorbent volume including any inaccessible pore volume, V_a is the adsorbed phase volume and V_f is the bulk fluid (gas) phase volume. We can now define the following reference quantity,

$$m_{\text{ref}} = \rho_H(V_{\text{total}} - V_{\text{solid}}). \tag{3.9}$$

This is the mass of gas phase hydrogen that would occupy the *dead volume* of the adsorption system, $V_{\text{total}} - V_{\text{solid}}$, at a density, ρ_H, in the absence of adsorption. This can be defined through the use of a reference gas, which is typically helium, using the assumption that it does not interact with the adsorbent during the reference volume measurement. This was termed the *direct route* for dead space determination by Rouquerol et al. [12]; the alternative is the *indirect route*, which involves the experimental measurement of V_{total} and the calculation of V_{solid} using the mass of the sample and an independent sample density determination measurement. In the absence of any experimental error, which is an unlikely scenario, this would have the same value as the directly determined volume. In both cases, the determined volume defines the position of $x = 0$ in Fig. 3.4, which is known as the *Gibbs dividing surface* (see Sect. 6.2.1.2).

The excess adsorption can then be obtained from,

$$m_{\text{excess}} = m_{\text{system}} - m_{\text{ref}} \tag{3.10}$$

where m_{system} is the total mass of hydrogen present in the adsorption measurement system. In this expression we do not distinguish between hydrogen in the non-adsorbed gas phase and the hydrogen in the adsorbate phase, as is the case in a practical gas adsorption measurement. To calculate the total adsorption we can then apply Eq. 3.7,

$$m_{\text{tot}} = m_{\text{excess}} + \rho_H V_a \tag{3.11}$$

using an assumed value for either V_a or $\rho_H V_a$. One simple approach is to assume that,

$$V_a = V_{\text{pore}} \tag{3.12}$$

where V_{pore} is the estimated, theoretical, or independently determined pore volume. This assumes that the volume of the adsorbed phase does not change as a function of pressure, and also relies on the accurate determination of the pore volume of the material. This assumption, termed the *total pore volume approximation* by Murata et al. [13], has been used widely in recent work on hydrogen storage by metal–organic frameworks, to the point that total adsorption is sometimes defined using this assumption.[9]

Following the approach of Murata et al. [12], the assumptions can be separated into two types: the first assumes a constant adsorbed phase volume and the second assumes a constant phase density. The total pore volume approximation mentioned above is an example of the former; another is the *effective thickness approximation*, which uses the molecular diameter of the adsorbate and the Specific Surface Area (SSA) of the adsorbent. This approximation is not recommended for microporous hydrogen storage materials because the concept of SSA is inappropriate for microporous solids.[10]

In the constant density approach, a constant value for the mean adsorbate phase density, $\langle \rho_a \rangle$, at a particular temperature, is used. A value for V_a is not required in these expressions but it can be calculated, if necessary, from,

$$V_a = \frac{m_{\text{tot}}}{\langle \rho_a \rangle}. \tag{3.13}$$

The various constant density methods then simply differ in the expressions used for the mean phase density, $\langle \rho_a \rangle$. The *critical density approximation* uses,

$$\langle \rho_a \rangle = \rho_c \tag{3.14}$$

where ρ_c is the critical density of the adsorbate. The *liquid density approximation* uses either,

$$\langle \rho_a \rangle = \rho_{\text{liq}} \tag{3.15}$$

---

9 See Murray et al. [14], a review on hydrogen storage by MOFs, in which the total adsorption is defined as the amount of hydrogen contained within the boundaries formed by the faces of the framework crystals, although this is only the case if the total pore volume approximation is used.

10 The measured SSA depends on the diameter of the adsorbate molecule and the chosen method. In addition, a 'surface layer', which does not physically form on the internal surfaces of micropores, would be three dimensional in the smallest pores. A material consisting exclusively of perfect slit-shaped pores would perhaps be an exception, but this is an idealised scenario that is unlikely to occur in practice. See Sect. 5.2.4 for further discussion of the applicability of the concept of surface area for microporous materials.

where ρ_{liq} is the liquid density of the adsorbate, or,

$$\langle\rho_a\rangle = \frac{\rho^b_{\text{liq}}}{(\exp[\alpha(T - T_b)])} \tag{3.16}$$

where ρ^b_{liq} is the liquid density at the boiling temperature, T_b, and α is a thermal expansion coefficient for superheated liquid. The *van der Waals constant approximation* uses,

$$\langle\rho_a\rangle = \frac{M}{b} = \frac{MRT_c}{8P_c} \tag{3.17}$$

where b is the van der Waals constant, M is the molar mass, R is the universal gas constant, T_c is the critical temperature and P_c is the critical pressure. In these constant density approximations, ρ_H is expressed in terms of the fugacity, f,

$$\rho_H = \frac{Mf}{RT}. \tag{3.18}$$

Using Eq. 3.13, the total or absolute adsorption, as a function of fugacity, is then given by,

$$m_{\text{tot}}(f) = m_{\text{excess}}(f) + \left(\frac{Mf}{RT}\right)\left(\frac{m_{\text{tot}}}{\langle\rho_a\rangle}\right). \tag{3.19}$$

And therefore,

$$m_{\text{tot}} = \frac{m_{\text{excess}}(f)}{\left(1 - \frac{Mf}{RT\langle\rho_a\rangle}\right)}. \tag{3.20}$$

In the treatment up to now we have assumed that we know the total volume of the adsorption system and we therefore know the number of moles of hydrogen present. We will now look at the gravimetric case, in which we do not know the volume of the system, but instead reference the derived quantities to the mass and density of the sample (T. J. Mays 2008, private communication). The experimentally measured mass, m_{exp}, in a gravimetric measurement is dependent on the mass of the adsorbent or solid, m_s, the mass of the adsorbed phase, m_{tot}, and the buoyancy correction (see Sects. 4.2.1.1 and 6.6.2), which is the product of the density of the hydrogen at the measurement temperature and pressure, and the displaced volume, V_{disp}, and so is given by $\rho_H V_{\text{disp}}$. Therefore,

$$m_{\text{exp}} = m_{\text{solid}} + m_{\text{tot}} - \rho_H V_{\text{disp}}. \tag{3.21}$$

The displaced volume is a combination of the volume of the solid, V_{solid}, including any inaccessible pores, and the volume of the adsorbed phase, V_a, and so,

$$m_{\text{exp}} = m_{\text{solid}} + m_{\text{tot}} - \rho_H(V_{\text{solid}} + V_a). \tag{3.22}$$

Therefore,

$$m_{\text{tot}} = m_{\text{exp}} - m_{\text{solid}} + \rho_H V_{\text{solid}} + \rho_H V_a. \tag{3.23}$$

If we now combine this with Eq. 3.7 we obtain,

$$m_{\text{excess}} = m_{\text{exp}} - m_{\text{solid}} + \rho_H V_{\text{solid}}. \tag{3.24}$$

Now, $V_{\text{solid}} = \frac{m_{\text{solid}}}{\rho_{\text{solid}}}$, where ρ_{solid} is the known or independently determined density of the sample, and so,

$$m_{\text{excess}} = m_{\text{exp}} - m_{\text{solid}} + \rho_H \left(\frac{m_{\text{solid}}}{\rho_{\text{solid}}} \right). \tag{3.25}$$

If we then substitute Eq. 3.13 into Eq. 3.7 we get,

$$m_{\text{excess}} = m_{\text{tot}} - m_{\text{tot}} \left(\frac{\rho_H}{\langle \rho_a \rangle} \right) \tag{3.26}$$

and so,

$$m_{\text{tot}} = \frac{m_{\text{excess}}}{\left(1 - \frac{\rho_H}{\langle \rho_a \rangle} \right)}. \tag{3.27}$$

Combining this with Eq. 3.25 we obtain,

$$m_{\text{tot}} = \frac{m_{\text{exp}} - m_{\text{solid}} + m_{\text{solid}} \left(\frac{\rho_H}{\rho_{\text{solid}}} \right)}{\left(1 - \frac{\rho_H}{\langle \rho_a \rangle} \right)}. \tag{3.28}$$

This is the equivalent of Eq. 3.20 but m_{tot} is expressed using the experimental parameters from a gravimetric measurement. If we assume that $\langle \rho_a \rangle > \rho_H$, which must be the case if adsorption occurs, then we calculate the total adsorbed quantity with a constant density approximation. The approximation then just depends on the choice of $\langle \rho_a \rangle$. It can be seen that the lower the assumed adsorbed phase density the greater the calculated m_{tot}. An unrealistically low value of $\langle \rho_a \rangle$ will therefore lead to an overestimated total adsorbed quantity, and hence an overestimated reversible hydrogen storage capacity. The practical problem is clearly that $\langle \rho_a \rangle$ is unknown. A typical assumed value is the liquid density of hydrogen, 0.07 g cm^{-3}, although it is not clear whether this assumption is physically reasonable at supercritical temperatures.[11]

[11] See Fig. 6.2, Sect. 6.6.2, for an example of this correction applied to hydrogen adsorption data for an activated carbon.

3.1.2 Long-Term Cycling Stability

The *long-term cycling stability* of a material defines its ability to retain its reversible storage capacity during repeated hydrogen charge and discharge cycles. The 2015 US DOE storage targets specify an ability to undergo 1,500 hydrogenation/dehydrogenation cycles. Factors affecting the cycling stability of a material can be complex and will vary greatly between different material types. The problem of poor cycling stability is more significant for hydrides than it is for porous adsorbents, because metal hydrides invariably undergo physical and chemical degradation during prolonged hydrogen cycling. This is true even for the compounds that are the most resistant to degradation. If a host hydrogen adsorbent, on the other hand, is thermally stable, and therefore does not decompose at the operating or hydrogen cycling temperature, the cycling stability of the material, using pure hydrogen, is likely to be high. If the hydrogen contains gas phase impurities, preferential adsorption of the contaminants will reduce the capacity. The affinity for any given material towards gaseous impurities in the hydrogen fuel, however, is discussed in Sect. 3.1.3. Therefore, we shall concentrate here on the cycling-induced degradation of metal hydrides.

The different metal hydride degradation mechanisms can be classified as either *intrinsic* or *extrinsic*. The former primarily refers to *disproportionation*, the process by which a ternary or higher hydride decomposes into more thermodynamically stable decomposition products, but defect formation also appears to play an important role; extrinsic mechanisms, meanwhile, are those induced by impurities in the gas phase hydrogen, and can include poisoning and corrosion of the surface [6, 15]. Extrinsic degradation mechanisms relate to the presence of impurities in the gas phase, which is addressed next, and so we will concentrate in this section on disproportionation and the effects of defect formation.

As stated above, disproportionation is the process by which a ternary or higher hydride decomposes into more thermodynamically stable decomposition products. For example, in the $LaNi_5$–H system, LaH_2 + Ni is more thermodynamically stable than $LaNi_5H_x$. Therefore, providing the metal atoms are sufficiently mobile within the host matrix, segregation into pure Ni and La hydride will tend to occur during the hydrogen cycling process, or there will at least be some separation into Ni-rich and La-rich regions. The mobility of the metal atoms, however, is crucial in governing the extent to which this process occurs and so the effects are partially dependent on the temperature at which hydrogen cycling is performed. Although disproportionation occurs during cycling, Sandrock et al. [16] found that neither hydrogen pressure cycling nor thermal cycling in a hydrogen atmosphere are required, providing the material is held in the ordered hydride (β) phase at sufficient temperature. They observed significant disproportionation of $LaNi_5$ when it was held in the β phase at 180°C (453 K). For this material, the degree of phase separation can be reduced by partial substitution of the Ni. It has been suggested that the mobility of the host metal atoms is reduced by the presence of these substituent atoms in the lattice [17]. An example is the presence of the larger

Al atom in the 3 g lattice position in the $LaNi_{5-x}Al_x$ structure. This site is normally only occupied by Ni in pure $LaNi_5$, but it has been proposed that the Al prevents, or at least reduces, the mobility of the Ni through the B component sub-lattice. This reduces the disproportionation of the host into LaH_2 and Ni, because the Ni cannot diffuse so readily. Both $LaNi_{5-x}Sn_x$ and, to a lesser extent, $LaNi_{5-x}Al_x$ show greatly improved long term cycling stability over that of the parent $LaNi_5$ compound. In a study by Lambert et al. [17] another AB_5 compound, $La_{0.9}Gd_{0.1}Ni_5$, which showed an 80% capacity loss over 10,000 cycles, was found by X-ray powder diffraction to contain a Ni phase whereas in hydrogen-cycled $LaNi_{4.8}Sn_{0.2}$, which showed less than 15% capacity loss after more than 10,000 cycles, pure Ni was not observed. Therefore, from this evidence, disproportionation appears to be the main origin of the loss of capacity with extended hydrogen cycling in these materials. In a subsequent study by Bowman et al. [18], $LaNi_{4.8}Sn_{0.2}$ was shown to have an improved resistance to degradation, in comparison to pure $LaNi_5$, by a factor of 20. They detected nanocrystalline LaH_x and a Ni metal phase in cycled $LaNi_5$ and $LaNi_{4.9}Sn_{0.1}$, which supports disproportionation as the primary degradation process. A significantly lower amount of a Ni metal phase was detected in the cycled $LaNi_{4.8}Sn_{0.2}$ sample, but no LaH_x phase was identified. A similar process is observed in other intermetallic hydrides. For example, Lee and Lee [19] attributed the cycling-induced degradation of an AB_2 Laves phase intermetallic, $Zr_{0.9}Ti_{0.1}Cr_{0.9}Fe_{1.1}$, to the formation of a more stable crystalline Cr-rich hydride phase.

The degradation of the hydrogen storage capacity of metal hydrides also appears to be linked to the formation of dislocations during initial activation and extended cycling, although the complex relationship between the crystallography and microstructure of the materials, and the degradation behaviour is not fully understood [20]. However, the wide variation in dislocation densities between pure $LaNi_5$ and a number of its partially substituted derivatives, following hydrogen cycling, has been identified by analysis of the anisotropic neutron and X-ray powder diffraction peak broadening exhibited by these materials [15, 21–24]. High vacancy concentrations in these and other intermetallic hydrides have also been identified by Positron Annihilation Spectroscopy (PAS) [25]. Compounds such as $LaNi_{5-x}Sn_x$ and $LaNi_{5-x}Al_x$, which exhibit the highest long term hydrogen cycling stabilities, appear to contain very low dislocation densities in comparison to pure $LaNi_5$. Given that disproportionation relies on metal atom mobility [16] and that dislocation cores are known to provide enhanced solid state diffusion paths[12] [26], it seems likely that the suppression of dislocation formation could play a significant role in the reduction in disproportionation and hence the enhanced long term cycling stability of some of the intermetallic hydrides, although this point has not been widely discussed in the literature.

The long term cycling stability of the complex hydrides has not been studied to the same extent as that of the intermetallics, due primarily to the shorter history

[12] So-called *pipe diffusion*.

of hydrogen storage research into these materials. While impressive extended cycling studies of the performance of intermetallic hydrides have been published,[13] similar studies have not yet been carried out on their complex hydride counterparts. A number of shorter term studies have, however, been performed and indicate that a number of different degradation mechanisms exist. These include the loss of the reversible storage capacity of lithium amide due to ammonia evolution [28] and the evolution of boranes in the case of borohydrides [29]. $TiCl_3$-doped sodium alanate, meanwhile, appears to exhibit good cycling stability over the course of 100 cycles [30]. In a recent review, Chandra [31] covered both the intrinsic and extrinsic degradation of hydrides and presented some provisional results on $LiNH_x$, showing the cycling stability over 1,100 cycles. The system lost approximately 25% of its initial capacity over the first 500 cycles, and had lost 2.53 wt% of the initial reversible capacity of approximately 5.6 wt% after 1,100 cycles. However, this practical cycling test used industrial hydrogen of a minimum 99% purity, containing impurities at the following approximate levels: 32 ppm of water, 10 ppm of oxygen, 400 ppm of nitrogen, 10 ppm of hydrocarbons, 10 ppm of carbon monoxide, and 10 ppm of carbon dioxide. Therefore, the degradation observed may have been more influenced by the gaseous impurities present (see Sect. 3.1.3) than the intrinsic cycling stability of the material.

Nevertheless, the important points to be noted are, firstly, that the cycling stability due to intrinsic degradation of a given material is strongly dependent on its physical and chemical properties; secondly, that long term cycling stability is a crucial performance criterion for hydrogen storage applications; and, thirdly, that this property can be practically investigated by determining the equilibrium hydrogen sorption properties before, during and after the hydrogen cycling process. There is no standard approach to describing the cycling stability of a material but a capacity loss in wt% or as a percentage of the maximum reversible capacity, over any given number of cycles, both appear to give a good indication of the resistance of a material to degradation. The English translation of a Japanese Industrial Standard (JIS) glossary of terms for hydrogen absorbing alloys defines the *degree of degradation* as 'the degree of decrease of the initial storage capacity of hydrogen absorbing alloys due to the number of repetitions or elapsed time' and the *cyclic durability with hydrogen absorption and desorption* as 'the change of the moving quantity of hydrogen depending on the number of cycles when hydrogen and hydrogen absorbing alloy are repeatedly reacted' [32]. In addition, the determination of the cycling stability of hydrogen absorbing alloys is covered by a separate JIS [33].

[13] See, for example, the study of the performance of three AB_5 compounds by Wanner et al. [27] in which each sample was subjected to approximately 95,000 thermally induced hydrogen absorption/desorption cycles.

Table 3.1 The impurities found in hydrogen fuel produced using different production methods or produced from different feedstocks [35]

Feedstock	Potential impurities
Crude oil	CO, NH_3, H_2S, HCN
Gasolines	Hydrocarbons, aldehydes
Diesels	Mercaptans
Natural gas	CO, NH_3, H_2S, HCN, hydrocarbons, mercaptans
Methanol/DME	CO, odorants, alcohols
Biomass	Cations, aldehydes, alcohols, formic acid, NH_3, H_2S, HCN
Water electrolysis	Anions, cations

3.1.3 Gaseous Impurity Resistance

Another important property is the resistance of a material to gaseous impurities in the hydrogen fuel supply. As mentioned in Sect. 1.5, hydrogen fuel can contain a variety of contaminants, and hydrogen purity specifications for fuel cell testing purposes include varying levels of contamination by CO, CO_2, H_2O, H_2S, NH_3, O_2, hydrocarbons, formaldehyde, formic acid and halogenates [34, 35]. Table 3.1 shows some of the common impurities found in hydrogen produced from different feedstocks [35]. Accurate hydrogen sorption measurements should ideally use the highest purity hydrogen possible. This will typically involve research grade hydrogen, perhaps combined with filtration if deemed necessary (see Sect. 6.2.1). However, although this will help ensure accurate characterisation of the storage capacity of a material, assessment of the reaction of a material to gaseous impurities is also essential in assessing its practical capability as a hydrogen store.[14]

In the case of microporous adsorbents this will primarily involve the tendency of the material to preferentially adsorb impurity species but could also involve the breakdown of the material. However, multicomponent gas adsorption measurements are challenging and the control of the partial pressures of various constituents in a gas mixture at impurity levels is difficult, although premixed gases can be used. It is possible to calculate the preferential adsorption of various contaminant adsorptives using the isotherms of individual species, using methods such as the Ideal Adsorbed Solution Theory (IAST) [36]. It is also worth noting that different adsorbents will have differing affinities for certain contaminant species. So, for example, hydrophilic zeolites are more likely to be affected by moisture contamination than a hydrophobic carbon. In the case of low temperature adsorptive storage, the temperature of operation will ensure that any moisture will have been frozen from the supply gas; however, if adsorbents are developed for ambient temperature operation, this will not be the case.

[14] If a good understanding of the hydrogen sorption properties of a material is to be obtained, its performance using high purity hydrogen should be determined initially before the controlled introduction of impurities. The performance using high purity hydrogen will then act as a baseline measurement against which future comparisons can be made.

Relatively few studies have been performed specifically on the effect of gaseous impurities on the hydrogen storage capabilities of microporous adsorbents. An example, however, is the work of Amankwah and Schwarz [37], who examined the effect of nitrogen as an impurity on the storage capacity of activated carbon using gravimetric measurements. They found that the hydrogen capacity of the super-activated carbon AX-31M was reduced by the presence of nitrogen at a level of 500 ppm. They found that the reduction reached 30% at pressures above approximately 5.0 MPa at a working temperature of 150 K, although the effect was lower at 195 K with an average reduction of 10% throughout the pressure range up to 5.0 MPa.

Although studies of the effects of gaseous impurities on the hydrogen adsorption capacity of materials for storage purposes are scarce, many porous adsorbents are used in gas separation processes [38] and therefore useful insights can be gleaned from studies in this research field. As an example, in a review of the use of activated carbon for gas separation and storage, Sircar et al. [39] show the preferential adsorption of CO_2 over H_2 for a range of activated carbons and 5A zeolite, although the carbons used are predominantly mesoporous and are therefore unlikely to be considered as effective hydrogen storage media. The data are also for higher temperatures (303 K) than those considered for adsorptive hydrogen storage; however, it can be seen that the preferential adsorption of CO_2, present as an impurity, by these materials will affect the hydrogen storage capacity at near ambient temperatures. Sircar et al. [39] also describe the purification of hydrogen using Pressure Swing Adsorption (PSA), involving the removal of CO_2, CH_4, CO and N_2 from a H_2-rich gas stream. The beds used in such a process[15] contain an activated carbon and 5A zeolite, which are chosen because they preferentially adsorb all four species mentioned above over H_2; the selectivities for the carbon are highest for CO_2, followed by CH_4, CO and then N_2, and for the zeolite, CO_2, CO, CH_4 and N_2. The temperature is again high in comparison to the low temperatures at which these materials store significant amounts of hydrogen but their tendency to adsorb these species in preference to hydrogen is the important point and this shows that preferential gaseous impurity adsorption could significantly reduce the hydrogen storage capacity of porous adsorbents. The use of a porous carbon to remove C_4H_{10}, C_3H_8, C_2H_6 and CH_4 from a hydrogen stream is also described, showing again the preference for the adsorption of other species over hydrogen.

Similarly, metal–organic frameworks have more recently been proposed for use in separation applications and some of the adsorption data from this literature can be used to provide useful information on their interaction with potential contaminants. Li et al. [40] recently reviewed selective gas adsorption by MOFs and a number of studies of the difference between the uptake of hydrogen and other gases by MOFs can be found in the literature [41–44]. However, the work so far understandably appears to concentrate on the interaction of different adsorptives

[15] Further details can be found in Yang [38].

Table 3.2 A summary of some example studies of the effects of gaseous impurities on hydrogen storage by metal hydrides

Host material	Impurities	Reference
$LaNi_5$, FeTi, $Fe_{0.85}Mn_{0.15}Ti$	CO, O_2, H_2O	[45]
$LaNi_5$, $Fe_{0.85}Mn_{0.15}Ti$	N_2, CH_4, CO, CO_2, O_2, H_2S, NH_3, C_2H_2, CH_3SH	[46]
$LaNi_5$, FeTi	N_2, CH_4, CO, CO_2, H_2S	[47]
$LaNi_5$, $LaNi_{4.7}Al_{0.3}$	CO	[48]
$LaNi_5$, $LaNi_{4.7}Al_{0.3}$	O_2	[49]
$LaNi_{4.7}Al_{0.3}$ (untreated and fluorinated)	CO	[50]
$LaNi_{4.9}Al_{0.1}$	CO	[51]
$LaNi_5$, $LaNi_{4.7}Al_{0.3}$, $MmNi_{4.5}Al_{0.5}$	CO, O_2	[52, 53]
$LaNi_5$	N_2, CO, CO_2, O_2, H_2S	[54]
$LaNi_{5-x}Al_x$ ($x = 0, 0.1, 0.3, 0.5$)	O_2, H_2O	[55]
$Fe_{0.85}Mn_{0.15}Ti$, $Fe_{0.85}Ni_{0.15}Ti$	CO	[48]
$Ti_{0.98}Zr_{0.02}Cr_{0.05}V_{0.43}Fe_{0.09}Mn_{1.5}$	N_2, CH_4, CO, CO_2, O_2 H_2O	[56]
Mg	N_2, CO, CO_2, O_2	[57]
Mg_2Ni	CO	[48]
Mg–V composite (5 at.% V)	N_2, CO_2, O_2	[58]

with the materials best suited to separation applications, rather than the multi-component adsorption behaviour of the most suitable materials for hydrogen storage. The hydrogen storage literature has also focused on the search for new high capacity adsorptive storage materials. It is likely, however, that the gaseous impurity resistance of the best performing storage materials will be the subject of future research, as this is crucial to their practical application in hydrogen storage.

In contrast, there have been a number of studies on the effects of various gaseous impurities on the performance of interstitial metal hydrides, some of which are listed in Table 3.2. The interaction of different impurities with these materials results in a range of effects. Sandrock and Goodell [46] proposed four types of interaction that are responsible for extrinsic degradation. The first is *poisoning*, which results in a rapid loss of hydrogen storage capacity without a decrease in the kinetics of the unaffected portion of the sample. Examples of poisoning given by Sandrock and Goodell [46] are the effects of CH_3SH (methyl mercaptan) on the hydrogen storage properties of $LaNi_5$ and Ti(Fe, Mn), and the effect of H_2S on $LaNi_5$. The second effect is *retardation*, which reduces the kinetics of hydrogen absorption but does not reduce the storage capacity. An example is the effect of NH_3 on $LaNi_5$. *Reaction* is the third type of interaction in which the capacity is reduced by bulk corrosion of the alloy. This is exemplified by the effects of long term cycling of $LaNi_5$ in oxygen-contaminated hydrogen. A fourth effect, which is defined as *innocuous*, does not damage the surface but decreases the sorption kinetics due to inert gas blanketing. This effect is seen in the hydrogen sorption properties of Ti(Fe, Mn) and $LaNi_5$ determined using hydrogen contaminated with CH_4, C_2H_4 or N_2. The JIS glossary of terms [32], meanwhile, defines *poisoning* more generally as 'a phenomenon in which a hydrogen absorption characteristic is deteriorated by adsorption on an alloy surface or reaction on the alloy of substances other than hydrogen such as water which are supplied together with hydrogen gas'.

We shall now briefly consider a specific example of the contamination of a material by a gaseous impurity. Goodell [49] studied the effects of oxygen on the cycling stability of $LaNi_5$ and $LaNi_{4.7}Al_{0.3}$. For $LaNi_5$ the degradation was found to be dependent on the oxygen content, with tested impurity levels of 0.03, 0.1 and 0.5%. The highest concentration caused a significantly greater decrease in the hydrogen storage capacity over the course of nearly 300 cycles, in comparison to the lower levels, and the degradation was also greater at higher temperatures (358 K compared to 298 K). Although this general summary is true, the response to the presence of oxygen was complex, with three observed stages: initial retardation, recovery and oxidation decay. The $LaNi_{4.7}Al_{0.3}$ alloy showed greatly improved long term cycling resistance in tests with 0.1% oxygen levels. The partial substitution of Ni by Al therefore appears to enhance the resistance of the $LaNi_5$-based materials to both intrinsic and extrinsic degradation.

In the case of the complex hydrides, moisture can break down the host material itself but, as for intrinsic degradation, studies on the effects of gaseous impurities on the complex hydrides are less common, although some of the practical aspects of hydrogen storage in sodium alanate have been addressed in the literature [59]. See Sect. 3.1.2 for a brief description of the work of Chandra [31] on the degradation of lithium amide and imide using industrially pure hydrogen.

From a practical point of view, if a material in a solid state storage system is affected or damaged by gaseous impurities, the hydrogen delivered to the store can be filtered or purified to avoid this consequence. It can be argued therefore that poor gaseous impurity resistance can be addressed in the design of the storage unit. However, purification inevitably adds to the cost and bulk of the system and so it is preferable for the material itself to exhibit a high impurity resistance to reduce potential problems that must otherwise be considered in the engineering of the storage unit. Any susceptibility to gas phase contamination is therefore a significant disadvantage and must be taken into account in the assessment of the suitability of a material for reversible hydrogen storage.

3.1.4 Ease of Activation

Most hydrogen sorbing materials require *activation*, of one kind or another, which is the process by which a sample is prepared for the reversible adsorption or absorption of hydrogen.[16] In the case of microporous adsorbents, this process

[16] The definition of *activation* in the JIS glossary of terms is 'a promotion of reaction for absorption and desorption of hydrogen absorbing alloys' [32].

essentially involves the removal of any environmental adsorbates from the internal pore structure and the external surfaces of the solid. If a material has been synthesised using wet chemistry methods any solvents used during the synthesis process must also be removed. Generally speaking, both environmental contaminants and remnants from the synthesis process can be removed by exposing the sample to vacuum at elevated temperatures, although samples can also be flushed through with an inert gas, which can also help remove pre-adsorbed species. Therefore, in the case of adsorbents, activation is a fairly straightforward process and generally involves heating and evacuating the adsorbent. On an experimental level this process involves monitoring either the weight change during degassing in a gravimetric system or the decrease in the pressure above the vacuum pump system in volumetric apparatus (see Sect. 6.5.1).

In the case of hydrides the situation is more complex [60]. Practically speaking, activation can involve the exposure of a material to a pressure of hydrogen at a particular, possibly elevated, temperature or the extended temperature and pressure cycling of a material to achieve practical and repeatable reversible absorption and desorption behaviour. As with microporous materials, the surface must first be cleaned to prepare for the initial reaction with hydrogen. However, rather than simply adsorbing molecular hydrogen onto its internal surface, the material will typically undergo significant changes in its physical properties. In the case of any brittle host compound this will involve the decrepitation of the bulk host material into a fine powder, with the result that fresh metallic surfaces are exposed for the dissociation of molecular hydrogen, allowing the subsequent absorption of atomic hydrogen. Although the precise mechanisms involved in the initial activation of a metallic absorber are not yet fully understood, the surface segregation of elemental metallic particles has been proposed as part of the process in a number of cases, including the segregation of metallic Ni in the case of $LaNi_5$ and Mg_2Ni, and metallic Fe in the case of TiFe. The segregated metallic particles are likely to act as an effective catalyst for the dissociation of molecular hydrogen. Although this seems to be the case in $LaNi_5$ and, to a lesser extent, Mg_2Ni there is contradictory evidence in the case of Fe-containing compounds. For TiFe, which is more challenging to activate than the AB_5 compounds, an alternative mechanism has been proposed, whereby initial bulk hydride phase formation results in the creation of cracks in the surface oxide layer. These cracks then allow the easier passage of hydrogen into the TiFe bulk. According to Manchester and Khatamian [60] this is the more likely mechanism for the activation process in this case. Other possible mechanisms involve the dissolution of the surface oxide layer and the enhancement of the permeability of the oxide layer due to the defect structure of the TiO_2 on the surface.

However, regardless of the exact mechanism involved, the important point is that the activation process is governed by the particular properties of the hydride-forming compound and this is an important practical consideration. A complicated and prolonged activation procedure can limit the practical use of a material or add economic constraints to its widespread use. Adsorbents do not have this disadvantage as they simply require degassing. With regard to the activation of recently proposed or developed absorbing materials, as with prolonged cyclic stability testing, many of these have not yet been studied thoroughly, and the detailed mechanisms will undoubtedly be the subject of future research.

3.2 Thermodynamic Properties

The thermodynamic properties of a storage material govern its operating temperature and pressure range. They must be considered in conjunction with the reversible storage capacity and the kinetic hydrogen sorption properties of a material, but are a fundamental measure of the hydrogen storage properties of a particular host, whether an adsorbent or an absorber. For many years, research in the hydride field has focused on the prediction of the formation and decomposition enthalpies of new hydrides through empirical, semi-empirical and first principles methods. Meanwhile, recent research on adsorbents for hydrogen storage has concentrated on increasing adsorption enthalpies in existing materials, in order to increase the temperatures at which a porous material will reversibly adsorb or desorb molecular hydrogen. In this section we will look at each of these properties in turn. From a practical point of view, the amount of heat generated by the hydrogen sorption process is another important thermodynamic consideration, as emphasised in a recent review by Eberle et al. [29]. In an adsorption system operating at low temperatures, the result will be the evaporation of the coolant, which presents an engineering challenge and would add to the economic cost of this hydrogen storage method. An absorption system, meanwhile, requires a heat exchanger to deal with the significant amount of heat generated while recharging the store in a short period of time.

3.2.1 Enthalpy of Adsorption

The enthalpy of adsorption provides a measure of the strength of the interaction between a molecule and the adsorbent surface or pore structure. The temperature at which hydrogen will physisorb onto a flat surface is low and of no practical use for storage purposes. Micropores result in the overlap of the potential fields from opposing pore walls and a subsequent increase in the density of the adsorbed hydrogen at any given temperature and pressure. Furthermore, recent work has included the investigation of the possibility of increasing the adsorption enthalpy of hydrogen by metal–organic frameworks through the interaction of hydrogen with exposed metal sites in the framework structure (see Sect. 2.1.3).

The enthalpy of adsorption can be defined in a number of ways. Rouquerol et al. [12] identify several definitions, including the transformed molar surface excess enthalpy, transformed differential surface excess enthalpy, transformed differential enthalpy, transformed integral molar enthalpy, differential surface excess enthalpy, and the differential enthalpy of adsorption. It is the last of these that is typically determined experimentally from the measurement of two or more isotherms at different, but not greatly separated, temperatures using the *isosteric method*. The differential enthalpy of adsorption, measured and determined for a particular uptake, or 'surface coverage', is then known as the *isosteric enthalpy of adsorption*,[17] ΔH_{iso}.

There are a number of ways of determining ΔH_{iso} and the result will depend to a certain extent on the chosen method. In each case, the pressure at which a fixed amount of hydrogen is adsorbed at different temperatures is required. One method is to apply the following expression,

$$\Delta H_{iso} = -\frac{RT_1T_2}{T_2 - T_1}\ln\left(\frac{P_2}{P_1}\right) \tag{3.29}$$

where T_1 and T_2 are two closely spaced measurement temperatures, P_1 and P_2 are the pressures at which a given quantity of hydrogen is adsorbed, and R is the universal gas constant [12, 61]. 10 K is a typical temperature difference, so that $T_2 - T_1 = 10$ K, and, due to the convenience of using liquid nitrogen and liquid argon, $T_1 = 77$ K and $T_2 = 87$ K are often chosen. In order to interpolate between data points it is necessary to fit the adsorption data to an appropriate adsorption equation (see Sect. 3.4.1). The method employed by Panella et al. [61] used the Langmuir equation to fit adsorption data that showed some scatter at higher pressures. Values of -3.8 and -4.5 kJ mol^{-1} H_2 were obtained for MOF-5 (IR-MOF-1) and Cu-BTC, respectively, with an estimated error of ±0.8 kJ mol^{-1} H_2. Alternatively, if isotherms are available at a series of temperatures, a plot of the natural log of the pressures at a fixed coverage or fixed absolute adsorbed quantity can be plotted against $1/T$, to produce a van 't Hoff isochore. The gradient gives ΔH_{iso}, according to the van 't Hoff relation (see Eq. 3.1 and Sect. 3.2.2.1).

Chen et al. [62] recently compared three methods of determining the isosteric enthalpy of adsorption for hydrogen and deuterium adsorption by a mixed zinc and copper metal–organic framework.[18] Two of these used virial equations to fit the adsorption data and the third used the Langmuir–Freundlich equation (see Sect. 3.4.1). The first method involves fitting the following virial equation to the adsorption isotherm [12],

$$\ln\left(\frac{n}{P}\right) = A_0 + A_1 n + A_2 n^2 + \cdots \tag{3.30}$$

where n is the amount adsorbed, P is the pressure and A_0, A_1 and A_2 are constants. The fitted parameters from this expression are then used for the interpolation and the van 't Hoff relation applied.

---

[17] This is also known as the 'isosteric heat', although this terminology is less precise and so 'isosteric enthalpy of adsorption' is preferable.

[18] $Zn_3(BDC)_3[Cu(Pyen)]$, where H_2BDC = 1,4 benzenedicarboxylic acid and $PyenH_2$ = 5-methyl-4-oxo-1,4-dihydro-pyridine-3-carbaldehyde.

The second method used a different virial equation, which was introduced by Czepirski and Jagiełło [63] and has since been used to fit low pressure (sub-ambient) hydrogen adsorption data for a range of materials, including carbon nanotubes [64, 65], activated carbon [65] and metal–organic frameworks [66, 67]. The equation is given by,

$$\ln(P) = \ln(n) + \left(\frac{1}{T}\right)\sum_{i=0}^{m} a_i n^i + \sum_{j=0}^{n} b_j n^j \tag{3.31}$$

and,

$$\Delta H_{\text{iso}} = -R\sum_{i=0}^{m} a_i n^i \tag{3.32}$$

where P is the pressure, n is the amount adsorbed, T is the temperature, R is the universal gas constant, and a_i and b_j are temperature independent empirical parameters.

The third method used the Langmuir–Freundlich equation (see Eq. 3.47 and Sect. 3.4.1) to fit the adsorption isotherm data, and a modification of the Clausius–Clapeyron equation. They conclude that the first method provided the most accurate description of the hydrogen and deuterium adsorption isotherms and so the calculation of ΔH_{iso} using this method was favoured.

Another quantity of interest is the *differential enthalpy of adsorption at zero coverage*, ΔH_{zero}. For a heterogeneous adsorbent the differential enthalpy will decrease as a function of surface coverage[19] because at low surface coverages, or rather low loadings, the adsorbate will interact initially with the most active surface sites, or enter the most narrow pores. The differential enthalpy of adsorption at zero coverage can be determined from the variation in the Henry's law constant with temperature. Henry's law describes the linear relationship between uptake and pressure at low surface coverage, and is given by,

$$n = k_H P \tag{3.33}$$

where k_H is known as the *Henry's law constant*. The differential enthalpy of adsorption at zero coverage is then given by [12],

$$\Delta H_{\text{zero}} = RT^2\left(\frac{\partial(\ln k_H)}{\partial T}\right). \tag{3.34}$$

If the data are fitted using the virial equation (Eq. 3.30), and the plot is linear, only the first two terms are required. From Eq. 3.33, $k_H = n/P$ and therefore

---

[19] An increase in the isosteric enthalpy of adsorption can occur due to increased adsorbate–adsorbate interactions [12] but this does not apply to supercritical hydrogen adsorption in microporous materials.

$k_H = \exp(A_0)$ at $n = 0$. The A_0 values obtained from the fits to the virial equation at two closely separated temperatures can therefore be used to calculate ΔH_{zero}.

As for the case of hydrogen absorption discussed in Sect. 3.2.2, the pressure at which a particular quantity of hydrogen is adsorbed depends also on the entropy change, ΔS, as a result of adsorption. In hydrogen absorption ΔS can be attributed primarily to the loss of disorder during the transition of the hydrogen from its gaseous molecular state to its atomically absorbed state and this can be considered to be relatively constant. In the case of adsorption the reduction in the disorder is lower because the hydrogen remains in molecular form and retains some rotational freedom [68]. Garrone et al. [69] recently analysed experimentally determined ΔH and ΔS data for a number of zeolites and concluded that there is a clear nonlinear correlation between these values, and so the entropy in adsorbent systems should not necessarily be considered constant. Using Grand Canonical Monte Carlo (GCMC) simulations, however, Bhatia and Myers [70] found that for porous carbons, with a pore dimension in the range 0.755–1.76 nm, ΔS was approximately constant with a value of $-8R$. This may be due to the fact that for the model carbons, unlike zeolites, the enthalpy is not affected by the presence of different framework cations or other chemical heterogeneity. It is worth noting, however, that the value determined for cylindrical silica pores of dimension 0.65 nm was $-8.95R$, reflecting the increased confinement of the hydrogen in pores of this size, and so differences in the ΔS values for materials with different pore dimensions can still be observed, regardless of the surface chemistry.

In the same study, Bhatia and Myers [70] also analysed the thermodynamic requirements for adsorptive gas storage and derived the following expression for the optimum enthalpy of adsorption for a store with an operating pressure between P_1 and P_2,

$$\Delta H^0_{\mathrm{opt}} = T\Delta S^0 + \frac{RT}{2}\ln\left(\frac{P_1 P_2}{P_0^2}\right) \tag{3.35}$$

where T is the temperature, ΔS^0 is the entropy change relative to the standard pressure, P_0 (1 bar), and R is the universal gas constant. This assumes Langmuir adsorption behaviour (see Sect. 3.4.1) and leads to an optimum enthalpy of 15.1 kJ mol^{-1} for ambient temperature adsorptive hydrogen storage. This value was calculated using upper and lower operating pressures of 30 and 1.5 bar (3.0 and 0.15 MPa), respectively, and an entropy change of $-8R$ (-66.5 J mol^{-1} K^{-1}). If different values are used for these parameters the optimum enthalpy will change. Garrone et al. [69] suggest that, due to the enthalpy–entropy correlation, the optimum enthalpy is in the range -22 to -25 kJ mol^{-1} based on the extrapolation of their collated data.

In addition, Bhatia and Myers [70] derived the following expression for the optimum operating temperature for an adsorbent with particular values of ΔH^0 and ΔS^0,

$$T_{\mathrm{opt}} = \frac{\Delta H^0}{\left[\Delta S^0 + \frac{R}{2}\ln\left(\frac{P_1 P_2}{P_0^2}\right)\right]}. \tag{3.36}$$

For the same operating pressures and standard entropy change, using an enthalpy of adsorption of 5.8 kJ mol^{-1}, an optimum operating temperature of 114.4 K was calculated. If we use the P_1 and P_2 values from the current 2015 US DOE criteria, which are 100 and 3 atm (10.1 and 0.3 MPa), respectively, the optimum temperature rises to 135.5 K.

In conclusion, the thermodynamic parameters are crucial to the assessment of the suitability of a material for the purpose of adsorptive hydrogen storage, and their determination is therefore critical for the characterisation of the performance of these materials. Careful consideration must be given to the chosen method of determination because different approaches can yield different results.

3.2.2 Enthalpy of Hydride Formation or Decomposition

The principle thermodynamic property of interest in hydride characterisation is the enthalpy of hydride formation or decomposition, ΔH. It can be seen from Eq. 3.1 that the plateau pressure for any given hydride can be expressed as a function of both the enthalpy, ΔH, and the entropy, ΔS, of the hydriding or dehydriding process. Both quantities can be determined from pressure–composition isotherms using van 't Hoff plots, which are described below. As mentioned in the previous section, ΔS can be considered relatively constant in comparison to the enthalpy term because, while ΔH varies greatly between different hydride compounds, the entropy change is dominated by the entropy of hydrogen gas that is lost upon hydrogenation[20] (ΔS_{gas} = 130 J K^{-1} mol^{-1} H_2 at ambient temperature) and hence will not vary significantly between different materials. The other contributions, which include the vibrational entropy of the atomic hydrogen in the solid, the modifications induced in the vibrational spectrum of the host structure, the entropy change due to the difference in the electronic heat capacity between the hydride and the empty host, and the configurational entropy of the hydride, are relatively small in comparison [71].[21] The important point to note is that ΔH determines the general operating temperature and pressure of a hydride compound, and if this value is beyond a particular range it will be either too stable or too unstable for practical storage purposes.[22]

---

[20] This argument is stronger in the case of hydrides partly because there is a larger range of formation enthalpy values for hydrogen absorption compared to the adsorption enthalpies for physisorption, which typically range from −4 to −12 kJ mol^{-1} (adsorption is always exothermic). In contrast, to use somewhat extreme examples, according to Buschow et al. [71], the enthalpy of hydride formation can range from −225 to +52 kJ mol^{-1} for the binary hydrides (hydrogen absorption can be both endo- and exothermic). Therefore, a relatively constant ΔS can be ignored when determining whether a given hydride is likely to form at practical temperatures and pressures.

[21] Further discussion of this is given by Buschow et al. [71].

[22] The kinetically stabilised hydrides mentioned briefly in Sect. 2.2.3 are a possible exception.

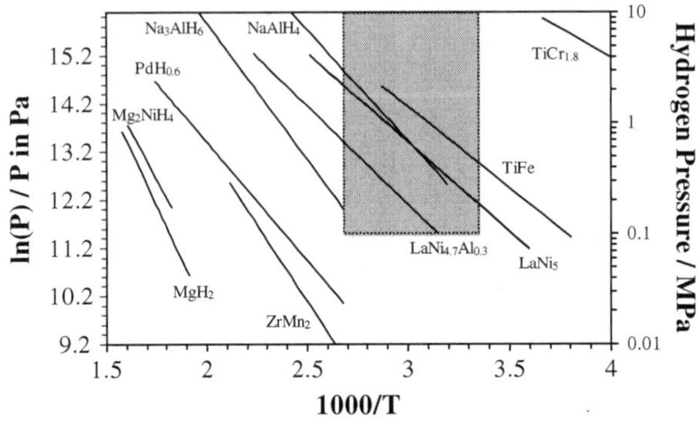

Fig. 3.5 Van 't Hoff plots for a number of hydrides, with the main region of interest for storage applications indicated by the *shaded box*, which marks the approximate temperature range of ambient to 100°C (373 K) and the pressure range from ambient to 100 bar (10 MPa). If the hydrogen stoichiometry is not indicated, the plots can be assumed to represent the mid-plateau pressures of the respective pressure–composition isotherms, measured during desorption or hydride phase decomposition. The data were plotted using enthalpy and entropy values from Sandrock [6], Bogdanović et al. [72], and the Sandia National Laboratories (US) Metal Hydride Properties database (http://hydpark.ca.sandia.gov/PropertiesFrame.html, accessed 30th July 2010) [73]

It is also worth noting that the enthalpies of hydride formation and decomposition determined for any given system will be different due to the hysteresis typically observed between the hydrogen absorption (hydride formation) and desorption (hydride decomposition) plateau pressures, which is discussed below. However, the differences are not particularly significant compared to those between stable and unstable compounds.

3.2.2.1 Van 't Hoff Plots

The enthalpy of hydride formation or decomposition is commonly calculated from pressure–composition isotherms using a graph of the natural log of the plateau pressure versus inverse temperature, which is commonly known as a *van 't Hoff plot*. Following Eq. 3.1, the slope of the line gives the enthalpy term and the intercept gives the entropy. In addition to this calculation, the van 't Hoff plots of a number of hydrides presented in the same figure allows the comparison of their suitability for use as hydrogen storage compounds. Figure 3.5 shows a number of plots for different hydrides, with the primary region of interest for hydrogen storage applications clearly indicated. For hydride characterisation purposes, the enthalpy calculated from such a plot can be readily compared to the value determined using calorimetry (see Sect. 5.1).

3.2.2.2 Plateau Pressure Hysteresis

The pressure of the plateau in the absorption isotherm for any given hydride is, with few exceptions,[23] always higher than the pressure of the plateau for desorption, for any given temperature, but the degree of hysteresis varies between compounds. Its origin is not fully understood, but it is a fundamental property of metal hydride systems and a number of models or explanations have been proposed [74]. It has been observed empirically that during desorption more repeatable or reproducible results are obtained compared to absorption and, according to Wang et al. [75], it is therefore sometimes wrongly assumed that the desorption isotherm represents equilibrium. However, this is not the case, and the absorption and desorption plateaus both represent metastable states [74]. The degree of plateau pressure hysteresis can be expressed, thermodynamically, either as the *uncompensated heat*, q', given by,

$$q' = \frac{1}{2} RT \ln\left(\frac{P_a}{P_d}\right) \tag{3.37}$$

where P_a and P_d are the plateau pressures in the absorption and desorption isotherms, respectively, measured at a temperature, T, and R is the universal gas constant; or as *entropy production*, $\Delta_i S$, given by,

$$\Delta_i S = \frac{1}{2} R \ln\left(\frac{P_a}{P_d}\right). \tag{3.38}$$

In practical terms, plateau pressure hysteresis will only cause a problem if it is very significant, to the point of preventing hydrogenation at the charging pressure of the hydrogen store, or dehydrogenation at its delivery pressure. However, minimisation of hysteresis is the ideal scenario, and its existence is another factor to consider in the characterisation of hydrides for storage applications.

3.3 Kinetic Properties

The kinetic hydrogen sorption properties of a storage material determine the rate at which a practical store can be charged and discharged. If a material cannot take up and release hydrogen in a practical timeframe then it will be of limited use in a reversible onboard storage unit. In this section we will cover some of the properties that characterise the kinetics of hydrogen sorption and desorption. We will firstly look briefly at hydrogen adsorption in which the kinetics of hydrogen uptake and release tend to be relatively rapid. The kinetics of hydrogen absorption will then be covered. We will discuss the latter in more depth, partly because

---

[23] One example is fully activated $Pd_{0.85}Ni_{0.15}$ alloy, which exhibits no hysteresis [75].

the rate of hydride formation and decomposition is more of an issue in practical terms,[24] but also because of the greater complexity of the hydrogen absorption process.

3.3.1 Hydrogen Adsorption

In general terms, diffusion through porous media can take a number of forms, and the dominant mechanism depends on the pore dimension and mean free path of the diffusing molecules [36]. In the continuum regime (see Sect. 6.1.4) diffusion can take place via *Poiseuille flow*, in which the diffusion is driven by the pressure gradient, or by *molecular diffusion*, which is also known as *continuum diffusion*. The latter occurs when there are two species present in the absence of a pressure gradient, and can therefore be ignored for our purposes. In the free molecule regime, the primary mechanism is *Knudsen diffusion*, which is induced by the collision of the gas molecules with the walls of the pore. A fourth mechanism is *surface diffusion*, in which adsorbed molecules diffuse between different surface adsorption sites. Surface diffusion[25] becomes more significant with decreasing pore diameter and so can be considered the primary hydrogen diffusion mechanism in microporous materials [62]. The surface diffusion of hydrogen on carbon nanotubes, carbon nanohorns and carbon black has been studied using Quasi-Elastic Neutron Scattering (QENS) (see Sect. 5.4.1), and diffusion coefficients of the order of 10^{-8} m^2 s^{-1} have been determined at temperatures below 70 K [77–80].

Another process which must occur before diffusion through the pore network can proceed is the entry of the hydrogen molecules into the pores (see, for example, the discussion by O'koye et al. [81] regarding the adsorption of other supercritical adsorptives in carbon molecular sieves). However, practically speaking, the entire hydrogen adsorption process is very rapid. For example, Zhao et al. [82] found that, during gravimetric isotherm determination of hydrogen uptake by microporous activated carbons, equilibrium was achieved within 2 min and was therefore too rapid for accurate measurement using the gravimetric method. Measurable isobaric adsorption kinetics were recently

[24] As an example, MgH_2 shows practically useful gravimetric and volumetric hydrogen storage capacities but the slow kinetics of hydrogen uptake and release mean that much of the research into MgH_2 and Mg-based hydrides for storage purposes has focused on the enhancement of the rates of hydrogen absorption and desorption by these materials; for example, by mechanical milling of the hydride, either to modify the microstructure and particle morphology or to mix the material with an appropriate additive (see Sects. 2.2.3 and Sect. 2.2.4.1).

[25] According to Ruthven [76] the terminology can vary in the literature and so this is also known as *micropore diffusion*, *configurational diffusion* and *intracrystalline diffusion*.

observed by Chen et al. [62] for a MOF exhibiting a high isosteric enthalpy of adsorption; however, the kinetic data showing the uptake of hydrogen and deuterium occurring over a timescale of 60 and 20 min at 77 and 87 K, respectively, were measured at pressures of 0.4 and 0.5 kPa, which are considerably lower than those of practical interest for storage applications. It therefore appears that neither entry into the pores nor surface diffusion effects limit the overall kinetic hydrogen adsorption process within timescales readily observable using macroscopic adsorption measurement methods at pressures relevant to storage applications, and we can conclude that slow hydrogen adsorption kinetics for microporous media is not likely to be an issue from a practical standpoint.

3.3.2 Hydrogen Absorption

A common approach to the analysis of reaction kinetics is to assume that there is a rate-limiting, or rate-determining, step [83]. There is some ambiguity in the meaning of this term but in hydride kinetics it generally means the slowest process in the hydrogen absorption or desorption reaction and therefore the partial reaction step that limits the rate at which the reaction can proceed. Although an understanding of the processes involved in the overall hydriding or dehydriding reaction is not necessarily required in order to assess the storage capabilities of a material, it is nevertheless worthwhile emphasising the complexity of the hydrogen absorption and desorption process for hydrides, relative to molecular hydrogen adsorption and desorption by porous solids, and to therefore explain the difficulty in practically quantifying the sorption kinetics using a single parameter, such as a diffusion coefficient. Furthermore, an understanding of the processes that can inhibit the rapid absorption and desorption of hydrogen can help in the improvement of the performance of a particular material, and so it is important to consider the microscopic processes related to the sorption reaction during the characterisation process. Later in this section we will then introduce some of the parameters that can help characterise the kinetic performance of a metal hydride.

The macroscopic kinetics of hydrogen absorption or desorption by hydrides are affected by a number of different microscopic processes [84, 85]. If we consider the case of interstitial hydrides, the first stage is the transport of the molecular hydrogen through the gas phase to the surface region. The second stage involves the physisorption of molecular hydrogen onto the surface. The hydrogen must then dissociate into atomic hydrogen and chemisorb. The fourth stage then involves the penetration of the chemisorbed atomic hydrogen through the surface of the material, followed by diffusion between interstitial sites in the lattice, which initially takes place through a disordered solid solution (α) phase. Providing the concentration of hydrogen in the material is sufficient, upon further hydrogen absorption, an ordered hydride (β) phase will form, or nucleate. The final stage of the absorption process is then the formation of the hydride[26] at its interface with the solid solution phase and the movement of this interface through the material.

At the mesoscopic level, hydrogen uptake by a single particle can be viewed as a shrinking core process, although alternative models exist.[27] In the *shrinking core model*, a β phase hydride film forms on the surface of a predominantly α phase and typically spherical particle. As the reaction proceeds, the α phase then constitutes a shrinking spherical core as the reaction front moves through the particle [89, 90]. The majority of the uptake by an interstitial hydride occurs during this process (hydride phase nucleation and growth). However, the other processes must continue in order for further hydrogen to be available at the solid solution–hydride phase interface to allow further hydride phase growth, and so any of the microscopic processes described above could potentially be rate-limiting regardless of the bulk hydrogen concentration. For example, the rate at which further hydride phase precipitates nucleate could be limited by the rate at which hydrogen penetrates the surface or the sub-surface layer and the process would then be surface-limited. However, if molecular hydrogen adsorption and dissociation on the surface is very rapid, as is hydrogen diffusion through the surface layer and bulk, but hydride phase formation and growth is relatively slow, the hydrogen uptake kinetics will be of the form expected from an appropriate nucleation and growth or shrinking core model (see Sect. 3.5.3). The rate-limiting process in this case is then hydride phase formation at the α/β interface. Alternatively, if the adsorption and

[26] The formation and growth of the hydride phase is not straightforward and can take many forms. The observation of the initial stages of hydride formation has been performed for some bulk hydride samples [84], but is experimentally challenging. There have also been a large number of surface studies of hydrogen adsorption on nearly perfect single crystal surfaces [86]. However, extrapolation of these results to real samples and elevated hydrogen pressure suffers from what has become known in surface science and catalysis as the materials, complexity and pressure gaps [87, 88]. The first two relate to the difference between near perfect single crystal samples and real (polycrystalline and heterogeneous) materials, and the latter to the difference between the nature of gas–solid interactions in Ultra-High Vacuum (UHV) environments and gas–solid interactions at the elevated pressures encountered in real applications.

[27] An alternative view is that the β phase nucleates randomly throughout the bulk of the particle. This tends to be called the *nucleation and growth model*, although in both cases the hydride phase must initially nucleate and then grow. According to Bloch and Mintz [84], in pure metals, nuclei form initially at the locations of the highest hydrogen concentration and the lowest activation energy for nucleation. These locations will include bulk discontinuities such as grain boundaries and defects. In detailed studies, different groups of nuclei have been observed, each with different relative rates of nucleation and growth, so that some groups rapidly form a large number of nuclei that grow slowly, whereas others slowly form a small number of nuclei that grow rapidly. The growth of the nuclei can also be anisotropic in terms of crystallographic direction. Hydride formation follows the hydrogen concentration gradient in the host metallic lattice. Hydrogen tends to accumulate under the surface passivation layer, the protective oxide coating discussed in Sect. 3.1.4, and so this is where the nucleation begins. If there is rapid hydrogen diffusion through grain boundaries, nucleation can then occur at greater depths. In the case of powder samples, however, for which the surface area-to-volume ratio is large, the surface region is the most likely location for hydride nucleation.

dissociation remains rapid, as above, but the hydride phase formation and growth is faster than hydrogen diffusion through the bulk and the surface layer, the hydrogen uptake kinetics will follow a diffusion model (see Sect. 3.5.2). In this case, hydrogen diffusion is the rate-limiting step. A number of the expressions that can describe these different individual processes are given in Sect. 3.5. Fromm [85] treated this topic in detail and provided an introduction to chemical reaction kinetics in the context of the absorption and desorption of hydrogen. The author emphasises the likelihood of more than one process being rate-limiting throughout the course of the hydrogen absorption process and describes a model that can be used to analyse scenarios involving the transition from one rate-limiting step to another at any stage of the reaction. Further discussion of this approach and the analysis of macroscopic hydrogen absorption and desorption kinetics is given in Sect. 3.5.

However, if we assume the existence of a single rate-limiting step, the absorption or desorption process can be parameterised, providing the experimentally determined hydrogen uptake or release can be described adequately by at least one of the appropriate analytical expressions. It should also be noted that the experimental conditions and experimental artefacts can greatly affect the results of a kinetic measurement. These include non-isothermal conditions due to the evolution of heat during the reaction, potential surface contamination and poisoning, and further decrepitation of the sample. A number of authors (see Wang and Suda [91] and references therein) have developed experimental apparatus, in particular specific sample reactors, for the determination of the intrinsic reaction kinetics of hydrogen–metal systems. There are two main reasons for this. Many hydrides, including most, if not all, of the proposed hydrogen storage compounds, absorb and desorb hydrogen when they are in powder form. Powders, by their nature, have low thermal conductivity in comparison with their bulk or single crystal equivalents. Meanwhile, the hydrogen absorption reaction for hydrides of practical interest is strongly exothermic. Conversely, hydrogen desorption by the same compounds is strongly endothermic. Therefore, any hydrogen absorption or desorption results in significant temperature excursions from the initial bed temperature and the powder form of hydride beds do not exhibit good heat transfer properties which might otherwise aid the dissipation of the generated heat. Experimental set-ups designed for the determination of isothermal kinetic measurements therefore seek to minimise these excursions through the use of thermal ballast or heat sinks. Although temperature excursions are seen in the case of adsorption, both as a result of the expansion of gas into measurement cells and the enthalpy of adsorption, these deviations are not as extreme as those seen in the hydride case. These experimental difficulties are another reason why attributing the reaction to a single rate-limiting partial step based on the fitting of an analytical expression to data is at best questionable.

Nonetheless, regardless of the acknowledged difficulties, in the following sections we will cover three parameters that can be applied to kinetic hydrogen absorption and desorption processes. The first is the activation energy, which expresses the energy barrier of any activated kinetic process. The second is the hydrogen diffusion coefficient, which expresses the rate at which atomic hydrogen diffuses through the host lattice. The third is the apparent rate of hydrogen absorption or desorption, which can be used to parameterise kinetic rates without attributing the macroscopic behaviour to any particular microscopic process and without describing the temperature dependence of the absorption or desorption. Some of the models that can describe the kinetics for different partial reaction steps and processes will be covered in Sect. 3.5.

3.3.2.1 Activation Energy

An activated process requires a particular amount of energy before it can proceed, and the rate at which the process occurs will change with temperature. The magnitude of this energy barrier is known as the *activation energy*, E_a. If the rate constant, k, of an activated kinetic process follows the Arrhenius relation,

$$k = A \exp\left[-\frac{E_a}{RT}\right] \tag{3.39}$$

where A is known as the pre-exponential factor, R is the universal gas constant and T is the temperature, then E_a can be determined from an Arrhenius plot of ln k versus $1/T$.

There are a number of different methods used for activation energy determination. One approach used in a number of recent studies of hydrogen desorption is the Kissinger method, which was originally proposed for the analysis of differential thermal analysis data [92]. In this method the thermal desorption spectra (see Sect. 4.3.2) at a series of different heating rates, β, are determined. Each spectrum will show a maximum in the desorption rate at a particular temperature, T_{max}. These two parameters can then be used to determine E_a using the following equation,

$$\ln\left(\frac{\beta}{T_{max}}\right) = -\frac{E_a}{RT_{max}} + \ln(k_0) \tag{3.40}$$

where k_0 is a constant. A plot of $\ln(\beta/T_{max})$ against $1/T_{max}$ therefore has a negative slope of E_a/R.

Other ways to determine the activation energy include the method due to Flynn and Wall, and Osawa (see Flynn [93]), and the approach of Gao and Wang [94], which uses an expression, derived from the Johnson–Mehl–Avrami equation (see Sect. 3.5.3), that was developed to determine the activation energy of crystallisation in metallic glasses. The latter was applied recently by Yang et al. [95] for the determination of the activation energy of a mixed lithium amide, lithium borohydride and magnesium hydride material. A number of these and other models are described by Brown [96] and references for a range of methods are given by Starinck [97, 98]. In view of the competing processes involved in hydrogen desorption, values determined using these methods should be considered apparent activation energies.

3.3.2.2 Hydrogen Diffusion Coefficient

Hydrogen diffusion between interstitial sites in a lattice can occur via a number of mechanisms and the dominant mechanism depends primarily on temperature [99]. Below a certain threshold temperature the processes are primarily quantum mechanical in nature. At the lowest temperatures, coherent tunnelling dominates. Then, as the temperature increases, phonon-assisted tunnelling becomes the primary mechanism. As the temperature increases yet further, the hydrogen atoms begin to behave as classical particles that migrate via over-barrier jumps induced by thermal excitations. At the highest temperatures, the hydrogen atoms behave like those in a gas, undergoing free motion through the lattice [99]. At the elevated temperatures of interest for hydrogen storage the quantum behaviour of hydrogen atoms described above for the lower temperature regimes can be ignored and they can be considered classical particles.

The rate at which hydrogen diffuses through a material, regardless of the actual mechanism, can be characterised by its diffusion coefficient. There are two types, namely the *chemical diffusion coefficient*, D_{chem}, and the *tracer diffusion coefficient*, D_t. The former describes the bulk diffusion of particles or atoms in the presence of a concentration gradient and the latter, the diffusion of individual, distinguishable particles. D_{chem} can be measured via bulk or macroscopic measurements and D_t, by microscopic measurement techniques such as Quasi-Elastic Neutron Scattering (QENS) and Nuclear Magnetic Resonance (NMR) (see Sect. 5.4). From a practical point of view we are more interested in D_{chem}, as this describes the motion of hydrogen through a material as a hydride store is charged or discharged, although D_t is still crucial as it helps characterise the rate at which the interstitial hydrogen can move through the lattice. D_{chem} and D_t will generally have different values, but can be related using the following expression,

$$D_t = \frac{D_{\text{chem}} k_B T}{c(\partial \mu / \partial c)} \tag{3.41}$$

where k_B is the Boltzmann constant, T is the temperature, c is the hydrogen concentration and μ is the chemical potential [100, 101].

The chemical diffusion coefficient can be determined by fitting data to a solution to the diffusion equation for the uptake of hydrogen by the sample, under isothermal isobaric conditions, using the assumption that the process is hydrogen diffusion-limited. There are many solutions to the diffusion equation, each dependent on the chosen boundary conditions. The solutions for diffusion into a sphere and a plane sheet or slab are covered in Sect. 3.5.2.

3.3.2.3 Apparent Rates of Absorption or Desorption

A practical way of expressing the rate at which hydrogen is absorbed and desorbed, without assuming any particular underlying mechanism, is through the use of an apparent rate of absorption and desorption. A number of different ways of defining this have been used. Goodell et al. [102] used the two parameters $T_{1/2}$ and $R_{1/2}$ to parameterise the kinetics of a number of interstitial metal hydrides. $T_{1/2}$ is the time the reaction takes to reach half completion, expressed in minutes, and $R_{1/2}$ is the reaction rate at the time $T_{1/2}$, expressed in H/M min^{-1}. The JIS glossary of terms [32], meanwhile, defines the hydriding rate as 'the hydriding amount per unit time' and the dehydriding rate as 'the dehydriding amount per unit time'. A JIS on the measurement of hydriding kinetics [103] then defines an *80% reaction time*, t_{80}, and an *80% reaction rate*, w_{80}. These express the time it takes for the hydrogen absorption or desorption to reach 80% of the equilibrium value, and the quantity of hydrogen absorption or desorption at the 80% reaction time divided by the 80% reaction time, respectively. According to the JIS [103], t_{80} is expressed in seconds and w_{80} in units of s^{-1}.

It can be seen from Fig. 2.3 that another approach used in the literature is to determine the average desorption rate between the times at which the desorbed hydrogen quantity is at 20 and 80% of its maximum. A potential problem with this approach would occur if it is used to compare materials with significantly different initial desorption rates, because this would result in a significant offset in the 20% desorption time. However, a reasonable assessment of the kinetic data should allow the potential for the distortion of the apparent rates, due to this effect, to be identified. Furthermore, this problem is more likely to be an issue in the comparison of absorption, rather than desorption, rates because different materials could potentially show sigmoidal-type behaviour during absorption in which a long incubation time of significant practical concern would be excluded from the calculated absorption rates. The result could be that materials exhibiting similar calculated absorption rates may practically take significantly different periods to charge. Attention should therefore be paid to the shape of kinetic profiles when determining how best to express apparent hydrogen absorption and desorption rates.

3.4 Isotherm Models

In this section we will cover some of the models that can be used to fit experimental equilibrium hydrogen uptake data. In the case of adsorption, these include some of the isotherm models that have been proposed for the adsorption of different gases and vapours of practical interest in a range of applications, as well as for fundamental studies. For absorption, we cover a number of models that have been proposed for absorption and desorption isotherms for metal–hydrogen systems. In this case, the models must account for the hydride phase formation and decomposition processes that occur during hydrogen absorption and desorption, as well as the heterogeneity of real host compounds.

3.4.1 Supercritical Hydrogen Adsorption

There are numerous adsorption isotherm models available to describe the interaction of gases with solid materials, and many of these have been covered extensively in books on adsorption measurement and theory [10, 12, 36, 38, 104, 105]. The simplest is the *Langmuir equation*, which describes monolayer adsorption on an ideal surface and is given by,

$$\theta = \frac{bP}{1 + bP} \tag{3.42}$$

where θ is the surface coverage, P is the pressure and b is the adsorption coefficient. The adsorption coefficient is related exponentially to the positive value of the energy of adsorption, E, so that,

$$b = K \exp\left(\frac{E}{RT}\right) \tag{3.43}$$

where K is proportional to the ratio of the relative rates of adsorption and desorption of the adsorbate molecules.

The Langmuir equation was originally derived from a kinetic model in which the gas–solid system is in dynamic equilibrium and the ratio of the relative rates of adsorption and desorption is constant for any given pressure. The model assumes that an adsorbed molecule occupies a localised adsorption site, that each site can be occupied by only one adsorbate molecule, that the energy of adsorption for each site is equal and that there are no adsorbate–adsorbate interactions. It can also, however, be derived from thermodynamics and from statistical mechanics [104]. In the latter case, the coefficient b is expressed in terms of the internal partition functions for the molecules in the adsorbed and gas phases.[28] Despite its limitations, the Langmuir isotherm equation has been used recently to fit supercritical hydrogen adsorption data [61, 107].

The linear form of the Langmuir equation, which is given by,

$$\frac{P}{n} = \frac{1}{n_m b} + \frac{P}{n_m} \tag{3.44}$$

where n is the amount adsorbed and n_m is the monolayer capacity, so that $\theta = n/n_m$, has also been used recently for supercritical hydrogen adsorption [61, 108].[29] However, the theoretical basis for the Langmuir expression does not seem satisfactory for the case of supercritical hydrogen adsorption on heterogeneous

---

[28] Fowler [106] derived the Langmuir equation using essentially the same assumptions that the atoms or molecules are adsorbed on definite adsorption sites, only one adsorbate atom or molecule can be accommodated on each site and that the energy of each is unaffected by the presence of neighbouring adsorbate atoms or molecules.

[29] Note that n in this case should be the absolute adsorbed quantity (see Sect. 3.1.1.3).

microporous solids[30] and its form does not practically appear to describe the overall uptake particularly well.

A number of empirical isotherm equations have been proposed to describe the adsorption of gases on heterogeneous surfaces. One approach, following the Langmuir equation, is to assume there are N independent regions, each of constant adsorption affinity, and that the Langmuir equation applies to each of these, so that,

$$C_\mu = \sum_{j=1}^{N} C_{\mu,j} \frac{b_j P}{1 + b_j P} \tag{3.45}$$

where C_μ is the total uptake and $C_{\mu,j}$ is the uptake by region j, which has an adsorption coefficient b_j. In other words, the adsorbent has a patchwise surface topography, with the Langmuir equation applicable to each patch and the patches consisting of only one type of adsorption site possessing a single adsorption energy. However, even in the case of $N = 2$, known as the *dual Langmuir isotherm*, this expression contains four fitting parameters and a number of empirical relations with only three have been proposed that better represent adsorption on heterogeneous surfaces. One of these combines the Langmuir isotherm with the empirical *Freundlich isotherm* [109].

The Freundlich expression was originally proposed as an empirical relation but it can be derived theoretically from the model described above, using the assumption that the adsorption energy distribution is described by an exponential. This gives the following result,

$$C_\mu = KP^{1/n} \tag{3.46}$$

where C_μ, as above, is the concentration of the adsorbed species, and K and n are empirical parameters. Obviously, for $n = 1$ this relation is linear; for $n > 1$, the isotherm becomes increasingly non-linear with increasing n. However, for $n \neq 1$, this expression does not show Henry's law behaviour (Eq. 3.33) in the low pressure region[31] and also does not reach a finite limit at high pressures.[32] Nonetheless, the Freundlich equation can be used to fit experimental data over a limited (intermediate) pressure range [36].

---

[30] In addition to the assumption of a homogeneous surface, another fundamental assumption in the derivation of the Langmuir equation is that the enthalpy of adsorption does not vary with the amount of gas adsorbed [104]. This has been shown experimentally not to be the case for supercritical hydrogen adsorption on a number of microporous adsorbents, which further supports the inapplicability of the Langmuir equation.

[31] Henry's law describes a linear uptake with pressure at low surface coverage, due to the absence of competition for adsorption sites and the non-interaction of neighbouring adsorbate molecules or atoms. The latter is expected to be the case for very low adsorbate concentrations (see also Sect. 3.2.1).

[32] The adsorbed quantity should saturate because it cannot increase indefinitely.

The combination of Eqs. 3.42 and 3.46, derived by Sips [109], assumes Langmuir isotherm behaviour but with a near Gaussian distribution of site energies, so that,

$$C_{\mu} = C_{\mu s}\frac{(bP)^{1/n}}{1 + (bP)^{1/n}}. \tag{3.47}$$

This expression reduces to the Langmuir equation for $n = 1$ and to the Freundlich equation at low pressure; as a consequence, it also cannot describe Henry's law behaviour at low pressures. The value of n can be considered as a measure of the level of homogeneity shown by the sample and is normally greater than one, with a larger value indicating a more heterogeneous surface. The *Sips equation*, also known as the *Langmuir–Freundlich isotherm*, has been shown to fit experimental adsorption data effectively and has been used recently to fit supercritical hydrogen adsorption data. See, for example, the report by Kaye and Long [110] on hydrogen adsorption by dehydrated Prussian blue analogues.

Two other well known expressions used to describe adsorption on heterogeneous surfaces are the *Tóth* and the *Unilan* equations. Unlike the Sips equation, both of these show Henry's law behaviour in the low pressure regime and saturate at elevated pressures. The first of these describes the adsorption in terms of another exponent, t,

$$C_{\mu} = C_{\mu s}\frac{bP}{\left[1 + (bP)^{t}\right]^{1/t}} \tag{3.48}$$

where b and t are specific to a given adsorbate–adsorbent system. This expression reduces to the Langmuir equation for $t = 1$. Usually, for a heterogenous adsorbent, $t < 1$. As with n for the Sips equation, t is said to represent the heterogeneity of the system. Lin et al. [111] recently used the Tóth equation, in conjunction with the Soave–Redlich–Kwong (SRK) equation of state (see Sect. 6.1.1.2), to represent the total, or absolute, hydrogen adsorption in the conversion of experimentally determined excess adsorption isotherms for a series of three metal–organic frameworks (NOTT-101, NOTT-102 and NOTT-103). The values of t obtained were 0.71, 0.56 and 0.66, respectively, indicating a variation in the heterogeneity of each material, assuming that the model is applicable in this case.

The Unilan equation, which derives its name from the combination of a *uni*form distribution and *Lan*gmuir local isotherm, assumes patchwise surface topography, local Langmuir behaviour and a uniform distribution of site energies. However, this equation has a more complicated form than both the Tóth and the Langmuir–Freundlich expressions and is yet to be applied, to the author's knowledge, to supercritical hydrogen adsorption data. Other examples of isotherm expressions that have been applied to the case of supercritical hydrogen adsorption include the virial equations covered briefly in the context of the determination of the isosteric enthalpy of adsorption in Sect. 3.2.1.

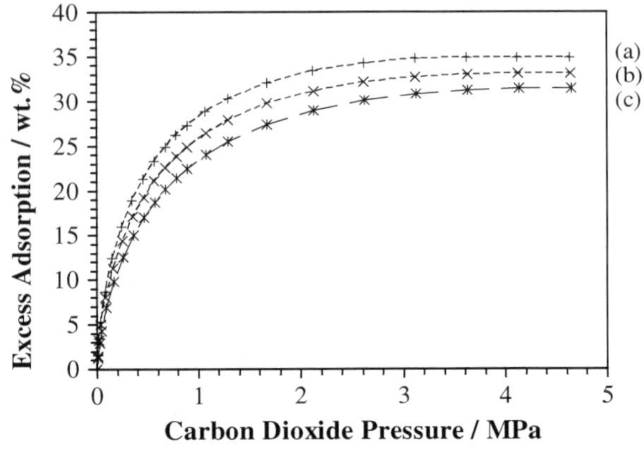

Fig. 3.6 Carbon dioxide adsorption isotherms for a commercial activated carbon sample measured at temperatures below and above the critical temperature of CO_2, T_c = 304.1 K. The data shown are for (a) T = 299 K, (b) T = 308 K and (c) T = 318 K. Unpublished data, reproduced with permission from Hiden Isochema Ltd

Apart from the Langmuir equation, the isotherm equations described above apply to adsorption on heterogeneous surfaces, whether porous or non-porous. For hydrogen storage, we are specifically interested in adsorption by microporous solids, as well as adsorption at supercritical temperatures. At sub or near-critical temperatures, the adsorption process for a microporous solid is governed by micropore filling, rather than monolayer formation followed by a transition to multilayer adsorption. This is not necessarily the case for supercritical adsorption in micropores, but the measurement of gas adsorption below and above the critical temperature by microporous solids has shown that there is no abrupt change in the form of the adsorption observed during the transition from sub-critical to supercritical conditions [36]. This can be seen in the adsorption of carbon dioxide both above and below its critical temperature by a microporous carbon, which is shown in Fig. 3.6. Therefore, equations based on a micropore filling mechanism, such as the Dubinin–Radushkevich (DR) and Dubinin–Astakhov (DA) expressions, can be empirically applied to supercritical gas adsorption by microporous materials.

The DA equation is one of a series of expressions, due to Dubinin and co-workers, based on Polanyi's potential theory, which views adsorption in terms of the *adsorption potential* of the material. For any given adsorbate–adsorbent system, the adsorption potential can be determined empirically by measuring its temperature-independent *characteristic curve*. This expresses the fractional filling of the micropore volume, θ, in terms of A, the Polanyi adsorption potential and E, the characteristic energy of the system,

$$\theta = \exp\left[-\left(\frac{A}{E}\right)^n\right] \tag{3.49}$$

where n is an empirical constant. This is a generalisation of an earlier proposal by Dubinin and Radushkevich in which n = 2. In the DA equation, which tends to fit experimental data more satisfactorily, n has been reported to range from 1.2 to 6.[33] The value of n can be considered an indication of the heterogeneity of the adsorbent because the isotherm, expressed as a function of the adsorption potential or relative pressure, sharpens with increasing n. Therefore, as the isotherm is expected to be relatively sharp for a homogeneous solid, the lower the value of the exponent, the greater the heterogeneity.

The adsorption potential, A, is given by the following expression,

$$A = -RT \ln\left(\frac{P}{P_s}\right) \tag{3.50}$$

where P_s is the pseudo saturated vapour pressure, which replaces the real saturated vapour pressure, P_0, in the standard DA equation.

It can be seen that the use of these equations requires the calculation of P_s, a parameter of questionable physical significance. One method involves the extrapolation of the Antoine equation,

$$\ln(P_s) = A - \frac{B}{C+T} \tag{3.51}$$

where A, B and C are the empirically determined Antoine coefficients, T is the temperature in K and P is the pressure in bar. In the case of hydrogen $A = 3.543$, $B = 99.395$ and $C = 7.726$. These values are calculated from the data of van Itterbeek et al. [112] in the temperature range 21.01–32.27 K, as listed in the NIST Chemistry WebBook.[34] Another assumption, suggested by Dubinin, calculates P_s in terms of the critical pressure, P_c, and the critical temperature, T_c, and is given by,

$$P_s = P_c\left(\frac{T}{T_c}\right)^2. \tag{3.52}$$

A generalisation of this approach was proposed by Amankwah and Schwarz [113], so that,

$$P_s = P_c\left(\frac{T}{T_c}\right)^k \tag{3.53}$$

where k is specific to the adsorbate–adsorbent system. The authors applied this expression in conjunction with the DA equation for hydrogen adsorption on four microporous carbons and found that n, the exponent in the DA equation, had values in the range 1.67–1.79 and k was in the range 2.72–4.16. Poirier and Dailly [114–116] have recently applied the DA equation to hydrogen adsorption data for

[33] According to Do [36], values for a strongly activated carbon are typically between 1.2 and 1.8, and for zeolites between 3 and 6, although Rouquerol et al. [12] quote values between 2 and 6 from original work by Dubinin, and Lowell et al. [105] mention the range 2–5.

[34] http://webbook.nist.gov/chemistry/, accessed 13th January 2010.

a series of MOFs and used the Amankwah and Schwarz [113] expression for the pseudo saturated vapour pressure.

3.4.2 Hydrogen Absorption

Models for the isothermal uptake of hydrogen by metal hydrides are not applied as widely as those for the isothermal adsorption of gases. This is due mainly to the wide applicability of the equations discussed in the previous section, which can be applied to the adsorption of various species by heterogenous and microporous solids in a wide array of practically important applications, as well as fundamental adsorption studies. An isotherm model that can parameterise the reversible absorption and desorption of hydrogen by a metallic host is therefore more narrow in scope. However, a number of models have been proposed, although they have not been widely tested on different materials and they have not yet passed into common usage.

An early model of the pressure–composition dependence of hydrogen absorption by Pd was proposed by Lacher [117]. The model assumes that there are a definite number of interstitial sites in the metal lattice for the hydrogen to occupy and that the heat of absorption increases as the sites become occupied. Hydride phase formation is considered to occur due to attractive interactions between the hydrogen atoms in the metal lattice and can be viewed as condensation of the absorbed hydrogen from a gaseous to a liquid-like state. Lacher assumed that the interaction was nearest-neighbour, and applied the Bragg–Williams (mean field) approximation [118] to derive the following expression for an absorption isotherm,

$$\ln P^{1/2} = \ln\left(\frac{\theta}{1-\theta}\right) - \frac{1}{k_B T}\left(\chi_0 - \frac{\chi_d}{2} + \theta\chi\right) + \ln A \tag{3.54}$$

where P is the hydrogen pressure, θ is the fractional site occupancy ($\theta = M_s/N_s$, where M_s is the number of hydrogen atoms and N_s is the number of interstitial sites), k_B is the Boltzmann constant, T is the temperature, $-\chi_0$ is the energy of a hydrogen atom in its lowest quantum state in an interstitial site, χ_d is the dissociation energy of a hydrogen molecule from it lowest quantum state, χ is a proportionality constant dependent on the critical temperature[35] ($T_{crit} = \chi/4k_B$) and ln A over a limited temperature range can be treated as a constant ($\log_{10}A = 2.3009$ in the work of Lacher).

A number of more complex models have been developed since this early work. Rees [119] formulated a model that accounted for the disorder in the parent lattice, assuming that either Frenkel or Schottky defects were present. Kierstead [120]

[35] Note that T_{crit} is the temperature of the critical point for the two phase co-existence region of the palladium–hydrogen system, not that of fluid phase hydrogen (see Sect. 3.1.1).

derived Lacher's expression thermodynamically and extended it to multiplateau isotherms so that each plateau corresponds to a different set of sites. Kierstead also applied the same approach to the Rees model and used it to fit experimental isotherms for $DyCo_3H_x$ [121]. Subsequent work by the same author included the improvement of the Lacher and Rees theories to better represent the temperature dependence of their description of hydrogen uptake behaviour [122], the generalisation of the multiplateau isotherms based on Lacher's theory [123], and the introduction of interactions between hydrogen atoms absorbed on different sites [124], which had been neglected in the previous models.

Bjurström et al. [125] took a different, empirical, approach. They reduced isotherms measured at different temperatures for the $LaNi_{4.79}Al_{0.21}H_x$ system into a single characteristic curve, analogous to the characteristic adsorption curves used in the Dubinin approach to adsorptive processes (see Sect. 3.4.1), using the free energy function $RT\ln(P/P_{X_{\mathrm{ref}}})$, where $P_{X_{\mathrm{ref}}}$ is a reference pressure for a given reference composition X_{ref}. A good fit to the data was achieved with the following six parameter expression for the equilibrium hydrogen-to-metal ratio,

$$\left(\frac{H}{M}\right)_{\mathrm{eq}} = a_1\frac{a_2x^{1/a_3}}{1+a_2x^{1/a_3}} + a_4\frac{a_5x^{1/a_6}}{1+a_5x^{1/a_6}} \tag{3.55}$$

where a_1 to a_6 are the six fitting parameters. A similar approach had been applied previously by Larsen and Livesay [126] to fit their equilibrium $SmCo_5H_x$ data. This expression does not describe Sieverts' law behaviour ($\sqrt{P} = K_S x$, where K_S is the Sieverts constant and x is the hydrogen-to-metal ratio [99]) at low pressures and fails in the critical region, but Bjurström et al. [125] argue that it is sufficiently accurate for engineering applications.

Fujitani et al. [127] modelled pressure–composition isotherms by fitting the α and β phase regions using the Lacher model, as applied by Evans and Everett [128], and by fitting the plateau region using a Gaussian distribution function to represent the plateau slope. They demonstrated the model by fitting data for $LaNi_{4.55}Al_{0.45}H_x$. Similarly, Zhou et al. [129] presented a model that represents the α, $\alpha + \beta$ and β phase regions separately but fits the sloping plateau region according to a 'slope factor', f_s, which expresses the gradient in the mixed phase region at a temperature, T, relative to the gradient at 298 K. They demonstrated their model by fitting data from two AB and five AB_5 compounds. The model was later applied to a series of AB_2 compounds [130, 131], and also modified to allow for curvature in the mixed phase region [132].

Lototsky and co-workers [133–136] have modified the Lacher model and Kierstead's multi-plateau version to incorporate a Gaussian distribution in the thermodynamic properties in the plateau region, in order to account for the inhomogeneity of real materials. The model appears to fit data for real alloys more accurately than the original Lacher and Kierstead models. Lexcellent and Gondor [137] recently presented a model, applicable to single plateau isotherms, which is based on the theory of hysteresis in open two-phase systems containing coherent interfaces developed by Schwarz and Khachaturyan [138, 139]. Lexcellent and

Gondor criticise the model of Lototsky and co-workers for not addressing the phase transition in the mixed phase region and for producing unconvincing fits to experimental data. However, in a recent article on PCI modelling, Payá et al. [140] point out that the Lexcellent and Gondor model predicts a linear dependence of ln *P* upon hydrogen content, which is inconsistent with published data. In their article, Payá et al. [140] compared a smoothed version of the Zhou et al. [129] model, with the Lototsky et al. [134] model and a polynomial fitting method, by fitting data for an AB_2 and an AB_5 compound. They conclude that the Zhou et al. [129] model provides the best fit to the data. The polynomial approach also provided good fits but non-physical oscillations are apparent in the fitted model in the plateau region; the authors therefore urge caution in using this method.

3.5 Kinetic Models

In this section we will cover models that can describe experimental kinetic hydrogen sorption and desorption data. Although the kinetics of the adsorption and desorption of molecular hydrogen by microporous materials are of practical interest, they are generally very rapid (see Sect. 3.3.1). This is a significant advantage for hydrogen storage applications but, firstly, provides less motivation for their study and analysis[36] and, secondly, makes their experimental determination challenging. On the other hand, the kinetics of hydrogen absorption and desorption by metal hydrides can be a significant problem and hence a significant disadvantage for hydrogen storage applications. Gaining a better understanding of their origin is therefore practically important, as is their measurement, and so we shall concentrate here on the kinetics of hydrogen absorption. As explained in Sect. 3.3, reaction kinetics are often analysed under the assumption that there is a single rate-limiting step. In the case of hydrogen absorption and desorption there are a number of processes involved, each of which potentially could be rate-limiting. These include surface penetration, hydrogen diffusion and hydride phase formation. In this section we will therefore cover some of the possible models for these three processes.

Firstly, however, we shall discuss some issues surrounding the analysis of kinetics, particularly with regard to the assumption of a single rate-limiting step, beginning with the work of Fromm and co-workers [85, 90, 141]. In their model of the hydrogen absorption and desorption process, Martin et al. [90] assumed a spherical particle shape and the shrinking core model (see Sect. 3.3.2), and presented expressions for five partial reaction steps: (i) physisorption of molecular hydrogen on the surface, (ii) the dissociation of the molecular hydrogen and subsequent atomic chemisorption, (iii) surface penetration by atomic hydrogen,

[36] In this regard, it is not a materials problem that requires a solution, although it is still of scientific interest.

(iv) hydrogen diffusion, and (v) hydride formation. Using their model, they analysed volumetrically determined kinetic data for $LaNi_{4.7}Al_{0.3}$ and a Ni-coated Mg sample, using Al powder in the reactor, for the $LaNi_{4.7}Al_{0.3}$ measurements, as a heat buffer in order to reduce the temperature excursions during the absorption and desorption processes. Schweppe et al. [141] subsequently presented a combined model that incorporates the physisorption, chemisorption, surface penetration and hydrogen diffusion partial steps, and allows for the possibility of different partial reaction steps being rate-limiting at different stages of the absorption or desorption processes. For example, a surface-related process could determine the initial absorption rate, with a subsequent transition to a diffusion-limited regime at longer times. Fromm and co-workers argued that the complex nature of the heterogeneous hydrogen–metal reaction means that the laws of first order reaction kinetics and the Johnson–Mehl equation (see Sect. 3.5.3) are not suited to the description of the kinetics of hydrogen absorption and desorption. The former describes the case of a single phase homogeneous reaction in which the reaction rate is proportional to the unreacted fraction, and the latter is often used to describe the kinetics of nucleation and growth processes.

In addition, prior to the work described above, the analysis of the hydrogen sorption kinetics of hydride powders was discussed in detail by Mintz and Zeiri [89]. They concluded that the determination of the rate-limiting step through the analysis of the sorption kinetics is possible but only under certain conditions. The three factors they identify as problematic to the application of sorption kinetics using a spherical particle model are the likelihood of a particle size distribution in the sample, variations in the shape of the powder particles, and a distribution in the time at which the reaction proceeds for each of the particles in the sample. Furthermore, in recent work on the interplay of diffusion and dissociation mechanisms in the hydrogen absorption process, Borgschulte et al. [142] concluded that, in most cases of hydrogen absorption, a single rate-limiting step does not exist.

When considering hydrogen sorption kinetics, temperature variation due to the exothermic or endothermic nature of both the absorption and the desorption process, is another factor that complicates an already complex picture. Any temperature variation is unlikely to be homogeneous and gradients will most likely be present. These temperature transients do not affect equilibrium uptake measurements to such an extent because the actual (equilibrium) measurement can and should be made after the material has returned to thermal equilibrium. In addition to the heat of reaction, we must also consider carefully the surface state of the hydriding material. When surface scientists study the interaction of gases with a sample they use a near-perfect surface in an Ultra-High Vacuum (UHV) environment. This means that both the crystallographic (*hkl*) plane of the surface is known and, due to the UHV conditions, there are minimal contaminants present in the gas phase. The sample will be nearly perfect with a relatively low level of defects. In contrast, laboratory hydrogen sorption measurements are typically performed using real polycrystalline materials with a complex surface structure. This surface, depending on the history of the material, will consist of oxides, hydroxides and many other surface functional groups. If a previously activated hydride sample has been exposed to air, varying amounts of surface oxidation may have occurred. This is known as *surface passivation*. Part of the hydrogenation reaction may therefore involve the reduction of the surface oxide layer. This will affect the absorption behaviour observed using bulk measurement techniques. Martin et al. [90] suggest that the Johnson–Mehl sigmoidal-type behaviour often observed is due to successive activation of the sample surface, which has been poisoned by impurities, rather than internal kinetics due to the nucleation and growth process. If this surface passivation effect is considered, combined with the possible presence of gaseous impurities, even at very low levels, in the hydrogen and their interference with the adsorption and dissociation process, possible temperature excursions and the highly defective nature of real surfaces, it can be seen that extreme caution should be exercised when attempting to draw conclusions from the form of the bulk kinetic uptake of hydrogen by materials based only on standard gas sorption measurements. Mintz and Bloch [143] make the point that although the measurement of hydrogen uptake kinetics for particular metals or alloys may provide important technological information regarding their performance in practical devices, including hydrogen storage systems, it is virtually impossible to extract meaningful basic information on the microscopic mechanisms involved in the reaction because the measured rates depend on so many extrinsic parameters, such as the morphology of the powder, the heat transfer, the reactor configuration, and so forth.

So, bearing in mind the uncertainties and difficulties in the measurement of hydrogen absorption and desorption kinetics, the problems associated with the assumption of a single rate-limiting step and the difficulty in analysing data when there is uncertainty in some of the properties and parameters, we shall now briefly look at some models that can be used assuming the rate-limiting steps are surface penetration, hydrogen diffusion and hydride phase formation.

3.5.1 Surface Penetration

The surface penetration step can be represented by the idealised surface barrier model [144, 145]. This has been used in different forms to describe the flux of hydrogen penetrating the surface in a number of combined hydrogen kinetic models [90, 141, 143, 146]. Following Pick et al. [144] and Davenport et al. [145], hydrogen penetration through the surface can be described by four fluxes, f_1 to f_4. f_1 is the flux from the gas phase onto the surface and f_2 is the flux from the surface to the gas phase. f_3 is the flux from the surface to the bulk and f_4 is the return flux from the bulk to the surface. f_1 is dependent on the flux of hydrogen molecules hitting the surface, Γ, the sticking coefficient, S, and the number of substrate atoms per unit area, N_s. The flux of hydrogen hitting the surface can be calculated from kinetic theory,

$$\Gamma = \frac{P}{\sqrt{2\pi m_H k_B T}} \tag{3.56}$$

where P is the hydrogen pressure, m_H is the mass of the hydrogen, k_B is the Boltzmann constant and T is the temperature. The assumption of second order Langmuir kinetics leads to the following dependence for the sticking coefficient, S, on the surface coverage, θ,

$$S = S_0(1-\theta)^2 \tag{3.57}$$

where S_0 is the sticking coefficient for the bare surface [147].[37] Therefore, f_1 is given by,

$$f_1 = \frac{2\Gamma S_0}{N_s}(1-\theta)^2 \tag{3.58}$$

Meanwhile, f_2 is given by,

$$f_2 = -K\theta^2 \tag{3.59}$$

with the rate constant, K, defined as,

$$K = K_0 \exp\left(-\frac{2E_D}{RT}\right) \tag{3.60}$$

where $K_0 \sim 10^{13}\ \text{s}^{-1}$, E_D is the surface energy per hydrogen atom, R is the universal gas constant and T is the temperature. f_3 is given by,

$$f_3 = -v\theta(1-x) \tag{3.61}$$

where x is the atomic fraction of hydrogen atoms in the bulk and the rate constant, v, representing the jump frequency from the surface to the bulk, is defined as,

$$v = v_0 \exp\left(-\frac{E_A}{RT}\right) \tag{3.62}$$

where $v_0 \sim 10^{13}\ \text{s}^{-1}$ and E_A is the activation energy for a jump from the bulk to the surface. Finally, f_4 is given by,

$$f_4 = \beta(1-\theta)x \tag{3.63}$$

where the rate constant, β, representing the jump frequency from the surface to the bulk, is defined as,

$$\beta = \beta_0 \exp\left(-\frac{E_B}{RT}\right) \tag{3.64}$$

where $\beta_0 \sim 10^{13}\ \text{s}^{-1}$ and E_B is the activation energy for a jump from the surface to the bulk.

---

[37] See Johansson et al. [148] for a discussion of the validity of this dependence, as well as comments on the variability of experimental results in Christmann [86].

By balancing these fluxes, the following two differential equations are obtained,

$$\frac{d\theta}{dt} = \frac{2\Gamma S_0}{N_S}(1-\theta)^2 - K\theta^2 - \nu\theta(1-x) + \beta(1-\theta)x \tag{3.65}$$

and,

$$\frac{dx}{dt} = \frac{\nu\theta}{N_l}(1-x) - \frac{\beta}{N_l}(1-\theta)x \tag{3.66}$$

where N_s is the number of metal atoms per unit area and N_l is the number of metal layers in the bulk. These equations do not have a general solution and so simplifying assumptions and approximations must be made to obtain a solution. Davenport et al. [145] discuss three different solutions; however, for the sake of brevity we will present the solution that Pick et al. [144] applied to Pd-coated Nb. Firstly, x is assumed to be small compared to 1 and so $(1 - x)$ is removed from the third term on the right-hand side of Eq. 3.65. Secondly, x and θ are assumed to be in quasi-equilibrium after an initial, very rapid, transient. Thirdly, $d\theta/dt$ is assumed to be very small compared to $N_l dx/dt$. Therefore,

$$\left(1+\frac{\beta x}{\nu}\right)^2 \frac{dx}{dt} = \frac{2\Gamma S_0}{N_s N_l} - \frac{K\beta^2 x^2}{N_l \nu^2}. \tag{3.67}$$

Pick et al. [144] then defined the following three variables,

$$y = \frac{x}{x_{\max}} \tag{3.68}$$

$$a = \frac{2\Gamma S_0}{N_s N_l x_{\max}} \tag{3.69}$$

$$b = \left(\frac{\beta}{\nu}\right) x_{\max} = \frac{\theta_{\max}}{1-\theta_{\max}} \tag{3.70}$$

and found, by integration, that for absorption,

$$\frac{(1-b)^2}{2}\ln(1+y) - \frac{(1+b)^2}{2}\ln(1-y) - b^2 y = at \tag{3.71}$$

and for desorption into vacuum,

$$\frac{1}{y} - \frac{1}{y_0} + 2b\ln\left(\frac{y_0}{y}\right) + b^2(y_0 - y) = at \tag{3.72}$$

where y_0 is the value of y at the start of the desorption process.

3.5.2 Hydrogen Diffusion

As we saw in Sect. 3.3.2.2, hydrogen diffusion can be described by the tracer or the chemical diffusion coefficient. The diffusion in terms of the latter for certain sample geometries can be described by various solutions to the diffusion equation [149]. Here we give the solutions for diffusion into a spherical and slab-like (plane sheet geometry) medium. These are the two solutions most relevant to the analysis of hydrogen diffusion in solids because a sphere can be used to approximate a powder particle and plane sheet geometry represents a foil or thin film.

If the surface concentration is constant, the mass change due to the diffusion of a substance into a sphere of radius, a, as a function of time, t, is given by,

$$\frac{M_t}{M_\infty} = 1 - \frac{6}{\pi^2}\sum_{n=1}^{\infty}\frac{1}{n^2}\exp\left(-\frac{Dn^2\pi^2 t}{a^2}\right) \tag{3.73}$$

where M_t is amount of the substance that has entered the sphere in a time, t, M_∞ is the amount after infinite time and D is the diffusion coefficient. The above solution is suitable for the conditions of an isobaric hydrogen sorption measurement.

For a plane sheet sample, with the surface concentration held constant on both sides, the mass change as a function of time is given by,

$$\frac{M_t}{M_\infty} = 1 - \sum_{n=0}^{\infty}\frac{8}{(2n+1)^2\pi^2}\exp\left(-\frac{D(2n+1)^2\pi^2 t}{4l^2}\right) \tag{3.74}$$

where l is the thickness of the sheet. In practice, for both solutions, only the first few terms of the summation are required to represent the uptake to sufficient accuracy.

3.5.3 Phase Transformation

The kinetics of solid state phase transformations can be described using a number of models and these can be used to describe hydride phase formation, subject to the caveats covered in the introduction to this section and in Sect 3.3. One of the best known is Avrami's model [150–152], which describes the transformation in terms of a time constant, k, and an exponent, n,

$$\alpha = 1 - \exp(-kt^n) \tag{3.75}$$

where α is the reacted fraction and t is the time. This model assumes random spatial nucleation and an exponential decrease in the nucleation rate. The value of n depends on the geometry (including the dimensionality) of the reaction. This model is used widely in materials science [153, 154], thermal analysis [96, 98] and many other fields, but has also been applied to metal hydrides [155]. It is also known variously as the Avrami–Erofeev [96], Johnson–Mehl–Avrami [155], Johnson–Mehl–Avrami–Kolmogorov (JMAK) [98] model, and other such variations [154]. It is commonly applied in its linear form,

$$kt = [-\ln(1-\alpha)]^{1/n}. \tag{3.76}$$

When expressed in terms of a mass uptake for the kinetic fitting of isobaric gravimetric kinetic measurements, it takes the form of Eq. 4.12 (see Sect. 4.2.1.2).

3.6 Summary

In this chapter we have discussed the various hydrogen sorption properties of materials that are of practical interest in the study of hydrogen storage. Firstly, we looked at the practical storage properties that can be assessed using the gas phase characterisation techniques that will be described in more detail in the following chapter. These included the reversible hydrogen storage capacity of both hydrides and adsorbents, the long term cycling stability, gaseous impurity resistance and the ease of activation. Where appropriate, we have discussed the origins of the effects that can lead to problems with regard to these properties in different materials. The thermodynamic properties, including the enthalpy of molecular hydrogen adsorption and the enthalpy of hydride formation or decomposition were then discussed. We briefly covered the kinetics of hydrogen adsorption, before discussing the kinetics of hydrogen absorption in more detail. The latter can be described using one of a number of characterisation parameters. We considered the activation energy, the hydrogen diffusion coefficient and the apparent rate of hydrogen absorption and desorption. The latter part of the chapter considered both equilibrium and kinetic models that can be used to describe experimental data. With regard to equilibrium properties, we covered both analytical adsorption isotherm equations and expressions that can describe hydrogen pressure–composition isotherms for interstitial hydrides. We concluded with kinetic models and focused on expressions for different stages of the hydrogen absorption process, including surface penetration, hydrogen diffusion and hydride phase formation. The methods used to determine many of the properties discussed above will be described in the next chapter.

References

1. Crabtree RH (2008) Hydrogen storage in liquid organic heterocycles. Energy Environ Sci 1:134–138
2. Graetz J, Reilly JJ (2007) Kinetically stabilized hydrogen storage materials. Scr Mater 56:835–839
3. Woodhead AP, Broom DP (2007) The activation of $LaNi_5$-H. Hiden Isochema Applications Article 125

4. Manchester FD, San-Martin A, Pitre JM (1994) The H–Pd (hydrogen–palladium) system. J Phase Equilib 15(1):62–83
5. Broom DP, Walton A, Book D, Benham MJ (2007) The accurate determination of the temperature dependence of hydrogen uptake by Na–X zeolite. Presented at the 15th international zeolite conference, Beijing, China, August 12–17th, 2007
6. Sandrock G (1999) A panoramic overview of hydrogen storage alloys from a gas reaction point of view. J Alloy Compd 293–295:877–888
7. Broom DP (2008) Hydrogen sorption measurements on potential storage materials: experimental methods and measurement accuracy. EUR 23242 EN. Office for Official Publications of the European Communities, Luxembourg
8. Sing KSW, Everett DH, Haul RAW, Moscou L, Pierotti RA, Rouquérol J, Siemieniewska T (1985) Reporting physisorption data for gas/solid systems with special reference to the determination of surface area and porosity. Pure Appl Chem 57(4):603–619
9. Garberoglio G, Skoulidas AI, Johnson JK (2005) Adsorption of gases in metal organic materials: comparison of simulations and experiments. J Phys Chem 109:13094–13103
10. Keller JU, Staudt R (2005) Gas adsorption equilibria: experimental methods and adsorptive isotherms. Springer, New York
11. Moellmer J, Celer EB, Luebke R, Cairns AJ, Staudt R, Eddaoudi M, Thommes M (2010) Insights on adsorption characterization of metal–organic frameworks: a benchmark study on the novel *soc*-MOF. Microporous Mesoporous Mater 129:345–353
12. Rouquerol F, Rouquerol J, Sing K (1999) Adsorption by powders and porous solids: principles, methodology and applications. Academic Press, London
13. Murata K, El-Merraoui M, Kaneko K (2001) A new determination method of absolute adsorption isotherm of supercritical gases under high pressure with a special relevance to density-functional theory study. J Chem Phys 114(9):4196–4205
14. Murray LJ, Dincă M, Long JR (2009) Hydrogen storage in metal–organic frameworks. Chem Soc Rev 38:1294–1314
15. Nakamura Y, Oguro K, Uehara I, Akiba E (2000) X-ray diffraction peak broadening and degradation in $LaNi_5$-based alloys. Int J Hydrogen Energy 25(6):531–537
16. Sandrock GD, Goodell PD, Huston EL, Golben PM (1989) On the disproportionation of intermetallic hydrides. Z Phys Chem NF 164:S1285–S1290
17. Lambert SW, Chandra D, Cathey WN, Lynch FE, Bowman RC (1992) Investigation of hydriding properties of $LaNi_{4.8}Sn_{0.2}$, $LaNi_{4.27}Sn_{0.24}$ and $La_{0.9}Gd_{0.1}Ni_5$ after thermal cycling and aging. J Alloy Compd 187:113–135
18. Bowman RC, Luo CH, Ahn CC, Witham CK, Fultz B (1995) The effect of tin on the degradation of $LaNi_{5-y}Sn_y$ metal hydrides during thermal cycling. J Alloy Compd 217:185–192
19. Lee H-H, Lee J-Y (1993) The intrinsic degradation behaviour of the Laves phase alloy $Zr_{0.9}Ti_{0.1}Cr_{0.9}Fe_{1.1}$ upon temperature-induced hydrogen absorption–desorption cycling. J Alloy Compd 202:23–28
20. Bowman RC, Lindensmith CA, Luo S, Flanagan TB, Vogt T (2002) Degradation behavior of $LaNi_{5-x}Sn_xH_z$ ($x = 0.20$–0.25) at elevated temperatures. J Alloy Compd 330–332: 271–275
21. Wu E, Kisi EH, Gray EM (1998) Modelling dislocation-induced anisotropic line broadening in Rietveld refinements using a Voigt function. II. Application to neutron powder diffraction data. J Appl Crystallogr 31:363–368
22. Wu E, Gray EM, Cookson DJ (2002) Synchrotron powder diffraction line broadening analysis of dislocations in $LaNi_5$–H. J Alloy Compd 330–332:229–233
23. Černý R, Joubert J-M, Latroche M, Percheron-Guégan A, Yvon K (2000) Anisotropic diffraction peak broadening and dislocation substructure in hydrogen-cycled $LaNi_5$ and substitutional derivatives. J Appl Crystallogr 33:997–1005
24. Joubert J-M, Latroche M, Černý R, Percheron-Guégan A, Yvon K (2002) Hydrogen cycling induced degradation in $LaNi_5$-type materials. J Alloy Compd 330–332:208–214

25. Shirai Y, Araki H, Mori T, Nakamura W, Sakaki K (2002) Positron annihilation study of lattice defects induced by hydrogen absorption in some hydrogen storage materials. J Alloy Compd 330–332:125–131
26. Legros M, Dehm G, Arzt E, Balk TJ (2008) Observation of giant diffusivity along dislocation cores. Science 319:1646–1649
27. Wanner M, Friedlmeier G, Hoffmann G, Groll M (1997) Thermodynamic and structural changes of various intermetallic compounds during extended cycling in closed systems. J Alloy Compd 253–254:692–697
28. Hino S, Ichikawa T, Ogita N, Udagawa M, Fujii H (2005) Quantitative estimation of NH_3 partial pressure in H_2 desorbed from the Li–N–H system by Raman spectroscopy. Chem Commun 3038–3040
29. Eberle U, Felderhoff M, Schüth F (2009) Chemical and physical solutions for hydrogen storage. Angew Chem Int Ed 48:6608–6630
30. Schüth F, Bogdanović B, Felderhoff M (2004) Light metal hydrides and complex hydrides for hydrogen storage. Chem Commun 2249–2258
31. Chandra D (2008) Intermetallics for hydrogen storage. In: Walker G (ed) Solid-state hydrogen storage: materials and chemistry. Woodhead Publishing, Cambridge
32. Japanese Industrial Standards Committee (2007) Glossary of terms used in hydrogen absorbing alloys. JIS H 7003:2007 (E)
33. Japanese Industrial Standards Committee (2007) Method for measurement of hydrogen absorption/desorption cycle characteristic of hydrogen absorbing alloys. JIS H 7203:2007 (E)
34. United States Department of Energy (2005) FreedomCAR and fuel partnership: hydrogen storage technologies roadmap. United States Department of Energy, Washington
35. Borup R, Meyers J, Pivovar B, Kim YS, Mukundan R, Garland N, Myers D, Wilson M, Garzon F, Wood D, Zelenay P, More K, Stroh K, Zawodzinski T, Boncella J, McGrath JE, Inaba M, Miyatake K, Hori M, Ota K, Ogumi Z, Miyata S, Nishikata A, Siroma Z, Uchimoto Y, Yasuda K, Kimijima K, Iwashita N (2007) Scientific aspects of polymer electrolyte fuel cell durability and degradation. Chem Rev 107:3904–3951
36. Do DD (1998) Adsorption analysis: equilibria and kinetics. Imperial College Press, London
37. Amankwah KAG, Schwarz JA (1991) Assessment of the effect of impurity gases on the storage capacity of hydrogen on activated carbon using the concept of effective adsorbed phase molar volume. Int J Hydrogen Energy 16(5):339–344
38. Yang RT (1997) Gas separation by adsorption processes. Imperial College Press, London
39. Sircar S, Golden TC, Rao MB (1996) Activated carbon for gas separation and storage. Carbon 34(1):1–12
40. Li J-R, Kuppler RJ, Zhou H-C (2009) Selective gas adsorption and separation in metal–organic frameworks. Chem Soc Rev 38:1477–1504
41. Dybtsev DN, Chun H, Yoon SH, Kim D, Kim K (2004) Microporous manganese formate: a simple metal–organic porous material with high framework stability and highly selective gas sorption properties. J Am Chem Soc 126:32–33
42. Chen B, Ma S, Zapata F, Fronczek FR, Lobkovsky EB, Zhou H-C (2007) Rationally designed micropores within a metal–organic framework for selective sorption of gas molecules. Inorg Chem 46:1233–1236
43. Lee JY, Olson DH, Pan L, Emge TJ, Li J (2007) Microporous metal–organic frameworks with high gas sorption and separation capacity. Adv Funct Mater 17:1255–1262
44. Ma S, Sun D, Wang X-S, Zhou H-C (2007) A mesh-adjustable molecular sieve for general use in gas separation. Angew Chem Int Ed 46:2458–2462
45. Sandrock GD, Goodell PD (1980) Surface poisoning of $LaNi_5$, FeTi and (Fe, Mn)Ti by O_2, CO and H_2O. J Less Common Met 73:161–168
46. Sandrock GD, Goodell PD (1984) Cyclic life of metal hydrides with impure hydrogen: overview and engineering considerations. J Less Common Met 104:159–173
47. Block FR, Bahs H-J (1983) Investigation of selective absorption of hydrogen by $LaNi_5$ and FeTi. J Less Common Met 89:77–84

48. Eisenberg FG, Goodell PD (1983) Cyclic response of reversible hydriding alloys in hydrogen containing carbon monoxide. J Less Common Met 89:55–62
49. Goodell PD (1983) Cycling hydriding response of $LaNi_5$ in hydrogen containing oxygen as a minor impurity. J Less Common Met 89:45–54
50. Wang X-L, Iwata K, Suda S (1995) Effects of carbon monoxide on the hydriding reactions of the untreated and fluorinated $LaNi_{4.7}Al_{0.3}$ alloys. J Alloy Compd 231:829–834
51. Han S, Zhang X, Shi S, Tanaka H, Kuriyama N, Taoka N, Aihara K, Xu Q (2007) Experimental and theoretical investigation of the cycle durability against CO and degradation mechanism of the $LaNi_5$ hydrogen storage alloy. J Alloy Compd 446–447:208–211
52. Han JI, Lee J-Y (1989) The effect of CO impurity on the hydrogenation properties of $LaNi_5$, $LaNi_{4.7}Al_{0.3}$ and $MmNi_{4.5}Al_{0.5}$ during hydriding–dehydriding cycling. J Less Common Met 152:319–327
53. Han JI, Lee J-Y (1989) Influence of oxygen impurity on the hydrogenation properties of $LaNi_5$, $LaNi_{4.7}Al_{0.3}$ and $MmNi_{4.5}Al_{0.5}$ during long-term pressure-induced hydriding–dehydriding cycling. J Less Common Met 152:329–338
54. Schweppe F, Martin M, Fromm E (1997) Hydrogen absorption of $LaNi_5$ powders precovered with O_2, CO, H_2S, CO_2 or N_2. J Alloy Compd 253–254:511–514
55. Nishimura K, Sato K, Nakamura Y, Inazumi C, Oguro K, Uehara I, Fujitani S, Yonezu I (1998) Stability of $LaNi_{5-x}Al_x$ (x = 0 0.5) during hydriding and dehydriding cycling in hydrogen containing O_2 and H_2O. J Alloy Compd 268:207–210
56. Töpler J, Feucht K (1989) Results of a test fleet with metal hydride motor cars. Z Phys Chem NF 164:1451–1461
57. Pedersen AS, Larsen B (1993) The storage of industrially pure hydrogen in magnesium. Int J Hydrogen Energy 18(4):297–300
58. Bouaricha S, Huot J, Guay D, Schulz R (2002) Reactivity during cycling of nanocrystalline Mg-based hydrogen storage compounds. Int J Hydrogen Energy 27(9):909–913
59. Sandrock G, Gross K, Thomas G, Jensen C, Meeker D, Takara S (2002) Engineering considerations in the use of catalyzed sodium alanates for hydrogen storage. J Alloy Compd 330–332:696–701
60. Manchester FD, Khatamian D (1988) Mechanisms for activation of intermetallic hydrogen absorbers. Mater Sci Forum 31:261–296
61. Panella B, Hirscher M, Pütter H, Müller U (2006) Hydrogen adsorption in metal–organic frameworks: Cu-MOFs and Zn-MOFs compared. Adv Funct Mater 16:520–524
62. Chen B, Zhao X, Putkham A, Hong K, Lobkovsky EB, Hurtado EJ, Fletcher AJ, Thomas KM (2008) Surface interactions and quantum kinetic molecular sieving for H_2 and D_2 adsorption on a mixed metal–organic framework material. J Am Chem Soc 130:6411–6423
63. Czepirski L, Jagiełło J (1989) Virial-type thermal equation of gas–solid adsorption. Chem Eng Sci 44(4):797–801
64. Ansón A, Callejas MA, Benito AM, Maser WK, Izquierdo MT, Rubio B, Jagiello J, Thommes M, Parra JB, Martínez MT (2004) Hydrogen adsorption studies on single wall carbon nanotubes. Carbon 42:1243–1248
65. Ansón A, Jagiello J, Parra JB, Sanjuán ML, Benito AM, Maser WK, Martínez MT (2004) Porosity, surface area, surface energy, and hydrogen adsorption in nanostructured carbons. J Phys Chem B 108:15820–15826
66. Rowsell JLC, Yaghi OM (2006) Effects of functionalization, catenation, and variation of the metal oxide and organic linking units on the low-pressure hydrogen adsorption properties of metal–organic frameworks. J Am Chem Soc 128:1304–1315
67. Furukawa H, Miller MA, Yaghi OM (2007) Independent verification of the saturation hydrogen uptake in MOF-177 and establishment of a benchmark for hydrogen adsorption in metal–organic frameworks. J Mater Chem 17:3197–3204
68. van den Berg AWC, Otero Areán C (2008) Materials for hydrogen storage: current research trends and perspectives. Chem Commun 668–681

69. Garrone E, Bonelli B, Otero Areán C (2008) Enthalpy–entropy correlation for hydrogen adsorption on zeolites. Chem Phys Lett 456:68–70
70. Bhatia SK, Myers AL (2006) Optimum conditions for adsorptive storage. Langmuir 22:1688–1700
71. Buschow KHJ, Bouten PCP, Miedema AR (1982) Hydrides formed from intermetallic compounds of two transition metals: a special class of ternary alloys. Rep Prog Phys 45:937–1039
72. Bogdanović B, Brand RA, Marjanović A, Schwickardi M, Tölle J (2000) Metal-doped sodium aluminium hydrides as potential new hydrogen storage materials. J Alloy Compd 302:36–58
73. Sandrock G, Thomas G (2001) The IEA/DOE/SNL on-line hydride databases. Appl Phys A 72:153–155
74. Flanagan TB, Park CN, Oates WA (1995) Hysteresis in solid state reactions. Prog Solid State Chem 23:291–363
75. Wang D, Flanagan TB, Kuji T (2002) Hysteresis scans for Pd–H and Pd-alloy-H systems. Phys Chem Chem Phys 4:4244–4254
76. Ruthven DM (2006) Transport in microporous solids: an historical perspective. Part I: Fundamental principles and sorption kinetics. In: Conner WC, Fraissard J (eds) Fluid transport in nanoporous materials. Springer, Dordrecht
77. Narehood DG, Pearce JV, Eklund PC, Sokol PE, Lechner RE, Pieper J, Copley JRD, Cook JC (2003) Diffusion of H_2 adsorbed on single-walled carbon nanotubes. Phys Rev B 67:205409
78. Fernandez-Alonso F, Bermejo FJ, Cabrillo C, Loutfy RO, Leon V, Saboungi ML (2007) Nature of the bound states of molecular hydrogen in carbon nanohorns. Phys Rev Lett 98:215503
79. Ross DK (2008) Neutron scattering studies for analysing solid-state hydrogen storage. In: Walker G (ed) Solid-state hydrogen storage: materials and chemistry. Woodhead Publishing, Cambridge
80. Haas O-E, Simon JM, Kjelstrup S, Ramstad AL, Fouquet P (2008) Quasi-elastic neutron scattering investigation of the hydrogen surface self-diffusion on polymer electrolyte membrane fuel cell catalyst support. J Phys Chem C 112:3121–3125
81. O'koye IP, Benham M, Thomas KM (1997) Adsorption of gases and vapors on carbon molecular sieves. Langmuir 13:4054–4059
82. Zhao XB, Xiao B, Fletcher AJ, Thomas KM (2005) Hydrogen adsorption on functionalized nanoporous activated carbons. J Phys Chem B 109:8880–8888
83. Laidler KJ (1996) A glossary of terms used in chemical kinetics, including reaction dynamics. Pure Appl Chem 68(1):149–192
84. Bloch J, Mintz MH (1997) Kinetics and mechanisms of metal hydrides formation—a review. J Alloy Compd 253–254:529–541
85. Fromm E (1998) Kinetics of metal–gas interactions at low temperatures: hydriding, oxidation, poisoning. Springer, Berlin
86. Christmann K (1988) Interaction of hydrogen with solid surfaces. Surf Sci Rep 9:1–163
87. Stoltze P, Nørskov JK (1985) Bridging the 'pressure gap' between ultrahigh-vacuum surface physics and high-pressure catalysis. Phys Rev Lett 55(22):2502–2505
88. Freund HJ, Kuhlenbeck H, Libuda J, Ruppenrechter G, Bäumer M, Hamann H (2001) Bridging the pressure and materials gaps between catalysis and surface science: clean and modified oxide surfaces. Top Catal 15(2–4):201–209
89. Mintz MH, Zeiri Y (1994) Hydriding kinetics of powders. J Alloy Compd 216:159–175
90. Martin M, Gommel C, Borkhart C, Fromm E (1996) Absorption and desorption kinetics of hydrogen storage alloys. J Alloy Compd 238:193–201
91. Wang X-L, Suda S (1990) A dehydriding kinetic study of $LaNi_{4.7}Al_{0.3}$ hydride by a stepwise method. J Less Common Met 159:83–90
92. Kissinger HE (1957) Reaction kinetics in differential thermal analysis. Anal Chem 29(11):1702–1706

93. Flynn JH (1996) Early papers by Takeo Ozawa and their continuing relevance. Thermochim Acta 282(283):35–42
94. Gao YQ, Wang W (1986) On the activation energy of crystallization in metallic glasses. J Non Cryst Solids 81:129–134
95. Yang J, Sudik A, Siegel DJ, Halliday D, Drews A, Carter RO III, Wolverton C, Lewis GJ, Sachtler JWA, Low JJ, Faheem SA, Lesch DA, Ozolinš V (2008) A self-catalyzing hydrogen-storage material. Angew Chem Int Ed 47:882–887
96. Brown ME (2001) Introduction to thermal analysis: techniques and applications, 2nd edn. Kluwer Academic Publishers, Dordrecht
97. Starinck MJ (2003) The determination of activation energy from linear heating rate experiments: a comparison of the accuracy of isoconversion methods. Thermochim Acta 404:163–176
98. Starinck MJ (2004) Analysis of aluminium based alloys by calorimetry: quantitative analysis of reactions and reaction kinetics. Int Mater Rev 49(3–4):191–226
99. Fukai Y (2005) The metal–hydrogen system. Basic bulk properties, 2nd edn. Springer, Berlin
100. Ross DK (1992) The interpretation of coherent quasielastic neutron scattering experiments on lattice gases and similar systems. Physica B 182:318–322
101. Stonadge PR, Benham MJ, Ross DK, Manwaring C, Harris IR (1993) The measurement of concentration dependent diffusion coefficients in the solid-solution alloy Pd–Y. Z Phys Chem 181:S125–S131
102. Goodell PD, Sandrock GD, Huston EL (1980) Kinetic and dynamic aspects of rechargeable metal hydrides. J Less Common Met 73:135–142
103. Japanese Industrial Standards Committee (2007) Method for measurement of hydrogen absorption/desorption reaction rate of hydrogen absorbing alloys. JIS H 7202:2007 (E)
104. Brunauer S (1943) The adsorption of gases and vapors vol 1—physical adsorption. Princeton University Press, Princeton
105. Lowell S, Shields JE, Thomas MA, Thommes M (2004) Characterization of porous solids and powders: surface area, pore size and density. Springer, Dordrecht
106. Fowler RH (1935) A statistical derivation of Langmuir's adsorption isotherm. Proc Camb Philos Soc 31:260–264
107. Schimmel HG, Kearley GJ, Nijkamp MG, Visser CT, de Jong KP, Mulder FM (2003) Hydrogen adsorption in carbon nanostructures: comparison of nanotubes, fibers, and coals. Chem Eur J 9:4764–4770
108. Langmi HW, Book D, Walton A, Johnson SR, Al-Mamouri MM, Speight JD, Edwards PP, Harris IR, Anderson PA (2005) Hydrogen storage in ion-exchanged zeolites. J Alloy Compd 404–406:637–642
109. Sips R (1948) On the structure of a catalyst surface. J Phys Chem 16(5):490–495
110. Kaye SS, Long JR (2005) Hydrogen storage in the dehydrated prussian blue analogues $M_3[Co(CN)_6]_2$ (M = Mn, Fe, Co, Ni, Cu, Zn). J Am Chem Soc 127(18):6506–6507
111. Lin X, Telepeni I, Blake AJ, Dailly A, Brown CM, Simmons JM, Zoppi M, Walker GS, Thomas KM, Mays TJ, Hubberstey P, Champness NR, Schröder M (2009) High capacity hydrogen adsorption in Cu(II) tetracarboxylate framework materials: the role of pore size, ligand functionalization, and exposed metal sites. J Am Chem Soc 131(6):2159–2171
112. van Itterbeek A, Verbeke O, Theewes F, Staes K, de Boelpaep J (1964) The difference in vapour pressure between normal and equilibrium hydrogen. Vapour pressure of normal hydrogen between 20°K and 32°K. Physica 30:1238–1244
113. Amankwah KAG, Schwarz JA (1995) A modified approach for estimating pseudo-vapor pressures in the application of the Dubinin–Astakhov equation. Carbon 33(9):1313–1319
114. Poirier E, Dailly A (2008) Investigation of the hydrogen state in IRMOF-1 from measurements and modeling of adsorption isotherms at high gas densities. J Phys Chem C 112(3):13047–13052
115. Poirier E, Dailly A (2009) Thermodynamic study of the adsorbed hydrogen phase in Cu-based metal–organic frameworks at cryogenic temperatures. Energy Environ Sci 2:420–425

116. Poirer E, Dailly A (2009) Thermodynamics of hydrogen adsorption in MOF-177 at low temperatures: measurements and modeling. Nanotechnology 20(20):204006
117. Lacher JR (1937) A theoretical formula for the solubility of hydrogen in palladium. Proc R Soc Lond A 161:525–545
118. Tanaka T, Keita M, Azofeifa DE (1981) Theory of hydrogen absorption in metal hydrides. Phys Rev B 24(4):1771–1776
119. Rees ALG (1954) Statistical mechanics of two-component interstitial solid solutions. Trans Faraday Soc 50:335–342
120. Kierstead HA (1980) A theory of multiplateau hydrogen absorption isotherms. J Less Common Met 71:303–309
121. Kierstead HA (1980) Application of the Rees theory to multiplateau hydrogen absorption isotherms. J Less Common Met 75:267–271
122. Kierstead HA (1981) Enhancement of the Lacher and Rees theories of hydrogen absorption isotherms. J Less Common Met 77:281–285
123. Kierstead HA (1982) A generalized theory of multiplateau hydrogen absorption isotherms. J Less Common Met 84:253–261
124. Kierstead HA (1984) A theory of hydrogen absorption with interactions. J Less Common Met 96:141–152
125. Bjurström H, Suda S, Lewis D (1987) A numerical expression for the *P–C–T* properties of metal hydrides. J Less Common Met 130:365–370
126. Larsen JW, Livesay BR (1980) Hydriding kinetics of $SmCo_5$. J Less Common Met 73:79–88
127. Fujitani S, Nakamura H, Furukawa A, Nasako K, Satoh K, Imoto T, Saito T, Yonezu I (1993) A method for numerical expressions of *P–C* isotherms of hydrogen-absorbing alloys. Z Phys Chem Bd 179:S27–S33
128. Evans MJB, Everett DH (1976) Thermodynamics of the solution of hydrogen and deuterium in palladium. J Less Common Met 49:123–145
129. Zhou Z, Zhang J, Ge J, Feng F, Dai Z (1994) Mathematical modeling of the PCT curve of hydrogen storage alloys. Int J Hydrogen Energy 19(3):269–273
130. Fang S, Zhou Z, Zhang J, Yao M, Feng F, Northwood DO (1999) Two mathematical models for the hydrogen storage properties of AB_2 type alloys. J Alloy Compd 293–295:10–13
131. Fang S, Zhou Z, Zhang J, Yao M, Feng F, Northwood DO (2000) The application of mathematical models to the calculation of selected hydrogen storage properties (formation enthalpy and hysteresis) of AB_2-type alloys. Int J Hydrogen Energy 25:143–149
132. Feng F, Geng M, Northwood DO (2002) Mathematical model for the plateau region of P–C-isotherms of hydrogen-absorbing alloys using hydrogen reaction kinetics. Comput Mater Sci 23:291–299
133. Davidson DJ, Sai Raman SS, Lototsky MV, Srivastava ON (2003) On the computer simulation of the *P–C* isotherms of $ZrFe_2$ type hydrogen storage materials. Int J Hydrogen Energy 28:1425–1431
134. Lototsky MV, Yartys VA, Marinin VS, Lototsky NM (2003) Modelling of phase equilibria in metal–hydrogen systems. J Alloy Compd 356–357:27–31
135. Singh RK, Gupta BK, Lototsky MV, Srivastava ON (2004) On the synthesis and hydrogenation behaviour of $MmNi_{5-x}Fe_x$ alloys and computer simulation of the *P–C–T* curves. J Alloy Compd 373:208–213
136. Singh RK, Lototsky MV, Srivastava ON (2007) Thermodynamical, structural, hydrogen storage properties and simulation studies of P–C isotherms of (La, Mm) $Ni_{5-y}Fe_y$. Int J Hydrogen Energy 32:2971–2976
137. Lexcellent Ch, Gondor G (2007) Analysis of hydride formation for hydrogen storage: pressure–composition isotherm curves modeling. Intermetallics 15:934–944
138. Schwarz RB, Khachaturyan AG (1995) Thermodynamics of open two-phase systems with coherent interfaces. Phys Rev Lett 74(13):2523–2526
139. Schwarz RB, Khachaturyan AG (2006) Thermodynamics of open two-phase systems with coherent interfaces: application to metal–hydrogen systems. Acta Mater 54:313–323

140. Payá J, Linder M, Laurien E, Corberán JM (2009) Mathematical models for the *P–C–T* characterization of hydrogen absorbing alloys. J Alloy Compd 484(1–2):190–195
141. Schweppe F, Martin M, Fromm E (1997) Model on hydride formation describing surface control, diffusion control and transition regions. J Alloy Compd 261:254–258
142. Borgschulte A, Gremaud R, Griessen R (2008) Interplay of diffusion and dissociation mechanisms during hydrogen absorption in metals. Phys Rev B 78:094106
143. Mintz MH, Bloch J (1985) Evaluation of the kinetics and mechanisms of hydriding reactions. Prog Solid State Chem 16:163–194
144. Pick MA, Davenport JW, Strongin M, Dienes GJ (1979) Enhancement of hydrogen uptake rates for Nb and Ta by thin surface overlayers. Phys Rev Lett 43(4):286–289
145. Davenport JW, Dienes GJ, Johnson RA (1982) Surface effects on the kinetics of hydrogen absorption by metals. Phys Rev B 25(4):2165–2174
146. Bloch J (2000) The kinetics of a moving metal hydride layer. J Alloy Compd 312:135–153
147. Colonell JI, Curtiss TJ, Sibener SJ (1996) Coverage dependence of the kinetics for H_2 desorption from Rh(111). Surf Sci 366:19–28
148. Johansson M, Lytken O, Chorkendorff I (2007) The sticking probability of hydrogen on Ni, Pd and Pt at a hydrogen pressure of 1 bar. Top Catal 46(1–2):175–187
149. Crank J (1975) The mathematics of diffusion, 2nd edn. Oxford University Press, Oxford
150. Avrami M (1939) Kinetics of phase change. I. General theory. J Chem Phys 7:1103–1112
151. Avrami M (1940) Kinetics of phase change. II. Transformation–time relations for random distribution of nuclei. J Chem Phys 8:212–224
152. Avrami M (1941) Granulation, phase change, and microstructure kinetics of phase change. III. J Chem Phys 9:177–184
153. Smallman RE, Bishop RJ (1999) Modern physical metallurgy and materials engineering: science, process, applications. Butterworth-Heinemann, Oxford
154. Pineda E, Crespo D (1999) Microstructure development in Kolmogorov, Johnson-Mehl, and Avrami nucleation and growth kinetics. Phys Rev B 60(5):3104–3112
155. Grant D (2008) Magnesium hydride for hydrogen storage. In: Walker G (ed) Solid-state hydrogen storage: materials and chemistry. Woodhead Publishing, Cambridge

Chapter 4
Gas Sorption Measurement Techniques

In this chapter we cover the common laboratory techniques used to measure the sorption of gaseous hydrogen by materials. The techniques can be broadly separated into three categories: gravimetric, volumetric and Temperature-Programmed Desorption (TPD), which is also known as Thermal Desorption Spectroscopy (TDS). The gravimetric and volumetric techniques are used to measure sorption isotherms, which are plots of hydrogen content or uptake versus hydrogen pressure, or vice versa, at a fixed temperature, whereas the temperature-programmed or thermal desorption techniques are used to determine the amount of hydrogen desorbed from a sample as a function of temperature. All three techniques can also be used to determine the kinetics of the sorption and desorption processes, depending on the experimental set-up and the material being studied.

In principle, the gravimetric and volumetric methods are the same as the techniques routinely used to determine the adsorption of gases other than hydrogen for a broad range of other purposes, including the characterisation of porous materials [1, 2] and the determination of gas–solid equilibria for practical applications such as adsorption-based gas separation [3, 4]. However, some of the properties of hydrogen, primarily its low molar mass and susceptibility to leakage, and the above ambient pressures required for hydrogen storage applications mean that specialised measurement apparatus is required. Similarly, the temperature-programmed desorption techniques described here are, broadly speaking, the same as techniques used for a range of other applications, including the study of the thermal decomposition of solids [5], the analysis of catalytic materials [6] and desorption processes in surface or vacuum science [7, 8]; however, as before, the unique properties of hydrogen mean that specific apparatus is required. Commercial analytical equipment is available for this purpose and Table 4.1 shows a list of the main apparatus suppliers.

D. P. Broom, *Hydrogen Storage Materials*, Green Energy and Technology,
DOI: 10.1007/978-0-85729-221-6_4,

Table 4.1 A list of the main suppliers of commercial instrumentation suitable for the characterisation of the hydrogen sorption properties of reversible storage materials at elevated pressures

Supplier	Type of apparatus	Country of origin
Advanced Materials Corporation (AMC)	Volumetric	USA
Bel-Japan	Gravimetric/volumetric	Japan
Hiden Isochema Ltd	Gravimetric/volumetric/TPD-MS	UK
HyEnergy (Setaram)	Volumetric	USA/France
Micromeritics/Particulate Systems	Volumetric	USA
Rubotherm	Gravimetric/volumetric	Germany
Suzuki Shokan Co. Ltd	Volumetric	Japan
VTI (TA Instruments Corporation)	Gravimetric/volumetric	USA

4.1 Volumetric Techniques

The volumetric method is the most common of the measurement techniques described in this chapter. It can be implemented relatively easily using a gas handling system of known volume in which the temperature and pressure can be accurately measured. The amount of hydrogen sorbed by a sample is calculated using the real gas law,

$$PV = nZRT \tag{4.1}$$

where P is the pressure, V is the volume, n is the number of moles, Z is the gas compressibility, R is the universal gas constant and T is the temperature. Volumetric measurements can be performed in a number of ways but the manometric method, which is known in the metal hydride research community as the *Sieverts Method* or *Sieverts' Technique*, is the most common.[1] We shall therefore concentrate primarily on this method, but we will also briefly cover some of the alternative approaches to volumetric measurement.

[1] There are some variations and inconsistencies in the terminology used for the measurements described in this section. The measurement type used often in metal hydride research is essentially the same as the method used by adsorption equipment for the determination of the BET surface area and pore size distribution of porous solids. It is widely called the *volumetric technique*, although the measurement parameter that determines the sorbed quantity is principally the pressure and therefore it should be termed *manometry* [1] or the *manometric technique* or *method* [4, 9]. The equipment is also sometimes referred to as *PCT apparatus*, because the measured data determine the Pressure–Composition–Temperature (PCT) properties of a metal hydride system. This is a particularly imprecise term because these properties can be determined in a number of ways. Here we use the term *volumetric* to encompass all techniques that use the measurement of the volume of hydrogen, as opposed to a change in the sample mass, to determine hydrogen uptake because this is common terminology, but acknowledge that for manometric apparatus this is, strictly speaking, incorrect.

4.1.1 The Manometric (Sieverts) Method

The *manometric method* is the most widely used hydrogen sorption technique. It is practically more convenient to implement than gravimetric methods because a simple manually operated system can be built relatively easily from off-the-shelf gas handling components. It is also the most common technique for gas adsorption measurement, in general, because commercial BET surface area and pore size distribution determination instruments also operate using the same principle.

A schematic diagram of a basic manometric system is shown in Fig. 4.1 [10]. V_1 and V_2 are known volumes, and valves A and B control the hydrogen gas inlet and vacuum outlet, respectively, allowing the control of the hydrogen pressure in V_1. Valve C allows the introduction or removal of gas to or from V_2, and the pressure in V_1 is measured using the manometer. The sample, which is held in V_2, should be secured appropriately, although this will depend on the form of the sample (for example, a fine or coarse powder, foil, single crystal, and so forth). The thermostat or thermal bath can be any temperature-controlling system, including a liquid nitrogen dewar, a cryostat, a low temperature fluid bath or a furnace, with the sample temperature monitored appropriately. The main system temperature should be controlled and monitored, preferably with sensors in more than one location. The manometer in Fig. 4.1 represents one or more pressure measuring devices, depending on the hydrogen pressure ranges required. In a system designed for both low and high pressure measurement, this is likely to include separate gauges for different ranges. The vacuum pump can be of any type, although an oil-free system with an Ultra-High Vacuum (UHV) compatible pump is preferable, particularly if samples are to be degassed. For microporous materials high vacuum[2] is essential and so a turbomolecular pump backed by a membrane or scroll pump is recommended. An oil-free system is favourable because oil vapour can backstream into the system causing contamination; although oil vapour filters can reduce backstreaming, they are unlikely to eliminate it entirely. The hydrogen supply should be of very high purity (>99.999%). Many of these points are discussed in further detail in Chap. 6, and some of the practical issues regarding hydrogen compatibility of pressure systems are covered in Sect. 4.2.3.2.

To perform a simple, single step sorption experiment on an activated sample (see Sect. 3.1.4) in the apparatus shown in Fig. 4.1, valves B and C are first opened to evacuate V_1 and V_2. After a sufficient period, valves B and C are then closed. Valve A is opened, allowing V_1 to fill with hydrogen to an initial pressure P_i. Valve A is then closed and valve C subsequently opened to fill V_2. Any drop in pressure beyond that due to the volume difference between V_1 and $(V_1 + V_2)$ is then assumed to be the result of hydrogen uptake by the sample. So, assuming the

---

[2] The *high vacuum* regime can be defined as the range $\sim 10^{-3}$ to 10^{-8} mbar (10^{-1} to 10^{-6} Pa). Either side of this regime is *ultra-high vacuum* in the range 10^{-8} to 10^{-12} mbar (10^{-6} to 10^{-10} Pa), *medium vacuum* in the range 1 to 10^{-3} mbar (10^2 to 10^{-1} Pa) and *low vacuum* in the range atmospheric pressure to 1 mbar (10^5 to 10^2 Pa) [11].

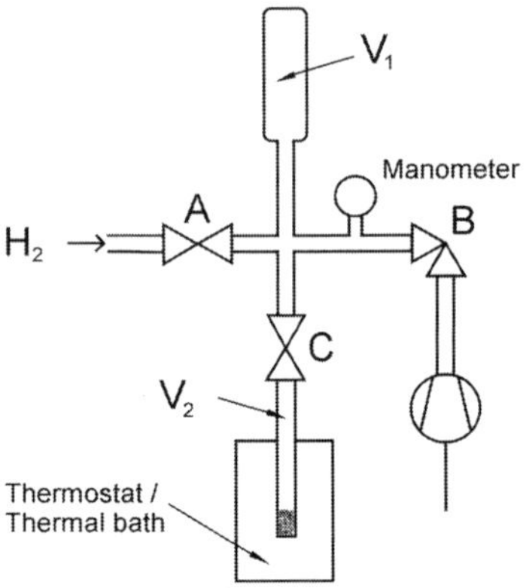

Fig. 4.1 A schematic diagram of a basic manometric sorption measurement system [10]

experiment is performed at a constant temperature, T, and that the final measured pressure is P_f, the number of moles sorbed is given by,

$$\Delta n = \frac{P_i V_1}{Z_{i,T} RT} - \frac{P_f (V_1 + V_2)}{Z_{f,T} RT} \tag{4.2}$$

where $Z_{i,T}$ and $Z_{f,T}$ are the hydrogen compressibilities at the measurement temperature T and the pressures P_i and P_f, respectively. To continue the measurement, the subsequent step then uses the values of Δn and P_f as a starting point. The repetition of the above procedure results in the measurement of a complete sorption isotherm, so that the total hydrogen uptake, after m sorption steps, is given by,

$$n_m = \sum_{j=1}^{m} \left[\frac{P_{f,j-1,T} V_2}{Z_{f,j-1,T} RT} + \frac{P_{i,j,T} V_1}{Z_{i,j,T} RT} - \frac{P_{f,j,T}(V_1 + V_2)}{Z_{f,j,T} RT} \right] \tag{4.3}$$

where $P_{i,j,T}$ and $P_{f,j,T}$ are the initial and final pressures at isotherm point j, $Z_{i,j,T}$ and $Z_{f,j,T}$ are the compressibilities of hydrogen at pressures $P_{i,j,T}$ and $P_{f,j,T}$, and T is the measurement temperature. The first term accounts for the residual gas phase hydrogen in the sample cell from the previous isotherm point. Note that for $j = 1$ this term is zero because the sample cell begins at vacuum, so $P_{i,0,T} = 0$, and Eq. 4.3 for the first step (or $m = 1$) then reduces to Eq. 4.2.

In this expression, V_2 is the *dead volume* of the sample cell; in other words, the free space that is not occupied by the sample or any containment material in the sample cell. This can be determined in two ways, using the methods called *direct* and *indirect* dead volume determination by Rouquerol et al. [1]. The direct method involves the measurement of the volume using a gas that is assumed to be non-interacting, which is typically helium, and the indirect method involves the subtraction of an estimated sample volume from the measured volume of the empty cell. For the direct method Eq. 4.3 is simply used with V_2 input as the dead volume of the sample cell determined using a helium calibration measurement on the sample cell with the sample in its activated state.[3] For the indirect method the expression becomes,

$$n_m = \sum_{j=1}^{m} \left[\frac{P_{f,j-1,T}\left(V_{\text{cell}} - m_s/\rho_s\right)}{Z_{f,j-1,T}RT} + \frac{P_{i,j,T}V_1}{Z_{i,j,T}RT} - \frac{P_{f,j,T}\left(V_1 + \left(V_{\text{cell}} - m_s/\rho_s\right)\right)}{Z_{f,j,T}RT} \right] \tag{4.4}$$

where m_s and ρ_s are the mass and density of the sample, respectively.

Equation 4.4 only applies to the situation in which the sample cell and the system are at the same temperature, T. This can be the case and some systems are immersed in a water bath, for example, to ensure complete isothermal conditions. However, it is common for measurements to be performed at temperatures at which it would be impractical to maintain the entire apparatus. Therefore the difference between the temperature of the main system, represented in the schematic in Fig. 4.1 by V_1, and the temperature of the sample cell, V_2, must be accounted for in the calculation. In reality, not all of V_2 will be at the sample temperature. The calculation can be made by assuming that there is a fixed dividing line at a point along V_2 at which the temperature changes from the sample temperature, T_{sample}, to the system temperature, T_{sys}. This point can be determined using calibration measurements and defined as the decimal fraction, f, of the sample cell volume that remains at T_{sys}. Therefore, fV_2 is at T_{sys} and $(1 - f)V_2$ is at T_{sample}. Equation 4.3 then becomes,

$$n_m = \sum_{j=1}^{m} \left[\begin{array}{l} \left(\frac{fP_{f,j-1,T_{\text{sys}}}V_2}{Z_{f,j-1,T_{\text{sys}}}RT_{\text{sys}}} + \frac{(1-f)P_{f,j-1,T_{\text{sample}}}V_2}{Z_{f,j-1,T_{\text{sample}}}RT_{\text{sample}}} \right) + \\ \frac{P_{i,j,T_{\text{sys}}}V_1}{Z_{i,j,T_{\text{sys}}}RT_{\text{sys}}} - \left(\frac{P_{f,j,T_{\text{sys}}}(V_1 + fV_2)}{Z_{f,j,T_{\text{sys}}}RT_{\text{sys}}} + \frac{(1-f)P_{f,j,T_{\text{sample}}}V_2}{Z_{f,j,T_{\text{sample}}}RT_{\text{sample}}} \right) \end{array} \right] \tag{4.5}$$

and Eq. 4.4 becomes,

$$n_m = \sum_{j=1}^{m} \left[\begin{array}{l} \left(\frac{fP_{f,j-1,T_{\text{sys}}}\left(V_{\text{cell}} - m_s/\rho_s\right)}{Z_{f,j-1,T_{\text{sys}}}RT_{\text{sys}}} + \frac{(1-f)P_{f,j-1,T_{\text{sample}}}\left(V_{\text{cell}} - m_s/\rho_s\right)}{Z_{f,j-1,T_{\text{sample}}}RT_{\text{sample}}} \right) + \\ \frac{P_{i,j,T_{\text{sys}}}V_1}{Z_{i,j,T_{\text{sys}}}RT_{\text{sys}}} - \left(\begin{array}{c} \frac{P_{f,j,T_{\text{sys}}}\left(V_1 + f\left(V_{\text{cell}} - m_s/\rho_s\right)\right)}{Z_{f,j,T_{\text{sys}}}RT_{\text{sys}}} + \\ \frac{(1-f)P_{f,j,T_{\text{sample}}}\left(V_{\text{cell}} - m_s/\rho_s\right)}{Z_{f,j,T_{\text{ssmple}}}RT_{\text{sample}}} \end{array} \right) \end{array} \right] \tag{4.6}$$

[3] Providing the empty cell volume is known this is also a helium pycnometry measurement of the sample volume and hence its density. For adsorption measurement, this defines the position of the Gibbs dividing surface.

If the system contains further volumes, Eqs. 4.5 and 4.6 gain further real gas law terms corresponding to the additional sections in the apparatus. Note that, strictly speaking, the decimal volume fraction should apply to the empty sample cell in both cases, so as not to shift the hypothetical dividing line between the two temperature zones. However, it is very important to apply modified versions of Eqs. 4.5 and 4.6 to the specific experimental set-up that is being used. The expressions presented here are intended to help explain the principle of the technique and provide a basis for the discussion in Chap. 6, but not to be used without critical evaluation of a particular piece of apparatus to which they are to be applied.

For hydrogen absorption, the quantity n_m is the total amount of absorbed hydrogen (Eq. 3.2 in Sect. 3.1.1.1) and for adsorption it is the excess adsorbed quantity (Eq. 3.7 in Sect. 3.1.1.3). In the case of adsorption measurement, V_2, measured using helium, is the reference volume, V_{ref}, defined in Sect. 3.1.1.3. A Japanese Industrial Standard (JIS) covers the determination of pressure–composition isotherms using Sieverts' method for hydrogen absorbing alloys in the temperature range 173–573 K, and in the pressure range 1 kPa to 5 MPa [12].

4.1.2 Other Volumetric Approaches

The manometric method described above is the most common approach to volumetric measurement but there are other types. In this section we will briefly describe the flowing and differential volumetric methods, the dynamic volumetric determination of PCIs and the variable volume hydrogenator proposed recently by Gray [13].

In the *flowing volumetric method*, the amount of hydrogen sorbed by a sample is determined through the measurement of the flow rate of hydrogen into the sample cell. In this case, a flow meter or a Mass Flow Controller (MFC) is positioned before the sample cell. If a mass flow meter is used with a standard manometric system, the integrated mass flow can be used to perform a kinetic measurement. If an MFC is used and controlled in such a way as to maintain isobaric conditions, the integral of the resulting flow rate as a function of time gives a measure of the isobaric absorption rate. The integral of the total mass flow signal gives the total absorbed quantity. An example is the flowing volumetric system presented by Poirier et al. [14], which operates up to atmospheric pressure. The authors presented data for hydrogen sorption by Pd and carbon nanotubes. The system has an MFC on the gas delivery line from the helium and hydrogen supplies, which is used to measure the amount of sorbed hydrogen. The result from this apparatus, which is designed only for low pressure measurement below 1 bar, is a single uptake point, rather than a full isotherm and so the reversibility of the sorption process cannot be established. This apparatus therefore seems relatively limited.

In the *differential volumetric method* two parallel reference volumes and sample cells are used. The differential pressure between the two sample cells is determined and used to calculate the gas uptake [1]. This method is not widely used for hydrogen sorption measurement but a differential system was reported by Blackman et al. [15] and used to measure the uptake of hydrogen by activated carbon and carbon nanofibres. It is argued that the system, which is an improved version of the unthermostated system used by Browning et al. [16], has advantages over Sieverts' method because differential pressure transducers have a higher accuracy than their absolute pressure measurement counterparts and also that errors associated with temperature fluctuations are minimised because each arm of the system is subject to the same thermal disturbances. However, the system measures only a single uptake point, rather than a full isotherm and so an isotherm must be constructed from a series of single point measurements. The system presented by Blackman et al. [15] also requires the entire apparatus to be maintained at 303 K. Presumably there is some scope for extending the operating temperature range but this will be limited by the operating temperature range of the pressure transducers and valves. A wider operating temperature range is possible by placing the sample cells in a separate thermostat, which was the approach taken by Zielinski et al. [17], who presented a more versatile differential system that operates up to 2000 psi (13.8 MPa). The instrument can determine full adsorption and desorption isotherms, and this was demonstrated by the measurement of hydrogen adsorption and desorption by a zeolite and an activated carbon, as well as methane adsorption on the same zeolite (Na-A). All measurements were performed near ambient temperature, in the range 273–323 K. They note that the differential method eliminates problems with the thermal effects of gas expansion (see Sect. 6.1.2) because the reference arm experiences the same expansion as the sample arm during gas dosing, and also emphasise the importance of considering helium adsorption effects during the dead space volume determination for microporous materials (see Sect. 6.2.1).

In the *dynamic volumetric method* hydrogen is delivered to a system of known volume at a fixed mass flow rate. If no sorption occurs then the pressure in the system will rise linearly, ignoring any non-linearity arising from the compressibility of the gas. Any deviation from the expected pressure increase is assumed to be due to sorption. Providing the sorption process is rapid enough and the mass flow rate, and hence the expected pressure increase, is slow enough, the system can be considered to be close to equilibrium and this method allows an entire isotherm to be measured relatively rapidly. This technique has been used to determine dynamic PCIs for a range of hydrides [18] and was also used in a commercial adsorption measurement system originally manufactured by Omicron Technology Corporation (USA), and subsequently by the Coulter Corporation (USA), which was known as the OMNISORP ([19]; J. D. Bullis 2009, private communication). The latter was used for the characterisation of microporous solids. Both Rouquerol et al. [1] and Lowell et al. [2] note the "unsurpassed" resolution of this approach, providing equilibrium conditions can be reached, and the former gives further details of the implementation of the method. In the case of hydrides, however,

Goodell et al. [18] found significant differences between isotherms determined statically and dynamically.

Another volumetric method was described recently by Gray [13]. The apparatus, named the *variable volume hydrogenator*, operates in a true volumetric mode in which the volume can be varied with a fixed quantity of hydrogen held within the system. A motor-driven piston allows the volume to be changed and the measurement of the hydrogen pressure for any given volume or temperature can be used to calculate the equilibrium hydrogen sorption properties of the loaded sample. This method eliminates the accumulative errors inherent in standard manometric measurements (see Sect. 6.5.4), but some additional practical complications are associated with the apparatus because of the need for a variable volume piston.

4.1.3 Kinetic Measurements

The absorption and desorption of hydrogen by the hydrides of interest for storage purposes are exothermic and endothermic reactions, respectively. Therefore, a temperature excursion is associated with any absorption or desorption that occurs. In order to determine the isothermal sorption kinetics of these materials a number of sample reactors have been developed and presented in the literature. The main approaches tend to involve the use of thermal ballast, such as Ni or Al powder, or double walled reactors (see, for example, Wang and Suda [20] and references therein). Another example of a container designed for kinetic measurements using the volumetric method was given in a Japanese Industrial Standard (JIS) [21], which shows a large copper sample container, designed to hold a relatively small amount of hydriding material in a flat, disc-shaped chamber 40 mm in diameter and less than 1 mm thick, in order to maximise heat dissipation.

4.2 Gravimetric Techniques

The gravimetric method was originally applied to the measurement of the adsorption of gases and vapours by materials (see, for example, McBain and Bakr [22]). More recently the technique has been used successfully to study the absorption of gases, including hydrogen [23–26], and it has now become a widely accepted method for the determination of both the adsorption and absorption of hydrogen. In this section we first describe the gravimetric method before covering the vacuum microbalances required for this measurement type. We then discuss the technical requirements for a hydrogen compatible system and also briefly describe an alternative gravimetric measurement method.

4.2.1 The Gravimetric Method

A gravimetric hydrogen sorption measurement system requires the following elements: an accurate (micro) balance, a vacuum system for sample degassing and system evacuation, a gas supply system that can control the hydrogen pressure in the balance chamber, and the ability to control the sample temperature. Once a sample has been mounted on the microbalance pan and the system evacuated, gaseous hydrogen is then introduced at a series of measurement pressures and the mass change measured at each step to determine the hydrogen content of the sample. Following the application of appropriate corrections, the measured weight as a function of hydrogen pressure is then used to plot an isotherm.

A schematic diagram of a basic gravimetric system is shown in Fig. 4.2 [10]. Valves A and B control the hydrogen gas inlet and vacuum outlet, respectively, allowing the control of the hydrogen pressure in the sample (microbalance) chamber, which is measured using the manometer. The sample is shown sitting in a holder suspended from a microbalance. The diagram implies the use of a compensating beam balance but this can also be a magnetic suspension balance in which the balance mechanism is isolated from the sample chamber (see Sect. 4.2.2). The thermostat or thermal bath can be any temperature-controlling unit, including a liquid nitrogen dewar, a cryostat, a low temperature fluid bath or a furnace, and the temperature of the system itself should be carefully controlled to ensure stability of the balance reading. The manometer represents one or more pressure measuring devices, depending on the hydrogen pressure ranges required. As for the manometric method, the vacuum pump can be of any type, although an oil-free system with UHV capability is preferable, particularly if samples are to be degassed. For microporous materials high vacuum is essential and so a turbo-molecular pump backed by a membrane or scroll pump is recommended. The hydrogen supply should be of very high purity (>99.999%), and the system must

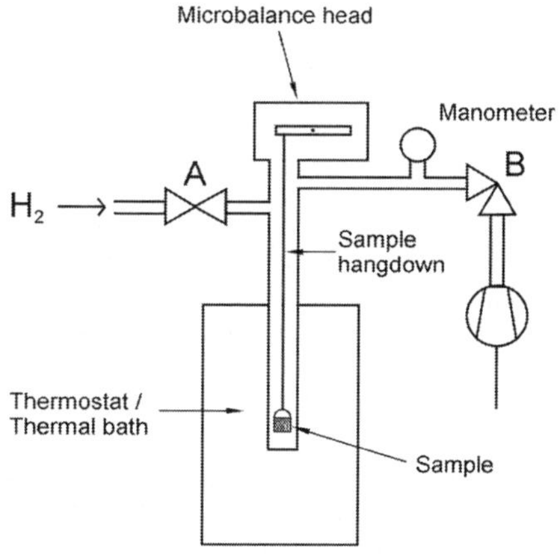

Fig. 4.2 A schematic diagram of a basic gravimetric sorption measurement system [10]

be compatible with hydrogen use as described in more detail in Sect. 4.2.3.2. These practical points are discussed further in Chap. 6.

To perform a simple, single step experiment on an activated sample in the apparatus shown in Fig. 4.2, valve B is first opened to evacuate the microbalance chamber for a sufficient period. The dry or empty sample mass[4] is then determined from the microbalance reading after this period. Valve B is then closed and valve A subsequently opened to allow the hydrogen pressure in the sample chamber to reach the pressure required for the single isotherm point. As the sample sorbs hydrogen, valve A can then be used to maintain a constant hydrogen pressure in the chamber. Once the weight has reached an appropriate "equilibrium", the hydrogen uptake can be determined from the sample weight, after the application of the buoyancy effect (Archimedes' Principle) corrections, using the dry or empty sample mass as a reference point. The achievement of sufficient equilibrium can be determined by defining an appropriate differential weight change threshold as a function of time or by fitting an appropriate function to the real time weight change. To extend this measurement to a full isotherm, the pressure is then increased and the weight monitored. After a sufficient period, the uptake can then again be determined from the buoyancy effect-corrected weight, with reference to the dry or empty sample reading. This process is then repeated until a full isotherm has been measured. At each point, the dry or empty sample mass is used as the reference point, thus avoiding the accumulative errors inherent in the volumetric measurement procedure.

4.2.1.1 Equilibrium Hydrogen Uptake Calculation

The measurement parameter of interest is the force, f_{sorb}, on the microbalance due to hydrogen sorption,

$$f_{\text{sorb}} = m_{\text{sorb}} g \tag{4.7}$$

where m_{sorb} is the mass of the adsorbed or absorbed hydrogen and g is the acceleration due to gravity. This measured force must be corrected for the buoyancy of the sample, which creates an upthrust, f_{buoyancy}, given by,

$$f_{\text{buoyancy}} = m_s g \left(\frac{\rho_H}{\rho_s} \right) \tag{4.8}$$

where m_s is the empty sample mass measured at vacuum, ρ_s is the sample density,[5] and ρ_H is the hydrogen density at the measurement temperature and pressure (see Sect. 6.1.1). Strictly speaking, in the case of adsorption, this term should also

[4] Empty in the sense of unloaded, unhydrogenated, or at the hydrogen loading required at the start of the measurement, including any trapped residual hydrogen.

[5] The (m_s/ρ_s) term therefore simply represents the volume occupied by the sample.

include the buoyancy effect from the adsorbed phase. In the case of absorption, the term will also include the increased sample volume due to hydrogen absorption and the sample density should be the density of the sample including the absorbed hydrogen.

In the absence of any other buoyancy effects, this produces a total force on the balance, f_{tot}, given by,

$$f_{\text{tot}} = f_{\text{sorb}} - f_{\text{buoyancy}} = m_{\text{sorb}}g - m_s g\left(\frac{\rho_H}{\rho_s}\right). \tag{4.9}$$

With knowledge of the sample mass and density, the mass of sorbed hydrogen can therefore be calculated.

In an actual measurement using a beam balance we must also account for the contributions from the other objects present in the gas phase. A counterweight is suspended from the reference arm and this contributes a negative force to the balance reading, along with the counterweight hangdown. On the sample side, a hangdown and the sample pan contribute positively. In the simplest arrangement, we must consider a sample hangdown of mass, m_{sh}, and density, ρ_{hd}, a counterweight hangdown of mass, m_{ch}, and the same density, ρ_{hd}, a counterweight of mass, m_{cw}, and density, ρ_{cw}, and the sample pan of mass, m_{sp}, and density, ρ_{sp}. Therefore, the buoyancy correction, f_{buoyancy}, becomes,

$$\begin{aligned} f_{\text{buoyancy}} = &\left(m_{sh}g\left(\frac{\rho_H}{\rho_{hd}}\right) + m_{sp}g\left(\frac{\rho_H}{\rho_{sp}}\right) + m_s g\left(\frac{\rho_H}{\rho_s}\right)\right) \\ &- \left(m_{ch}g\left(\frac{\rho_H}{\rho_{hd}}\right) + m_{cw}g\left(\frac{\rho_H}{\rho_{cw}}\right)\right) \end{aligned} \tag{4.10}$$

The first bracketed term in the above expression includes the items on the sample side of the balance and the second term includes those on the counterweight side. In this expression, for the sake of simplicity, we have assumed that the entire apparatus is at the same temperature and so both the counterweight and sample are suspended in hydrogen of gas density, ρ_H, which is the density of gaseous hydrogen at the measurement temperature and pressure. If the counterweight and sample are not at the same temperature, the appropriate gas densities for each term should be used. Additional terms must be included if any additional material, such as quartz wool to secure the sample, is present in the microbalance chamber.

The mass of the absorbed hydrogen is then determined using the definition of f_{buoyancy} from Eq. 4.10 and rearranging Eq. 4.9 to give,

$$f_{\text{sorb}} = f_{\text{tot}} + f_{\text{buoyancy}}. \tag{4.11}$$

At each hydrogen pressure we therefore apply a buoyancy correction to the total measured force, based on the parameters in Eq. 4.10 and the appropriate value of ρ_H. Note that, as the buoyancy is expected to create an upthrust on the balance with increasing hydrogen pressure, f_{buoyancy} is expected to be negative at elevated pressures. For the case of hydrogen adsorption, adding a buoyancy term to the first bracketed term in Eq. 4.10 for the adsorbed phase, using an assumed adsorbed phase density, gives a calculated total adsorbed quantity, following Eq. 3.28, rather than the experimental parameter, the excess adsorption.

4.2.1.2 The Approach to Equilibrium

During an isotherm measurement, at each hydrogen uptake point, the approach to equilibrium should be monitored as a function of time. This allows us to determine whether the state defined by Keller and Staudt [4] as *technical equilibrium* has been achieved. This is the point at which sufficient adsorption, absorption or desorption of hydrogen has taken place, so that the measurement can move on to the next isotherm point. The two approaches mentioned previously in the introduction to this section are those taken by Agarwal and Schwarz [27] and Benham and Ross [25], respectively. Agarwal and Schwarz [27] defined a threshold differential weight change of 0.8 μg min^{-1} for their adsorption measurements, corresponding to the achievement of a constant balance reading for a period of 5 min on a balance with a 4 μg resolution.[6] Benham and Ross [25], in their work on the gravimetric determination of hydrogen absorption by Pd, fitted the mass change in real time to Avrami's model,

$$m(t) = m_0 + \Delta m(1 - \exp(-kt^x)) \tag{4.12}$$

where *m(t)* is the mass as a function of time, t, m_0 is the sample mass at the start of the isotherm point, k is the time constant and x is an exponent. This function was originally proposed by Avrami to represent the kinetics of solid state phase transformations and has subsequently been used to represent a wide range of physical and chemical processes (see Sect. 3.5.3). For $x = 1$ it reduces to a simple exponential, which represents the Linear Driving Force (LDF) mass transfer kinetic model [28, 29]. For $x < 1$, it shows a steeper initial rise than the LDF model, and can therefore approximate solutions to the diffusion equation, which take the form of sums of exponentials (see Sect. 3.5.2). For $x > 1$, the function becomes sigmoidal which describes the kinetics of the nucleation and growth of new phase precipitates in a host matrix.

Providing the uptake is adequately represented by Eq. 4.12, an asymptotic uptake ($m_0 + \Delta m$) can be predicted. Therefore, the isotherm measurement can proceed to the next pressure point after a sufficient percentage of uptake has been reached. The asymptote can then be used as the uptake value for the current isotherm point. This approach has the advantage of optimising the measurement time and allowing the adjustment of the necessary waiting time for a particular isotherm point using the predicted time constant. This method is implemented by

---

[6] This is also the method of determining technical equilibrium described by Keller and Staudt [4].

the Intelligent Gravimetric Analyser (IGA) instruments currently manufactured by Hiden Isochema Ltd (UK).

4.2.2 Vacuum Microbalances

A vacuum microbalance is a highly accurate balance that can measure the weight of a sample suspended in vacuum and is the main component of a gravimetric sorption measurement system. There are a number of different types, including the spring balance, compensating beam balances, the magnetic suspension balance and the quartz crystal microbalance [30]. The use of vacuum microbalances for adsorption measurement has a long history [31], but the spring balance was one of the early types used for this purpose in the 1920s by McBain and Bakr [22]. It consists of a suspended helical spring made from fused silica with a sample holder attached below. Any sample weight change results in an extension or contraction of the spring. The magnitude of the change can then be determined from the displacement. Helical spring balances do not have the resolution of modern compensating beam balances and have now been largely superseded by the latter [1], although spring balances combined with modern displacement determination methods are still being developed [32].

Electronic compensating beam balances, such as those previously manufactured by Sartorius (Germany) and Cahn (USA), and currently produced by CI Electronics Ltd (UK), are the most widely used for the purpose of hydrogen sorption measurement. They consist of a pivoted beam, the displacement of which is compensated by the application of an electromagnetic signal. The strength of this signal gives a measure of the weight of the sample. More specifically, the CI Electronics compensating beam balance, which is currently used in Hiden Isochema IGA instruments, uses an infrared sensing system to detect the beam displacement. The balance beam, which is made from fine aluminium tubing, is then held in a null position using a galvanometer coil. The electromagnetic force required to maintain the null beam position is produced by passing a current through the coil. The current required to maintain the position is proportional to the applied force. Providing the balance is held in a well thermostatted enclosure, this allows the detection of mass changes of ± 0.1 μg.

The magnetic suspension balance, currently manufactured by Rubotherm (Germany), is a relatively recent advance in which the sample chamber is separated from the balance chamber [33]. The balance pan is attached to a permanent magnet and any change in the sample mass is detected through a non-magnetic, dielectric, wall by a second, coupled, magnet attached to the balance mechanism. This system has significant advantages over the compensating beam balances when performing measurements with corrosive species, which can attack the mechanism of the latter, but the disadvantages of the large buoyancy contribution of the permanent magnet assembly in the sample chamber and the requirement to re-zero the balance regularly throughout a measurement do not make it an obvious choice for hydrogen sorption measurement in which corrosion resistance is unnecessary and high accuracy is required due to the low molar mass of hydrogen. Further details of this balance type can be found in Keller and Staudt [4].

Quartz crystal microbalances use the frequency shifts of a quartz crystal due to an increase in its mass to determine the magnitude of the mass change, and can be used to measure hydrogen uptake by metal hydrides [34–36]. However, to perform the measurement the crystal is coated with a thin film of the hydrogen-absorbing material, which is then hydrogenated while its mass is monitored. As a consequence, their use is restricted to thin film samples and this technique has therefore not found widespread use in hydrogen sorption measurement, particularly for potential storage materials.

The commercial gravimetric systems that are currently available for hydrogen sorption measurement therefore use either a compensating beam balance (Hiden Isochema) or a Rubotherm magnetic suspension balance (Rubotherm/BEL Japan and VTI/TA Instruments).

4.2.3 High Pressure Systems

In order for a vacuum microbalance to be used in hydrogen sorption measurement it must be interfaced with a gas delivery system. Although low pressure (sub-ambient) measurements are of interest, for storage applications we are primarily interested in the above ambient performance of the material, and so it is crucial that the hydrogen uptake at pressures above ambient can be determined. The pressure system in which the microbalance is housed must have a method to accurately measure and control the pressure and must also be suitable for safe operation in pressurised gaseous hydrogen. We will now discuss both the pressure control system and the technical requirements for high pressure hydrogen-compatible operation.

4.2.3.1 Pressure Control

The pressure can be controlled in a number of ways. In the simplified description of the gravimetric method given in Sect. 4.2.1, valves A and B are described as controlling the pressure in the microbalance chamber. Simple open or shut valves will not perform this function accurately and so these should be valves that can open or close to varying degrees to precisely control the hydrogen pressure. In the Hiden Isochema IGA instruments, the equivalent of valves A and B are stepper motor-controlled bellows valves that allow PID control of the hydrogen pressure in conjunction with the manometer. An alternative is the control of the pressure using MFCs with feedback control. There are a number of possible practical solutions, but the important point is that the gas inlet and vacuum outlet of the microbalance chamber must be able to control the hydrogen pressure sufficiently. If accurate isobaric kinetic measurements are to be performed, a high degree of pressure control is essential.

4.2.3.2 Technical System Requirements

In order for a pressure system to be compatible with hydrogen it must be resistant to hydrogen leakage through joints and connections, as well as hydrogen permeation, embrittlement and attack of the construction material. If a system has been built to allow high vacuum operation then it is likely to conform to at least the first of these requirements. All metal seals are required and there should be minimal hydrogen leakage. Swagelok® VCR® components and valves, which are suitable for use with high pressure hydrogen and have high vacuum capability, for example, have a quoted helium leak rate $<4 \times 10^{-11}$ std $cm^3\ s^{-1}$. Hydrogen permeation, embrittlement and attack can be avoided by the selection of 316L stainless steel. The grade is important because the permeation rate of hydrogen through, for example, 403 and 316L stainless steel differs by orders of magnitude [37, 38]. The temperature of operation is also an important consideration. 316L stainless steel is known to be suitable for cryogenic use and so can also be used for low temperature adsorption measurements. At the other end of the scale, it can be used safely up to temperatures of 500°C (773 K). Above this temperature alternative specialist alloys may be suitable for use with hydrogen; however, hydrogen permeation and attack data are not readily available for many materials. The hydrogen embrittlement of steel is caused by a number of mechanisms. These include stress-induced hydride formation and cleavage, hydrogen enhanced localised plasticity, and decohesion [39]. Hydrogen embrittlement tends to occur at lower temperatures, near ambient, whereas hydrogen attack occurs at elevated temperatures. In the latter, high pressure bubbles form as a result of gaseous methane formation from the reaction of hydrogen with carbides or from water vapour formation from interaction with oxides [39].

Leak tightness can be determined using helium pressure hold leak testing. Providing high pressure helium testing is successful, the leak tightness can then be confirmed using high pressure hydrogen hold tests. If high pressure hydrogen is being used in a laboratory, a hydrogen safety sensing system should be used with appropriate fail safe mechanisms. The requirements can vary depending on the location and the physical size of the laboratory, but an extraction hood above the equipment will, for example, ensure that any hydrogen leakage will be safely removed from the vicinity of the equipment as hydrogen rises rapidly in air. The Lower Explosive Limit (LEL) for hydrogen is 4%. Safety sensors for flammable gases are typically set with two alarm settings at 10–20 and 40%LEL, and will therefore be set at a concentration of approximately 0.4–0.8 and 1.6% hydrogen in air. Another safety precaution is to ensure that the alarm system triggers slam-shut valves if these concentrations are met. However, any safety systems must be considered within the context of the particular laboratory and the relevant national safety regulations, and specific guidelines are beyond the scope of this book.

4.2.4 Other Gravimetric Approaches

So far we have focused on the standard gravimetric gas sorption measurement technique but to close this section we shall briefly describe an interesting alternative gravimetric approach for the rapid screening of potential hydrogen adsorbents, presented by Zielinski et al. [40]. Their straightforward and cost effective method involves measuring the mass change due to hydrogen pressurisation of a detachable pressure cell filled with adsorbent. The sample is first activated in separate apparatus before being weighed and transferred into a 180 g sealable pressure cell under argon. The cell is then attached to a gas dosing manifold and the sample degassed at ambient temperature to remove the residual argon from the cell. A pressure of hydrogen is applied and, after equilibration, the valve on the cell is closed and the cell disconnected from the manifold. Measurement of the mass of the cell then allows the total amount of hydrogen stored to be calculated. Note that this method does not just determine the amount of adsorbed hydrogen but also the amount held in the gas phase. It therefore provides an easy way of determining the benefit of the presence of the adsorbent in the cell, and hence a storage unit, versus pressurised gas phase hydrogen storage in the same volume. However, the determined value will depend to a certain extent on either the bulk or the tap density of the adsorbent (see Sect. 6.2.1), which is dependent on the manner in which the cell is loaded.

The method requires a relatively large sample but the minimum sample size depends on the uptake shown by the material and the pressure.[7] Activated carbon samples weighing around 1.7 g were used to demonstrate the determination of both methane and hydrogen storage, with impressive agreement shown with data measured using the differential volumetric measurement apparatus covered in Sect. 4.1.2. The method is recommended only for ambient temperature storage measurement because of the uncontrolled pressurisation that will occur if the cell is first cooled, pressurised and then allowed to warm prior to mass measurement, although presumably this would not be a problem providing the cell was rated to a sufficiently high pressure.[8]

4.3 Thermal Desorption

We now come to the third type of measurement, in which hydrogen desorption from a material is induced by the application of a thermal ramp. This measurement

---

[7] A greater uptake and higher pressure allows a smaller sample size.

[8] Careful calculations would be required before this was attempted because the resultant pressure rise due to an increase from 77 to 300 K, for example, is nearly fourfold ignoring any desorption from the sample. However, this might only limit the upper pressure of the measurement because 2–3 MPa would seem to be feasible, whereas 20 MPa would not because a 77 K measurement could potentially result in pressures approaching a kbar (100 MPa).

type has several analogues. It is similar to standard Thermogravimetric Analysis (TGA), which is widely used to study the thermal decomposition of solids [5], and Thermal Desorption Spectroscopy (TDS) or Temperature-Programmed Desorption (TPD), which is used widely in surface science [7, 8] and for the study of catalytic materials [6]. Standard TGA can be used to determine the amount of hydrogen released as a function of temperature by measuring the mass loss of the sample, providing a hydrogen loaded material can be mounted in the instrument. Otherwise a system in which the sample can be hydrogenated in situ must be used. TDS can be performed in a system equipped with a mass spectrometer in which the hydrogen is desorbed into vacuum or in a flowing configuration in which it is desorbed into an inert carrier gas stream. A more technically straightforward arrangement involves the desorption of the hydrogen into a calibrated volume while monitoring the subsequent pressure increase.

4.3.1 Thermogravimetric Analysis (TGA)

In TGA, the change in the weight of a sample as a function of temperature is measured. Standard TGA equipment differs from gravimetric sorption apparatus because it cannot typically be used to hydrogenate samples at different hydrogen pressures to determine isotherms. TGA equipment can, however, be used to desorb hydrogen from a material, or to decompose a sample to induce hydrogen release. TGA measurements are normally performed as a function of temperature; although for hydrides with slow desorption kinetics experiments can be performed by ramping the sample to a temperature at which hydrogen evolves and then maintaining a fixed temperature during the desorption process.

TGA apparatus consists of a microbalance onto which a sample can be mounted [5]. A carrier gas is flowed over the sample and the mass monitored as a function of temperature. The carrier gas is typically inert and at ambient pressure in standard systems, although both elevated and sub-ambient pressure operation are available commercially. To determine the nature of the decomposition or desorption products a mass spectrometer can be coupled to the apparatus using, for example, a heated capillary to provide the necessary pressure reduction and to transport the decomposition products effectively to the quadrupole analyser [41]. The pressure reduction is required as quadrupole mass spectrometers operate in the high vacuum regime and so, in order to analyse the composition, the gas must be sampled from the experimental pressure and reduced to levels below 10^{-4} mbar (10^{-2} Pa). The vacuum level depends on the detector being used. A Faraday detector operates at higher pressures or a lower vacuum than a Secondary Electron Multiplier (SEM); the latter, however, provides higher sensitivity for low analyte concentrations. The combination of mass spectrometry with TGA is known as TGA-MS. This can be particularly useful if the characterisation of decomposition products other than hydrogen is of interest; an example being the possible evolution of NH_3 during the decomposition of lithium amide and lithium hydride (see Sect. 2.3.2).

4.3.2 Thermal Desorption Spectroscopy

Thermal Desorption Spectroscopy (TDS) involves the desorption of hydrogen into either a vacuum or a flowing inert carrier gas. In a volumetric-type instrument, a known calibrated vessel of sufficient volume initially at vacuum can be used. The hydrogen is desorbed into the vessel and the increase in pressure used to determine the quantity of desorbed hydrogen. Hydrogen can also be desorbed into vacuum with the pressure, measured by a low pressure range gauge, used to plot a thermal desorption spectrum [42]. However, although this vacuum method allows the determination of the peak desorption temperature, it does not allow the quantitative determination of the desorbed hydrogen. The desorbed quantity can also be determined using a flow meter [43]. A more accurate method uses a mass spectrometer to monitor the hydrogen signal [44].

Von Zeppelin et al. [44] desorbed hydrogen into vacuum and used the mass spectrometer to determine the amount of desorbed hydrogen. The signal was calibrated using the response from a sample of known hydrogen content. In this case, hydrogenated PdGd alloy and TiH_2 was used. This kind of TDS is similar to the measurements commonly performed in surface science in UHV equipment and was the technique used in the report of Dillon et al. [45] on hydrogen storage in carbon nanotubes (see Sect. 2.1.1.2). Panella et al. [46] later reported TDS apparatus that can be used to study low temperature hydrogen desorption from 20 K. Subcritical temperatures are required to study the temperature-programmed desorption of adsorbed molecular hydrogen in order to retain the physisorbed hydrogen before the application of the thermal ramp. The desorption of hydrogen can also be calibrated using a controlled flow of a calibrant gas mixture (H_2/N_2 or H_2/He, for example). This method is used by the Hiden Isochema HTP1-S hydrogen desorption analyser and involves flowing the calibrant gas at three appropriate flow rates and using the mass spectrometer response as a calibration curve [47].

Regardless of the calibration method, the total desorbed hydrogen quantity is calculated from the integral of the calibrated thermal desorption signal,

$$N_{\text{tot}} = \int_{t_0}^{t_1} n(t)dt \tag{4.13}$$

where N_{tot} is the number of moles of desorbed hydrogen, $n(t)$ is the molar hydrogen desorption rate as a function of time, t, and t_0 and t_1 are the start and end times of the measurement, respectively [44].

TDS also allows the application of the Kissinger method, in which desorption spectra are measured at different thermal ramp rates, to allow the calculation of the activation energy of the desorption process (see Sect. 3.3.2.1). For example, Zlotea et al. [47] used this method to determine activation energies of 182 and 153 kJ mol^{-1} for MgH_2 and $Mg_{24}Y_5$-H, respectively.

4.4 A Comparison of the Techniques

Before concluding the chapter we will briefly discuss the comparative merits of the different techniques. Many of the accuracy issues discussed in Chap. 6 play an important role in the choice of technique and these can depend significantly on the material being studied, as well as the measurement temperature and pressure ranges of interest. Other aspects to consider, however, are the ease of implementation and use, the required sample throughput (i.e., measurement times) and the versatility. The information required from the measurement is also important. For example, if the determination of the high pressure behaviour, including the uptake as a function of pressure and temperature, is of significant importance, the volumetric and gravimetric techniques are more appropriate than temperature-programmed desorption, assuming that the thermal desorption is performed into either an inert carrier gas stream or into a vacuum. On the other hand, temperature-programmed methods are very useful for activation energy determination and TDS performed with a mass spectrometer can be used for the characterisation of the hydrogen desorption behaviour of small samples. This is particularly important if novel synthesis methods that produce only small sample quantities of the order of milligrams are being used. It also allows relatively rapid screening of different materials because the thermal desorption procedure is performed using a single, relatively rapid, temperature ramp.

With regard to the volumetric and gravimetric techniques, the former is easier to implement from a practical point of view. Gravimetry, on the other hand, provides higher accuracy at moderate pressures because the measurement error is not accumulative and high accuracy microbalances provide higher precision and accuracy in the determination of hydrogen uptake. Typical microbalances can detect weight changes of ± 0.1 μg, a quantity that is challenging to measure volumetrically, or manometrically, because it would require both a very small volume and high accuracy pressure measurement to achieve an equivalent level of sensitivity. However, the accuracy of the microbalance reading is counteracted at increasing measurement pressures by the increasing significance of the buoyancy corrections, which also increase with decreasing sample density. For adsorption measurement, both techniques are increasingly affected at higher pressures by our knowledge of the sample volume. This affects both absorption and adsorption measurement, but in the latter there is the added complication of the effects of possible helium adsorption during sample density or dead volume determination, which defines the position of the Gibbs dividing surface. For hydrides the sample volume problem still exists but the issue of helium adsorbing in micropores does not affect the measurement.

Generally speaking, volumetric measurements can be performed more rapidly than gravimetry. Measurements that use a microbalance need to be performed more carefully, in terms of pressurisation rates and the possibility of the physical disturbance of the apparatus, which will not affect volumetric measurements to the same extent. However, in the case of hydrides, isobaric hydride formation or synthesis is an intrinsically slower process than the approach used in manometric apparatus because, in the latter, the hydrogen absorption is driven by a pressure differential. Versatility can be defined in a number of ways, but it can include the ease with which a technique can be combined with other complementary characterisation techniques (see Chap. 5) and, in the case of hydrides, the possible variation in the sample size for activation purposes. In principle, with a large enough internal volume, large sample sizes can be activated in manometric apparatus, whereas the sample size is limited in gravimetry by the capacity of the microbalance. Both techniques can be combined with complementary characterisation techniques but the use of gravimetry for this purpose is complicated by the need to keep the sample independently suspended from the microbalance. Volumetric measurement is therefore more often used for the performance of in situ X-ray and neutron scattering experiments (see Sects. 5.3 and 5.4.1). Air or moisture sensitive sample handling is also easier to perform with volumetric apparatus because the sample does not need to be loaded onto a balance but instead can be easily loaded inside a glovebox, into a cell equipped with an isolation valve.

From the point of view of the information required from a measurement, the volumetric and gravimetric techniques both allow the determination of either the enthalpy of hydride formation or decomposition (see Sect. 3.2.2) or the isosteric enthalpy of hydrogen adsorption for microporous materials (see Sect. 3.2.1) via the measurement of a series of isotherms at different temperatures. Temperature-programmed techniques, on the other hand, allow the determination of the activation energy of the desorption process and, in the event of different peaks appearing in the desorption spectra, the occupation of different sorption energy sites.

In each case, the techniques described in this chapter have significant advantages compared to the others and the selection of the appropriate method depends on many of the points discussed above. The most thorough approach to the characterisation of hydrogen storage materials is to use a range of techniques, because good agreement between the measurements performed with different apparatus, if expected from the different hydrogenation routes (isobaric or isochoric), will provide validation of the results from each method and increase confidence in the determined hydrogen storage properties of the material. A thorough discussion of the experimental errors that can affect measurements of each type is presented in Chap. 6.

4.5 Summary

In this chapter we have covered the common laboratory gas sorption techniques that can be used to determine the hydrogen storage properties of materials. Firstly, we described volumetric methods, which determine the amount of hydrogen uptake by measuring the change in the pressure of hydrogen in a measurement system of a fixed known volume or by measuring the volumetric hydrogen flow rate. We focused on the manometric method as this is the most commonly used technique, and is the approach used in the majority of commercial instrumentation. Secondly, we discussed gravimetric measurement in which a vacuum microbalance mounted in a pressure-rated, hydrogen-compatible, vessel is used to determine the uptake of hydrogen by measuring the weight change of the sample due to hydrogen sorption. We have briefly covered the different microbalance types and the technical requirements for systems specified for pressurised hydrogen use. We then covered thermal desorption, or temperature-programmed, techniques in which desorption from a hydrogen loaded sample is driven by an applied temperature ramp, before briefly comparing the different techniques. Issues surrounding the accuracy of these techniques and sources of experimental error in hydrogen sorption measurement are covered in more detail in Chap. 6.

References

1. Rouquerol F, Rouquerol J, Sing K (1999) Adsorption by powders and porous solids: principles, methodology and applications. Academic Press, London
2. Lowell S, Shields JE, Thomas MA, Thommes M (2004) Characterization of porous solids and powders: surface area, pore size and density. Springer, Dordrecht
3. Yang RT (1997) Gas separation by adsorption processes. Imperial College Press, London
4. Keller J, Staudt R (2005) Gas adsorption equilibria: experimental methods and adsorptive isotherms. Springer, New York
5. Brown ME (2001) Introduction to thermal analysis: techniques and applications, 2nd edn. Kluwer Academic Publishers, Dordrecht
6. Bhatia S, Beltramini I, Do DD (1990) Temperature-programmed analysis and its applications in catalytic systems. Catal Today 7:309–438
7. Redhead PA (1962) Thermal desorption of gases. Vacuum 12(4):203–211
8. Carter G (1962) Thermal resolution of desorption energy spectra. Vacuum 12(5):245–254
9. Blach TP, Gray EM (2007) Sieverts apparatus and methodology for accurate determination of hydrogen uptake by light-atom hosts. J Alloy Compd 446–447:692–697
10. Broom DP (2008) Hydrogen sorption measurements on potential storage materials: experimental methods and measurement accuracy. EUR 23242 EN. Office for Official Publications of the European Communities, Luxembourg
11. Chambers A, Fitch RK, Halliday BS (1998) Basic vacuum technology, 2nd edn. Institute of Physics Publishing, Bristol
12. Japanese Industrial Standards Committee (2007) Method for measurement of pressure–composition–temperature (PCT) relations of hydrogen absorbing alloys. JIS H 7201:2007 (E)
13. Gray EM (2008) Reliably measuring hydrogen uptake in storage materials. In: Walker G (ed) Solid-state hydrogen storage: materials and chemistry. Woodhead Publishing, Cambridge
14. Poirier E, Chahine R, Tessier A, Bose TK (2005) Gravimetric and volumetric approaches adapted for hydrogen sorption measurements with in situ conditioning on small sorbent samples. Rev Sci Instrum 76:055101
15. Blackman JM, Patrick JW, Snape CE (2006) An accurate volumetric differential pressure method for the determination of hydrogen storage capacity at high pressures in carbon materials. Carbon 44:918–927
16. Browning DJ, Gerrard ML, Lakeman JB, Mellor IM, Mortimer RJ, Turpin MC (2002) Studies into the storage of hydrogen in carbon nanofibers: proposal of a possible reaction mechanism. Nano Lett 2(3):201–205

17. Zielinski JM, Coe CG, Nickel RJ, Romeo AM, Cooper AC, Pez GP (2007) High pressure sorption isotherms via differential pressure measurements. Adsorption 13:1–7
18. Goodell PD, Sandrock GD, Huston EL (1980) Kinetic and dynamic aspects of rechargeable metal hydrides. J Less Common Met 73:135–142
19. Pieters WJM, Gates WE (1984) Method and apparatus for determining the amount of gas adsorbed or desorbed from a solid. United States Patent 4489593
20. Wang X-L, Suda S (1990) A dehydriding kinetic study of $LaNi_{4.7}Al_{0.3}$ hydride by a step-wise method. J Less Common Met 159:83–90
21. Japanese Industrial Standards Committee (2007) Method for measurement of hydrogen absorption/desorption reaction rate of hydrogen absorbing alloys. JIS H 7202:2007 (E)
22. McBain JW, Bakr AM (1926) A new sorption balance. J Am Chem Soc 48(3):690–695
23. Lutz HM, Schmitt R, Steffens F (1978) A high-temperature, high-pressure microbalance for the determination of the hydrogen sorption characteristics of metal hydrides. Thermochim Acta 24:369–381
24. Feenstra R, Griessen R, de Groot DG (1986) Hydrogen induced lattice expansion and effective H–H interaction in single phase PdH_c. J Phys F Met Phys 16:1933–1952
25. Benham MJ, Ross DK (1989) Experimental determination of absorption–desorption isotherms by computer-controlled gravimetric analysis. Z Phys Chem NF 163:S25–S32
26. Bououdina M, Soubeyroux JL, Juen P, Mouget C, Argoud R, Fruchart D (1995) An apparatus for gravimetric analysis: its application to metal-hydrogen systems. J Alloy Compd 231:422–426
27. Agarwal RK, Schwarz JA (1988) Analysis of high pressure adsorption of gases on activated carbon by potential theory. Carbon 26(6):873–887
28. Chagger HK, Ndaji FE, Sykes ML, Thomas KM (1995) Kinetics of adsorption and diffusional characteristics of carbon molecular sieves. Carbon 33(10):1405–1411
29. O'koye IP, Benham M, Thomas KM (1997) Adsorption of gases and vapors on carbon molecular sieves. Langmuir 13:4054–4059
30. Gast T, Robens E (1996) Modern vacuumbalances. J Therm Anal 47:605–617
31. Kiefer S, Robens E (2008) Some intriguing items in the history of volumetric and gravimetric adsorption measurements. J Therm Anal Calorim 94:613–618
32. Trexler MD, Sanders TH, Singh PM (2006) Automation of a McBain-Bakr-type thermo-gravimetric analyzer using a digital image correlation technique. Rev Sci Instrum 77: 025103
33. Robens E, Gast T, Hoinkis E, Müller U (1989) Surface area, density and porosity measurements using the magnetic suspension balance. Fresenius Z Anal Chem 333:428–432
34. Frazier GA, Glosser R (1979) Phase diagrams of thin films of the palladium hydrogen system using a quartz crystal thickness monitor. J Phys D 12:L113–L115
35. Frazier GA, Glosser R (1980) Characterization of thin films of the palladium hydrogen system. J Less Common Met 74:89–96
36. Kulchytskyy I, Kocanda MG, Xu T (2007) Direct mass determination of hydrogen uptake using a quartz crystal microbalance. Appl Phys Lett 91:113507
37. Schefer RW, Houf WG, San Marchi C, Chernicoff WP, Englom L (2006) Characterization of leaks from compressed hydrogen dispensing systems and related components. Int J Hydrogen Energy 31:1247–1260
38. San Marchi C, Somerday BP, Robinson SL (2007) Permeability, solubility and diffusivity of hydrogen isotopes in stainless steels at high gas pressures. Int J Hydrogen Energy 32:100–116
39. Birnbaum HK (2001) Hydrogen embrittlement. In: Buschow KHJ, Cahn RW, Flemings MC, Ilschner B, Kramer EJ, Mahajan S, Veyssière P (eds) Encyclopedia of materials: science and technology. Elsevier, Amsterdam
40. Zielinski JM, McKeon P, Kimak MF (2007) A simple technique for the measurement of H_2 sorption capacities. Ind Eng Chem Res 46:329–335
41. Hatton PJ, Southward BWL (2003) Optimisation of the connection between TA-MS systems together with improved data interpretation for TA-MS applications. J Therm Anal Calorim 72:83–92

42. Stern A, Kreitzman SR, Resnik A, Shaltiel D, Zevin V (1981) Thermal desorption spectra of hydrogen from the bulk: ZrV_2H_x. Solid State Commun 40(8):837–841
43. Castro FJ, Meyer G (2000) A novel thermal desorption spectroscopy apparatus. Rev Sci Instrum 71(5):2131–2133
44. von Zeppelin F, Haluška M, Hirscher M (2003) Thermal desorption spectroscopy as a quantitative tool to determine the hydrogen content in solids. Thermochim Acta 404:251–258
45. Dillon AC, Jones KM, Bekkedahl TA, Kiang CH, Bethune DS, Heben MJ (1997) Storage of hydrogen in single-walled carbon nanotubes. Nature 386:377–379
46. Panella B, Hirscher M, Ludescher B (2007) Low-temperature thermal-desorption mass spectroscopy applied to investigate the hydrogen adsorption on porous materials. Microporous Mesoporous Mater 103(1–3):230–234
47. Zlotea C, Sahlberg M, Özbilen S, Moretto P, Andersson Y (2008) Hydrogen desorption studies of the $Mg_{24}Y_5$–H system: formation of Mg tubes, kinetics and cycling effects. Acta Mater 56(11):2421–2428

Chapter 5
Complementary Characterisation Techniques

We will now cover some of the techniques that can be used to complement the common laboratory hydrogen sorption characterisation methods described in Chap. 4. We focus on methods that have typically been used for the characterisation of hydrogen storage compounds and, in particular, have been used for comparison with data obtained using one or more of the sorption measurement techniques. We begin with thermal analysis and calorimetry, both of which can be used to determine the thermodynamic parameters associated with phase transitions in metal hydrides. In addition, thermal analysis can also be used to rapidly determine the temperature required for hydrogen absorption and desorption. We then cover the gas adsorption characterisation methods that are commonly used to characterise microporous adsorbents, in terms of surface area, pore volume and the pore size distribution. The experimental procedures used in these methods are essentially the same as the laboratory hydrogen sorption techniques described in Chap. 4, but the data analysis methods are more sophisticated than the analysis typically applied to hydrogen adsorption isotherms. We will therefore concentrate on the isotherm analysis methods, rather than experimental details. We then discuss X-ray and neutron powder diffraction, which are techniques that can be used to determine the structural changes that occur as a result of hydrogen absorption by hydrides, as well as the positions of adsorbed hydrogen molecules at suitably low temperatures in microporous adsorbents. Following this, we look at different types of spectroscopy, such as Inelastic Neutron Scattering (INS) and Nuclear Magnetic Resonance (NMR), which can be used to study the dynamics of hydrogen in materials, together with Variable Temperature Infrared (VTIR) spectroscopy, which can be used to determine the strength of the hydrogen-adsorbent interaction. We end the chapter with a brief discussion of some other techniques that have also been used in one way or another to complement sorption measurements on storage materials.

D. P. Broom, *Hydrogen Storage Materials*, Green Energy and Technology,
DOI: 10.1007/978-0-85729-221-6_5, © Springer-Verlag London Limited 2011

5.1 Thermal Analysis and Calorimetry

Thermal analysis and calorimetry can be used in a number of ways to complement hydrogen sorption measurements. *Thermal analysis* is a broad term that covers a number of experimental methods, including TGA and TDS or TPD, which were covered in the previous chapter in the context of the quantification of the hydrogen content of materials. Here we shall briefly consider two other thermal analysis techniques, Differential Thermal Analysis (DTA) and Differential Scanning Calorimetry (DSC), together with reaction calorimetry. We will focus primarily on the study of hydrides using these techniques.

In DTA, the difference in temperature between the sample and a reference material is measured while the same heating program is applied to both. Heat flux DSC is similar to DTA except that the heat flow between the sample and the reference material, rather than the temperature difference, is determined. In power compensated DSC, meanwhile, the sample and the reference material are heated independently but at the same temperature ramp rate. The difference between the independent supplies of power to each gives a measure of the heat flow (dq/dt) into the sample. In each case, the occurrence of a phase transition or a reaction involving the sample will result in a peak in ΔT or dq/dt as a function of time [1, 2]. In the case of hydride phase formation or decomposition, the integral of the signal will be related to the enthalpy of hydride formation or decomposition. The temperature at which a DSC peak occurs also provides valuable information regarding the temperature of desorption. This is useful, for example, when studying the effects of mechanical milling because the results of two samples under the same conditions will indicate the level of improvement due to the treatment. In addition, methods used for the determination of the activation energy from thermal desorption measurements, such as the Kissinger method (see Sect. 3.3.2.1), were originally developed for DSC and can therefore be used for the analysis of hydrogen absorption or desorption reactions. In these methods the peak of the DSC signal as a function of heating rate is used for the determination of an apparent activation energy.

Reaction calorimetry, meanwhile, measures the amount of energy consumed or released during any given chemical reaction. The technique can be implemented in a number of ways [3], including heat flux and power compensated variants, and it provides higher accuracy than DSC [4]. Calorimetry also allows the determination of the enthalpies of hydrogen solution through the course of a pressure-composition isotherm measurement, by combining a calorimeter with volumetric sorption apparatus, so that more detailed thermodynamic studies as a function of hydrogen content can be performed. The other thermal analysis methods can also be combined with other complementary characterisation techniques; this is often known as *hyphenation*, or the resultant combination as a *hybrid technique*.

DTA, DSC and reaction calorimetry have been used to characterise many metal hydride materials, including intermetallic AB [5], AB_2 [6], AB_3 [6, 7], AB_5 [8–12], A_2B_{17} [7] and AB_{12} [7] hydrides, amorphous alloys [13], ball-milled nanocrystalline complex transition metal hydrides [14, 15], and various other complex hydrides, including the alanates [15–18] and borohydrides [19–21].

An example of a detailed thermodynamic study is due to Murray et al. [10] who used differential heat conduction calorimetry to study the enthalpy of hydrogen solution for $CaNi_5$. This AB_5 intermetallic has a two plateau isotherm and a series of over 70 simultaneous pressure-composition and calorimetric enthalpy measurements were performed through a single absorption and desorption isotherm on two samples. Another example of hyphenation, meanwhile, was presented by Fernández et al. [4]. The authors combined DSC with TDS (Sect. 4.3.2) and applied the technique to PdH_x, $TiCr_{1.85}H_x$ and $ZrCr_2H_x$. A $TiH_{1.92}$ sample was used to calibrate the TDS signal and the simultaneous determination of the enthalpies of hydride phase decomposition and the desorbed quantity of hydrogen was therefore possible. However, it should be noted that the enthalpies determined at different (Ar) carrier gas rates for a particular material differed significantly. It was therefore necessary to perform measurements at a series of different flow rates and extrapolate a linear fit to obtain a "true" enthalpy. As a consequence, the enthalpy value for each material contained a significant estimated uncertainty. The DSC and MS signals for the $TiCr_{1.85}H_x$ sample are shown in Fig. 5.1, in which it can be seen that the signals show good agreement.

An alternative approach to the use of thermal analysis was presented recently by Rongeat et al. [15], who used high pressure DSC to determine the enthalpies of hydride formation and decomposition of a Ni-doped MgH_2 sample, ball milled Mg-Co hydride, and Ti-doped $NaAlH_4$. A series of DSC scans were performed at different hydrogen pressures in the range 3 bar (0.3 MPa) to 120 bar (12.0 MPa).

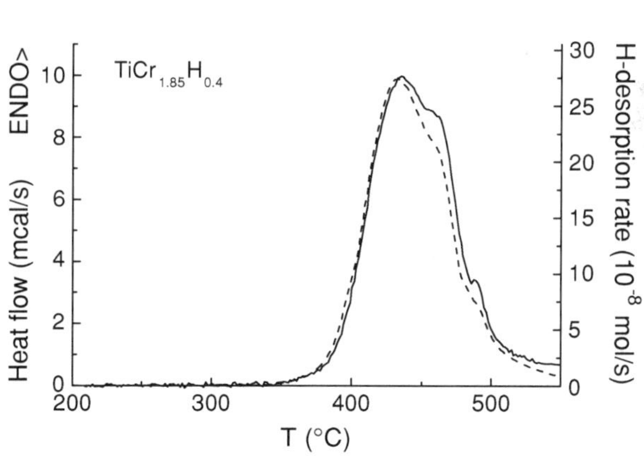

Fig. 5.1 A plot of the DSC and MS signals for hydrogen desorption from $TiCr_{1.85}H_{0.4}$, performed at a heating rate of 20°C min^{-1}. The *solid line* shows the DSC signal and the *dotted line* shows the MS signal. (Reprinted from the Journal of Alloys and Compounds, Volume 298, Simultaneous differential scanning calorimetry and thermal desorption spectroscopy measurements for the study of the decomposition of metal hydrides, pages 244–253 [4], Copyright 2000, with permission from Elsevier)

A van 't Hoff plot (see Sect. 3.2.2.1) was then constructed from the natural log of the pressure of each scan and the onset temperature of the endo- or exothermic peak. The authors argue that this provides a rapid alternative to the measurement of pressure-composition isotherms. A recent example of the application of different activation energy determination methods is the study by Zakotnik et al. [22] in which the Kissinger and Friedmann methods were used to determine the activation energy for hydrogen desorption from a Sm_2Co_{17}-based magnetic material using DSC.

5.2 Gas Adsorption

The measurement of gas adsorption is used widely for the characterisation of powders and porous materials. The most common method is surface area determination using the BET (Brunauer–Emmett–Teller) method, which typically involves the measurement of nitrogen adsorption at its boiling point, 77 K, to determine the Specific Surface Area (SSA) of a material. Other commonly used characterisation methods include the determination of the pore volume of porous solids from the measurement of nitrogen uptake, and the determination of the pore size distribution of porous materials using nitrogen, argon or carbon dioxide adsorption. In this section we will consider each of these methods in turn, before concluding with a discussion of the concepts of surface area and micropore volume, and their relation to hydrogen storage capacity. For more in-depth coverage of the methods covered here, interested readers are referred to the books specifically on the topic by Rouquerol et al. [23] and Lowell et al. [24].

5.2.1 *Surface Area Determination*

The most common way to measure surface area is to use the BET method. In this technique the adsorption isotherm for a particular subcritical adsorptive is measured, normally using volumetric apparatus, and the isotherm is analysed on the basis of a series of assumptions. The BET theory is an extension of the Langmuir theory of monolayer formation (Sect. 3.4.1) but accounts for the adsorption of further layers above the first. The theory assumes adsorption on a flat surface, with no limit on the number of adsorbed layers, so that there is an infinite number of layers at $P/P_0 = 1$. As in the Langmuir theory, it assumes that the surface is energetically homogeneous and that there is no interaction between neighbouring adsorbate molecules. At equilibrium, after monolayer formation, it is also assumed that the rate of condensation onto the first layer is equal to the rate of evaporation from the second layer, and that these properties are the same as for the adsorbate in the liquid state. These adsorption and desorption rates are dependent on the respective energies of adsorption and liquefaction, which are therefore assumed to be equal.

The above assumptions lead to the BET equation,

$$V = \frac{V_m CP}{(P_0 - P)\left[1 + (C-1)\left(\frac{P}{P_0}\right)\right]} \quad (5.1)$$

where V is the adsorbed phase volume, V_m is the volume of a complete monolayer, C is the BET constant, P is the pressure and P_0 is the saturation pressure of the adsorptive, so that P/P_0 is the relative pressure. The full derivation can be found in a number of books on adsorption [23–26]. The BET constant, C, in Eq. 5.1 expresses the magnitude of the adsorbate-adsorbent interaction and, in order for it to carry physical meaning, should have a positive value [27, 28]. If C has a low value, the BET equation describes a Type III isotherm, whereas for higher values of C, it describes Type II. There is not a distinct boundary between these two extremes, but instead a gradual transition; however, generally speaking, if $C < 1$ the isotherm will be Type III but if it is of the order of 100 or greater, the isotherm will be Type II.

The following linearised form of the BET equation is then used to determine the surface area of a material,

$$\frac{P}{V(P_0 - P)} = \frac{1}{V_m C} + \left(\frac{C-1}{V_m C}\right)\left(\frac{P}{P_0}\right). \quad (5.2)$$

BET analysis therefore involves plotting the left-hand side of this equation, $P/V(P_0 - P)$, against relative pressure, P/P_0, and determining the gradient, m, and intercept, c, of a straight line fit to the linear part of the resulting plot. The gradient will be $m = (C - 1)/V_m C$ and the intercept $c = 1/V_m C$, so that,

$$V_m = \frac{1}{(m+c)} \quad (5.3)$$

and,

$$C = \left(\frac{m}{c}\right) + 1. \quad (5.4)$$

The BET SSA, a_{BET}, can then be calculated from the following expression,

$$a_{BET} = V_m N_A \sigma \quad (5.5)$$

where N_A is the Avogadro constant and σ is the molecular area of the chosen adsorbate [29]. This is typically nitrogen, with σ assumed to have a value of 0.162 nm^2; however, a number of other adsorptives can be used where appropriate. The BET equation is commonly considered to be valid for data in the approximate range $0.05 < P/P_0 < 0.3$, but it is important to assess the linearity of the BET plot on a case-by-case basis; for microporous materials the range is considered to be lower than this (see Sect. 5.2.4). Furthermore, the chosen relative pressure region should be stated when reporting the results of BET surface area analysis.

5.2.2 Pore Volume Determination

The pore volume, expressed for example as $cm^3\ g^{-1}$, is an important parameter for a porous adsorbent. If adsorption by a porous material is due to a pore filling mechanism rather than Langmuir-type monolayer formation, the adsorption capacity of an adsorbent will be dependent on its specific pore volume rather than its SSA. The pore volume of a material can be determined by measuring the adsorption isotherm of a probe gas other than hydrogen. If the isotherm exhibits Type I behaviour then the pore volume can be determined from the saturation capacity, assuming that the adsorbate density is known. The assumption that the adsorbate has the same volume as an equivalent quantity of bulk liquid is known as the *Gurvich rule* [23]. Although confined fluids are known to behave in a different manner to bulk fluids [30], this rule seems to hold generally for subcritical adsorbates. See, for example, the recent work of Moellmer et al. [28] who found good agreement between the pore volumes of an indium-based MOF (*soc*-MOF) using the Gurvich rule applied to the adsorption of subcritical argon at 87.3 K, methane at 107.4 K and carbon dioxide at 273 K, with values of $0.47\ cm^3\ g^{-1}$, $0.45\ cm^3\ g^{-1}$ and $0.44\ cm^3\ g^{-1}$ obtained, respectively.

5.2.3 Pore Size Distribution Determination

There are many different methods available for the determination of the Pore Size Distribution (PSD) of a porous material using the analysis of gas or vapour adsorption [24]. Some of these are applicable to pore sizes that are not of specific interest for hydrogen storage applications. These include mesopore analysis methods based on the Kelvin equation of which the Barrett–Joyner–Halenda (BJH) method is one of the best known. However, a number of approaches apply to microporous adsorbents and they can be grouped into classical and modern methods. The former category, which includes the Dubinin–Radushkevich (DR), Dubinin–Astakhov (DA) and Horvath–Kawazoe (HK) methods, generally consider the adsorbed fluid as if it behaves in the same manner as the bulk fluid phase. Although the HK method allows for the attractive solid–fluid forces in narrow pores it does not describe the mechanism of pore-filling correctly[1] and also does not account for fluid layering near the pore walls. These classical methods tend to

[1] Micropore filling is considered to be a continuous process, but in the HK method it is assumed that pore filling occurs discontinuously at a specific pressure [24].

lead to underestimated pore sizes, compared to the results obtained using their modern counterparts. The modern methods include Density Functional Theory (DFT), and the Monte Carlo and Molecular Dynamics methods, and they allow the determination of the equilibrium density profiles of confined fluids and the subsequent calculation of equilibrium adsorption isotherms from these densities. As the Grand Canonical Monte Carlo (GCMC) and Non-Local DFT methods are considered to be the most accurate methods of PSD determination, we shall concentrate on these in this section.

In the DFT method the local fluid density, $\rho(\boldsymbol{r})$, at a position $\boldsymbol{r}$, is determined by minimising the corresponding grand potential, $\Omega[\rho(\boldsymbol{r})]$. Once the equilibrium form of $\rho(\boldsymbol{r})$ has been determined, the isotherm and other properties of the system can then be calculated. The grand potential is given by,

$$\Omega[\rho(\boldsymbol{r})] = F[\rho(\boldsymbol{r})] - \int d\boldsymbol{r}\rho(\boldsymbol{r})(\mu - V_{ext}(\boldsymbol{r})) \tag{5.6}$$

where F is the intrinsic Helmholtz free energy of the fluid (the Helmholtz energy in the absence of an external field), μ is the chemical potential and $V_{ext}(\boldsymbol{r})$ is the potential field due to the pore walls [30, 31].

The description of the fluid–fluid interaction is chosen to allow the reproduction of the bulk behaviour of the fluid. The parameters that describe the solid–fluid interactions, meanwhile, are obtained from fitting calculated isotherms to the standard (nitrogen) isotherm on a planar surface. It is assumed that the fluid is contained in individual pores of simple, typically slit-like or cylindrical, geometry. An individual carbon slit pore, for example, is represented by two infinite, parallel graphitic slabs separated by the pore width [24].

There is more than one type of DFT and they differ in the way that the fluid–fluid interactions are represented. Local DFT uses the Local Density Approximation (LDA), which omits short range correlations, whereas Non-Local DFT uses a Non-Local Smoothed Density Approximation (SDA) that includes short range correlations. The latter accounts for the oscillations of the density profile near to the pore walls [30, 32], and the advantage of Non-Local DFT is that it therefore provides a more accurate description of the behaviour of fluids confined in narrow pores in comparison with Local DFT.

The GCMC method, meanwhile, simulates the situation of an adsorbed fluid in equilibrium with a bulk fluid reservoir. Molecules are moved and rotated randomly, leading to different system configurations, which are then accepted or rejected using thermodynamic criteria. After many such moves, an average is taken and an equilibrium density profile is determined. This allows the subsequent calculation of an adsorption isotherm. GCMC is considered to be the most accurate method for the description of fluids confined in narrow pores, but is more computationally expensive than DFT.

Both the GCMC and DFT methods produce a set of isotherms calculated for a corresponding set of pore sizes in a given range for the adsorptive under consideration. The isotherms are calculated by integrating $\rho(\boldsymbol{r})$ for the fluid in a model pore of each size in the chosen range. The isotherm set is known as a *kernel*, and is used as the basis of the PSD determination. Such a kernel, $N(P/P_0,W)$, where P/P_0 is the relative pressure and W is the pore width (dimension), is then correlated with experimental data using the Generalized Adsorption Isotherm (GAI),

$$N(P/P_0) = \int_{W_{\min}}^{W_{\max}} N(P/P_0, W)f(W)dW \tag{5.7}$$

where $N(P/P_0)$ is the experimental isotherm for a relative pressure of P/P_0, and $f(W)$ is the pore size distribution function. This expression assumes that the total isotherm consists of a number of individual isotherms for given pore sizes multiplied by their relative distribution. The GAI is then solved numerically from a kernel produced using either GCMC or DFT. This is, generally, an ill-posed problem but stable solutions can be found using regularisation algorithms. These algorithms are numerical methods, or rather different implementations of the Regularisation Method, which can be used to solve the GAI regardless of the chosen kernel [33–36].

The accuracy of the results of PSD determination using these methods depends on the availability of kernels that are applicable to the adsorbate-adsorbent system of interest. The currently available methods also do not take into account the chemical and geometric hetereogeneity of pore walls in real materials. The calculated theoretical isotherms therefore exhibit steps due to layer transitions, which are not observed in experimental isotherms [24]. They are, however, considered to be the most accurate methods for PSD determination for both microporous and mesoporous media. Related work, due to Jagiello and co-workers, which is of considerable interest in the context of hydrogen storage, has investigated the use of hydrogen itself as a probe molecule for PSD determination [37–40]. Although not traditionally used for this purpose, the argument is that hydrogen can probe smaller pore sizes than argon, nitrogen and carbon dioxide. Jagiello et al. [38] used low (sub-ambient) pressure hydrogen adsorption data measured at 77 K to predict the uptake of hydrogen by three porous carbons up to a pressure of 100 atm (10.1 MPa) at temperatures of 87 and 298 K, with very good agreement shown between the predicted isotherms and experimental data measured gravimetrically in the elevated pressure regime. This process required the determination of additional high pressure kernels at the temperatures of interest. Jagiello et al. [39] advocate the combination of hydrogen and nitrogen adsorption for the PSD determination of carbons; the hydrogen probes smaller pores, which are either inaccessible to nitrogen molecules or lead to kinetic limitations at 77 K, and the nitrogen provides better sensitivity than supercritical hydrogen in the mesoporous regime. Similarly, Jagiello and Betz [40] demonstrate the combination of hydrogen and argon adsorption measurements for the same reasons, also using microporous carbons.

5.2.4 Discussion

With regard to hydrogen storage material characterisation, these methods are primarily of interest for the characterisation of the physical properties of microporous adsorbents.[2] In this section we discuss the validity of the concept of surface area for microporous materials and the relevance of the assessment of porous material characteristics in the context of hydrogen storage.

The BET method has been very successful and is widely used for surface area determination, even though the assumptions behind the theory are somewhat questionable. Another method that has been used recently for the characterisation of microporous materials in the hydrogen storage literature is the determination of the Langmuir surface area. However, the use of these methods tends to result in a significant difference in the calculated SSAs[3] because they use different models for the adsorption process. The Langmuir picture of adsorption considers pure monolayer formation, while the BET theory assumes the formation of multilayers. Type I adsorption behaviour, which is typically exhibited by microporous media, results from pore filling rather than monolayer or multilayer formation, and therefore the 'surface area' determined using either method is a measurement parameter, as opposed to a real physical property of the system. The value obtained for the surface area using either method will also depend on the adsorbate because different size probe molecules [29] will 'see' a different surface area (or micropore volume). The chosen value of the adsorbed molecular area for a specific probe species will also affect the result.

For the application of BET analysis to microporous materials, Rouquerol et al. [27] argue that it would more appropriate to replace the concepts of surface area or micropore volume with the terms *BET strong retention capacity*, *external surface area*, *micropore capacity* and *saturation capacity*.[4] The use of the concept of a single SSA for microporous materials, however, remains in common usage.

---

[2] SSA is an interesting physical property of metal hydrides because a larger surface area provides a greater area for the surface dissociation of molecular hydrogen. The surface area can be increased by the ball milling of materials, and also as a natural consequence of the decrepitation of brittle metal hydrides upon hydriding and dehydriding. A change in the surface area of a material can likely be associated with an improvement in the sorption kinetics if the process is surface-limited. However, it is not a property that can be linked in any direct manner to the determination of total hydrogen uptake or the measurement of isotherms; hence our focus here on microporous adsorbents rather than hydrides.

[3] See, for example, the data tabulated by Thomas [41] and the lower gradient in Fig. 5.3 compared to Fig. 5.2.

[4] Under these proposals, the *BET strong retention capacity* would consist of two components: the *micropore capacity* and the *external surface area*, the latter of which is the monolayer content on the non-microporous sections of the surface. The *saturation capacity* is then the amount adsorbed at saturation.

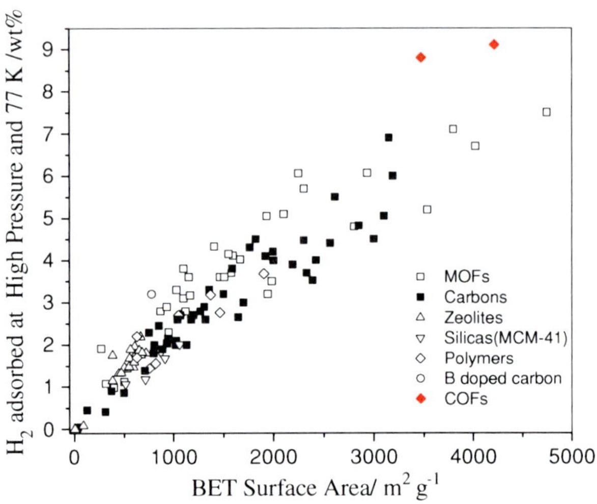

Fig. 5.2 A plot of hydrogen adsorbed (wt%) at saturation at 77 K against the BET SSA (m^2 g^{-1}) for metal–organic frameworks, microporous carbons, boron-doped carbons, MCM-41, zeolites, microporous polymers and covalent organic frameworks. (Thomas [41]—Reproduced by permission of the Royal Society of Chemistry)

The application of the BET method to microporous carbon has been the subject of a recent interlaboratory study (Round Robin exercise), which was reported by Silvestre-Albero et al. [42]. The nitrogen BET surface areas for two commercial activated carbons (Takeda 4A and 5A) were found to be 313–422 and 476–589 m^2 g^{-1}, respectively, thus demonstrating the potential variability in the results of surface area determination for microporous media, even when using both the same method and the same adsorbate. The authors concluded that the large discrepancies may be due to kinetic restrictions on the diffusion of nitrogen in the pore network and this means that measurements performed at 77 K are strongly dependent on the choice of equilibration time. Discussion of the range of applicability and the practical use of the BET method, in general, can be found in Rouquerol et al. [23] and Lowell et al. [24], and it is also covered by IUPAC guidelines [43], while Rouquerol et al. [27] discuss the validity of the method specifically for microporous materials. Although they conclude, as mentioned above, that the BET 'surface area' does not represent a real surface area they propose conditions that should aid the consistent application of the BET equation. These include ensuring a positive value of C and only using a relative pressure range in which $V(P_0 - P)$ increases continuously with P/P_0. The application of these criteria by the authors to argon adsorption data at 87 K for 13X (Na-X) zeolite led to the use of only the data in the range below $P/P_0 = 0.04$ for BET analysis.

With regard to the relationship between surface area, pore structure and hydrogen storage, Figs. 5.2 to 5.4 show data for various microporous materials, as tabulated and plotted by Thomas [41]. These plots seem to suggest, for the chosen materials, that the correlation between the BET surface area and hydrogen uptake (Fig. 5.2) is stronger than the correlation between pore volume and hydrogen uptake (Fig. 5.4). If we use the terminology of Rouquerol et al. [27] and assume perhaps incorrectly that, in highly microporous materials, the BET SSA is dominated by the micropore capacity, rather than the external surface area, this

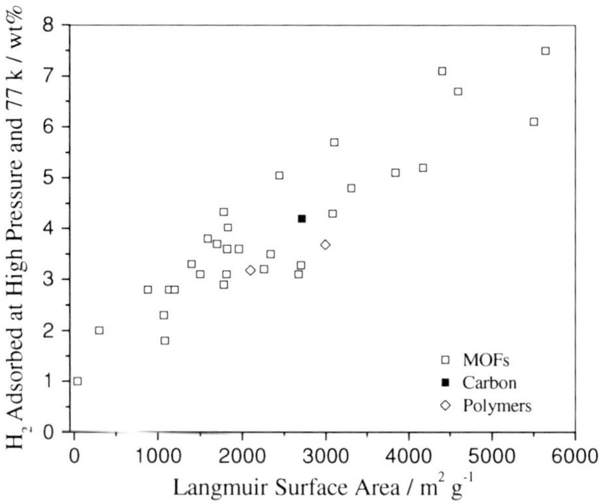

Fig. 5.3 A plot of hydrogen adsorbed (wt%) at saturation at 77 K against Langmuir surface area ($m^2\ g^{-1}$) for metal–organic frameworks, microporous carbons and organic polymers. (Thomas [41]—Reproduced by permission of the Royal Society of Chemistry)

observation means that the hydrogen uptake correlates more strongly with the micropore capacity than it does with the total pore volume. This sounds reasonable because any pores of a larger dimension will not contribute significantly to the overall hydrogen adsorption capabilities of a predominantly microporous material. It is notable that the greatest variation in hydrogen capacity versus pore volume in Fig. 5.4 is shown by carbons, which are most likely to have a broader pore size distribution. This could therefore explain the greater overall scatter in the presented pore volume data. This topic is, however, the subject of ongoing research. A number of studies have investigated the relationship between the pore structure characteristics of carbons and their hydrogen storage capabilities [44–48] with one suggestion being that the hydrogen storage capacity of activated carbons scales best with the CO_2 micropore volume determined from DR analysis of data measured at 273 K [44]. For some of the materials in this study, the CO_2 DR micropore volume was approaching only a third of the value of the total pore volume determined using N_2 at 77 K, while the trend for the CO_2 DR micropore volumes between the different materials more closely follows the trend shown by the surface area[5] than that shown by the total volume. Further detailed discussion of this and other related work [46, 48] is beyond the scope of this chapter, but we can perhaps tentatively conclude that the hydrogen storage capacity correlates reasonably well with micropore capacity, although this is more effectively determined using the BET method than the total pore volume determination methods used to obtain the data compiled in Fig. 5.4. A more thorough analysis would require the separation of these data into those determined for the pore volumes of different material types using each individual method, different adsorbates (e.g., N_2, Ar and CO_2) and also, potentially, different assumed densities

[5] The method of surface area determination was not defined in the article.

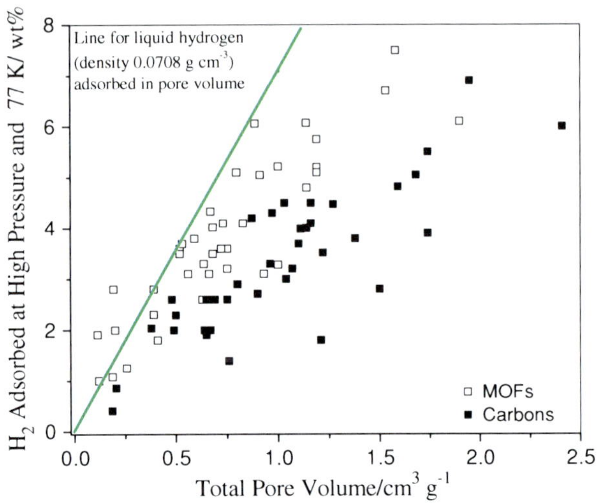

Fig. 5.4 A plot of hydrogen adsorbed (wt%) at saturation at 77 K against pore volume ($cm^3\ g^{-1}$) for metal–organic frameworks and microporous carbons. (Thomas [41]—Reproduced by permission of the Royal Society of Chemistry)

of the adsorbate in each case, because each of these factors could affect the quantitative determination of the pore structure characteristics and hence the conclusion. In related work, Gogotsi and co-workers [49, 50] argue that, for Carbide-Derived Carbons (CDCs), the pore dimension is critical to the hydrogen capacity. They show that, of two CDCs with differing SSAs, it is the material with the lower SSA but smaller pore size that adsorbs the most hydrogen at 77 K and 0.1 MPa. However, the maximum hydrogen uptake, or at least the uptake at an elevated pressure, was not determined in these studies. The results of Gogotsi et al. [49] also, however, showed a correlation with the pore volume, which was determined using Ar adsorption at 77 K, analysed using NLDFT, and the pore volume also followed the BET SSA. It would seem possible that the material with the smaller pore dimensions adsorbs more hydrogen at lower pressure but the material with the greater measured SSA adsorbs more hydrogen at saturation because it also has a greater total pore volume. In a subsequent report, Yushin et al. [50] showed plots of sub-nanometre micropore volume against low pressure hydrogen capacity that exhibited a much clearer correlation than the plot of the volume of pores > 1 nm.

With regard to MOFs, recent modelling work has interestingly suggested that the BET method can apply to these materials [51, 52]. Furthermore, other modelling work by Snurr and co-workers has indicated that hydrogen uptake by a metal–organic framework will scale with different characterisation parameters depending on the pressure regime [53]. At low pressures the uptake correlates with the heat of adsorption, at intermediate pressures it correlates with surface area and at higher pressures it correlates with the free volume (the available pore volume).[6]

[6] The three pressure regimes, in this case, are of the order of 0.01, 3 and 12 MPa, respectively.

If, as an initial indication, we are interested in the maximum potential hydrogen uptake this work also therefore suggests that it is the pore volume that is important, rather than the surface area. However, as discussed above, the pore volume must clearly also be considered in conjunction with the pore size, as larger pore meso- or even macroporous materials may have large pore volumes but will not exhibit high hydrogen uptakes because broader pores do not provide the enhanced adsorption potentials necessary for the effective physisorption of hydrogen.

In conclusion, if we consider the BET method as a quick and easy way to determine the micropore capacity, rather than the surface area, of a microporous material then it can be seen that it provides a useful way to assess the approximate amount of volume available for the effective adsorption of hydrogen at elevated pressures. Furthermore, the empirical observation that the hydrogen capacity scales approximately with BET surface area does not contradict the idea that, physically, the pore volume is actually more important; it is just, perhaps, a question of terminology. However, given that the apparently complex relationship between hydrogen adsorption behaviour and the enthalpy of hydrogen adsorption, the measured surface area, the total pore volume, the micropore volume and the pore dimension has not yet been conclusively determined for any material type, it seems likely that the interrelationship between the pore structure characteristics and the hydrogen storage capabilities of porous materials will be the subject of future research. Any further conclusions that can be drawn will greatly aid our understanding of hydrogen storage by this class of materials, and will also increase the usefulness of gas adsorption analysis as a complementary technique. In the meantime, it is clear that the gas adsorption characterisation methods covered in this section provide crucial complementary information regarding the potential hydrogen storage performance of newly synthesised microporous adsorbents, and that they will continue to play a pivotal role in the characterisation of porous hydrogen storage materials.

5.3 Powder Diffraction

Powder diffraction is a solid state characterisation technique that allows the determination of the crystallographic structure of materials. In both X-ray and neutron diffraction, the incoming beam is elastically scattered and produces peaks at angles defined by Bragg's law,

$$\lambda = 2d \sin \theta \tag{5.8}$$

where λ is the wavelength, 2θ is the scattering angle and d is the lattice spacing for a particular set of Miller (*hkl*) indices. Diffraction can be performed with single crystals or with powders. In the single crystal case, the sample is rotated to allow the reflection intensities from each lattice plane to be determined. In contrast, a powder consists of a large number of small crystallites that are randomly oriented with respect to each other. The result is that a sufficient number of crystallites are in the correct orientation for scattering from each aligned *hkl* plane to occur. All information regarding the atomic arrangement of a material is therefore contained in a single pattern.

The advantage of powder diffraction, or rather the reason why powder diffraction is often used, is that a single crystal is not required for its implementation. It is often the case for hydrogen-absorbing materials that single crystal samples are unavailable,[7] particularly for in situ studies of the microstructural or crystallographic effects of hydrogen absorption, and so powder diffraction must be used. In this section we will look at powder diffraction using both neutrons and X-rays, which interact with matter in fundamentally different ways. The X-ray interaction is governed by the scattering strength of the electrons of the scattering atoms. The strength of the scattering, expressed as a scattering length or scattering cross-section, is thus dependent on the number of electrons surrounding the nucleus, and so the larger the atom, generally speaking, the greater the X-ray scattering strength. X-ray scattering by light atoms, such as hydrogen, can therefore be overwhelmed by the scattering strength of heavier surrounding atoms. Neutron scattering, on the other hand, is governed by the interaction strength between the incoming neutrons and the nucleus of the scattering atom. As a consequence, the scattering strength of neutrons as a function of atomic mass varies in an unpredictable manner, and the scattering strengths of elements close in atomic mass can therefore be significantly different. This allows, under certain circumstances, the scattering of light atoms, such as hydrogen, to be clearly distinguished from scattering by heavy atoms. In addition, there can be very significant differences between the scattering strengths of different isotopes, which can often be exploited experimentally.

The use of X-ray and neutron powder diffraction methods for the solution of the structures of hydrides for hydrogen storage was covered recently by Hauback [54], and this introduction and review is recommended to interested readers. The topic is also discussed by Yvon [55], who provides a useful comparison of X-rays and neutrons. Although the author concludes that neutrons generally yield superior results to X-rays, a number of disadvantages of neutrons were highlighted. These include limited flux, and hence the need for large samples, the large background induced by incoherent scattering, strong disorder scattering from natural isotope mixtures, and strong thermal diffuse scattering from mobile hydrogen, and the large degree of neutron absorption by some natural isotope mixtures of interest for hydrogen storage. Despite the importance of powder diffraction for structure solution, we will concentrate here on the use of in situ methods for the study of the structural evolution of hydrides during the hydrogenation and dehydrogenation processes, because it is this measurement type that can most effectively

[7] Most of the materials of practical interest for storage applications decrepitate upon hydride phase formation, although some metallic hosts, such as Pd, do not.

complement hydrogen sorption measurement. We shall also, however, touch upon in situ neutron diffraction studies of adsorbed molecular hydrogen.

5.3.1 Neutrons

Neutron scattering, and hence neutron powder diffraction, experiments require a neutron source of sufficient flux. Unfortunately, this tends to mean a centralised experimental facility that produces either constant wavelength neutrons using a nuclear reactor or pulsed neutrons, from a so-called *spallation neutron source*. Despite its numerous advantages over X-ray powder diffraction, it is therefore not a standard laboratory technique and this is one of its main practical disadvantages.

Constant Wavelength (CW) neutron powder diffraction experiments are performed in a similar manner to laboratory X-ray powder diffraction, and so the technique therefore involves the determination of the angular spread of a scattered beam from an incoming monochromatic source. The resultant diffraction pattern is a plot of scattered intensity versus scattering angle, and so consists of Bragg peaks corresponding to each of the *hkl* planes in any crystalline phase or phases in the sample. The positions of the peaks will be dependent primarily on the crystallographic structure and the lattice parameters of the material. Their intensity will depend on a number of factors, including the scattering strength of the elements responsible for the scattering within the unit cell, in particular, but also microstructural details such as the size of the coherent diffracting domains.[8]

Pulsed neutron powder diffraction experiments, on the other hand, use an incoming polychromatic beam of neutrons and measure the diffraction pattern as a function of the neutron *Time-of-Flight* (TOF). This allows the measurement of an entire diffraction pattern from a measurement made at a single scattering angle. To illustrate this we can consider that the neutron wavelength, λ_n, is given by the de Broglie equation,

$$\lambda_n = \frac{h}{m_n v_n} \tag{5.9}$$

where h is Planck's constant, m_n is the neutron mass and v_n is the neutron velocity. If a neutron takes a time, t, to travel a distance, L, then $v_n = L/t$, and so,

$$\lambda_n = \frac{ht}{m_n L}. \tag{5.10}$$

By combining this expression with Bragg's law, Eq. 5.8, for a fixed instrumental path length, L, and scattering angle, 2θ, the plane spacing of a particular *hkl* plane, d_{hkl}, can be expressed as a function of the time-of-flight, t,

---

[8] A reduced domain size results in the broadening of diffraction peaks, which reduces the overall peak height.

$$d_{hkl} = \frac{ht}{2m_n L \sin\theta}. \tag{5.11}$$

The experimental advantage of measuring a diffraction pattern at only one scattering angle is that complex sample environment equipment [56] can surround most of the sample, if necessary, without affecting the performance of a measurement. In reality, to minimise counting time, modern TOF diffractometers use a bank of detectors each at a different 2θ position and the data measured at each position is summed to provide better counting statistics for a given data acquisition time.

Powder diffraction data are often analysed using the method originally proposed by Rietveld [57, 58] for the analysis of CW neutron data. However, the approach is now also widely used for TOF neutron powder diffraction, as well as laboratory and synchrotron X-ray powder diffraction, data analysis and involves the calculation of a full powder diffraction pattern from a crystallographic model consisting of one or more crystalline phases. Various parameters used in the model are then refined using least squares minimisation to improve the agreement between the calculated pattern and the experimental data. A number of software packages are available for the analysis of data using the Rietveld Method, of which GSAS [59] and FULLPROF [60] are two of the best known. These packages allow the refinement of models that include multiple phases. The lattice expansion and phase transitions that occur through the hydrogen absorption process can therefore be determined with this method, thus providing crucial complementary information. Using this technique the absorption of hydrogen by a material can effectively be 'seen', rather than just inferred from the volumetric or gravimetric measurement of hydrogen uptake.

A crucial point to note with regard to neutron powder diffraction studies of hydrogen storage media is the significant difference in the scattering behaviour of hydrogen and deuterium. Any element or isotope has a coherent and an incoherent scattering cross-section, σ_{coh} and σ_{inc}, respectively.[9] These two properties express how strongly the respective nucleus scatters neutrons. The coherent scattering is exploited in powder diffraction and contributes to the Bragg peaks, whereas the incoherent scattering contributes only to the background.[10] Neutron scattering by

[9] The scattering of neutrons by nuclei is dependent on their nuclear spin states. In a real scattering system there is a deviation in the scattering lengths of the nuclei from the mean value due to the differing spin states of each individual nucleus. The *coherent scattering* is the scattering that would occur if all of the nuclei in the scattering system had a scattering length equal to the mean value. In this case, scattered waves from each pair of nuclei are in phase, which results in interference effects. This provides us with information regarding the relative positions of the atoms, which is the reason that coherent scattering is exploited in neutron diffraction studies. However, in addition to the coherent scattering, in a real system there is a contribution that is due to the random distribution of the deviations of the scattering lengths from their mean value, which is known as the *incoherent scattering*.

[10] Incoherent neutron scattering methods are also used to measure hydrogen content at low concentrations in metals [61, 62], but this is beyond the scope of this discussion.

hydrogen is dominated by the incoherent contribution, but this is not the case for deuterium;[11] the latter should therefore be used for diffraction experiments on hydrogen-containing materials. This approach assumes that the isotope effect between hydrogen and deuterium is not significant; the validity of this assumption should, however, be assessed on a case-by-case basis.

In situ hydrogenation and dehydrogenation experiments can be performed, in which diffraction patterns are determined for a series of different hydrogen concentrations. This type of experiment allows intermediate concentrations of hydrogen to be maintained in the neutron beam, which would not otherwise be possible.[12] The hydrogen content can be measured volumetrically or gravimetrically during the experiment. Such studies require the use of a sample cell that can withstand the application of a hydrogen pressure but does not interfere significantly with the measured diffraction pattern. Examples include the use of a coated vanadium window or a so-called null matrix Ti-Zr alloy [64]. In the former case, vanadium contributes only an incoherent background whilst the coating prevents hydrogenation or hydrogen embrittlement and, in the latter case, the coherent scattering from the Ti and Zr in the window material cancel each other out.

In situ neutron powder diffraction studies have been carried out on a variety of materials, including both hydrides and microporous adsorbents. Example data from a recent in situ neutron powder diffraction study of a complex hydride is shown in Fig. 5.5, which was measured on OSIRIS at the ISIS neutron source (UK) [65]. The deuteration of lithium nitride is shown as a function of deuterium content for an applied deuterium pressure of 5 bar (0.5 MPa). A section of the measured diffraction pattern, as a function of time, can be seen in Fig. 5.5a. The (111) peak from LiD, the (101) peak from Li_3N, the (200) peak from $LiND_2$ and the (004) peak from $LiND_2$ are labelled. Rietveld refinement of models for each pattern, in which the relative phase fractions were refined together with other crystallographic parameters, allowed the molar phase fractions to be determined as a function of time (Fig. 5.5b).

In the case of microporous materials, the refinement of structural models using the Rietveld Method has allowed the positions of adsorbed hydrogen (or deuterium) molecules to be determined at low temperatures. This approach has been applied recently to a number of different MOFs [66–70]. The hydrogen (deuterium) positions are typically determined at temperatures in the range 3.5–5 K. In addition to studies of hydrides and microporous adsorbents, in situ neutron diffraction has also been applied to clathrates, in order to study the cage occupancy of hydrogen (deuterium) as a function of temperature [71].

[11] For hydrogen, σ_{coh} = 1.8 b and σ_{inc} = 80.2 b; for deuterium, σ_{coh} = 5.6 b and σ_{inc} = 2.0 b (1 b = 1 barn = 1×10^{-28} m^2) [63].

[12] The surface of hydrides can be poisoned in order to trap hydrogen in the sample but this approach is not as flexible as in situ hydrogenation and does not allow time-resolved studies of the hydrogenation or dehydrogenation processes.

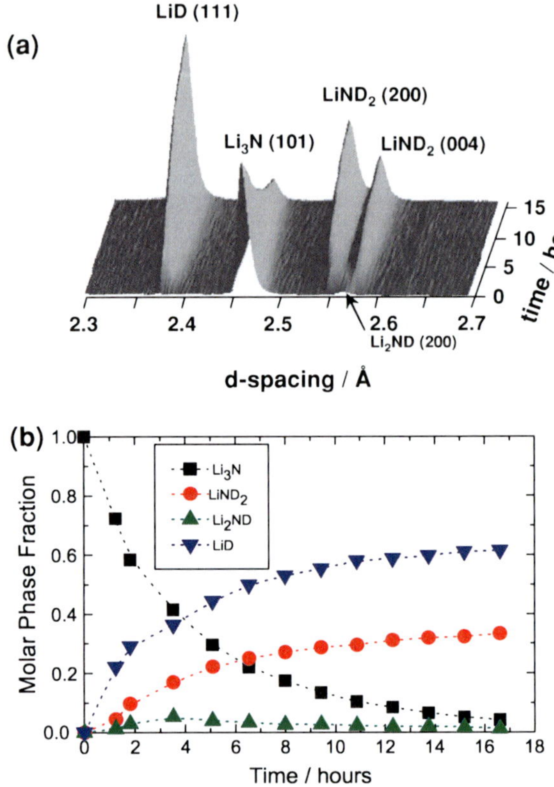

Fig. 5.5 In-situ neutron powder diffraction data showing the absorption of deuterium by Li_3N following the application of a deuterium pressure of 5 bar (0.5 MPa). **a** The time-dependent evolution of four of the Bragg peaks as a function of lattice spacing (d-spacing). **b** The molar phase fractions for Li_3N, $LiND_2$, Li_2ND and LiD as a function of time for the same kinetic process, as determined by Rietveld refinement of the diffraction patterns. (Bull et al. [65]—Reproduced by permission of the PCCP Owner Societies)

5.3.2 X-rays

Although the use of laboratory X-ray powder diffraction is widespread in materials characterisation and will be familiar to many materials scientists and chemists, there are two general types of the method that are worth emphasising with respect to the complementary characterisation of hydrogen storage materials. Firstly, in situ laboratory powder diffraction is particularly useful for the characterisation of hydrogen sorption by hydrogen-absorbing materials because the structural evolution, whether lattice expansion and contraction or phase transformation, of the material during the hydrogen absorption and desorption processes can be observed. The second is the use of synchrotron radiation, which provides both higher resolution, in terms of diffraction peak widths, and greater penetration depths than the laboratory equivalent. The improved resolution means that closely spaced neighbouring peaks are resolved more easily, allowing the analysis of complex structural details. It also means that complicated peak broadening effects can be more readily observed. Meanwhile, greater penetration depths mean more sophisticated sample environment equipment can be used for in situ studies. Significantly higher beam intensities also allow time-resolved studies because the time required to record a single diffraction pattern is much reduced, and synchrotron sources also enable the easy selection of a given wavelength due to the broad continuous distribution of beam intensity they produce as a function of wavelength [54]. Rietveld refinement, as discussed in the previous section, is commonly used to analyse X-ray powder diffraction data and can yield detailed information regarding the microstructure and phase composition of samples.

In situ laboratory X-ray diffraction has been used to study a range of hydrides including sodium alanates [72–74], Mg hydride-based materials [75, 76], La-Mg alloys [77], $Mg(BH_4)_2$ [19] and Pd nanoparticles [78, 79]. Examples of recent in situ synchrotron radiation work, meanwhile, include the study of the dehydrogenation of $LiBH_4$ [80], $Mg(BH_4)_2$ [20, 81], $Ca(BH_4)_2$ [81], $LiAlH_4$ [82], Ti and V-doped $LiAlH_4$ [83], $Mg(AlH_4)_2$ [84], AlH_3 [85], ball-milled Mg-based materials [75, 86–88] and other composites [89, 90]. The hydrogenation and dehydrogenation of AB_5 compounds has also been studied recently using synchrotron radiation by Joubert et al. [91] ($LaNi_5$) and Stange et al. [92] ($LaNi_{4.7}Sn_{0.3}$). As mentioned above, in situ synchrotron radiation studies typically exploit the high flux and high resolution of synchrotron radiation, which allows the performance of detailed time-resolved studies of the hydrogen absorption and desorption processes. In the case of the study by Joubert et al. [91], for example, data acquisition times of the order of 10 s allowed the formation of a transient γ phase to be observed during both the absorption and desorption processes, following the initial hydrogenation of the sample. As is the case for neutron powder diffraction, such in situ studies require the use of specialised sample cells. Gray et al. [93] recently presented an in situ synchrotron X-ray diffraction cell, which is suitable for hydrogen storage material studies. It was designed for use in transmission geometry with the powder sample held between two parallel plate beryllium windows. The cell allows sample thicknesses of 0.5 and 1.0 mm, and it can withstand a hydrogen pressure of up to 100 bar at a sample temperature up to 573 K. The use of the cell was demonstrated using the hydrogenation of $LaNi_5$ at 283 K following a step change in the hydrogen pressure. As an example of ex-situ studies of hydrides using synchrotron radiation, Černý et al. [94] reported a study of the anisotropic peak broadening observed in the powder diffraction patterns of intermetallic hydrides due to high dislocation densities. Synchrotron X-ray diffraction was used in this work due to its high resolution, and this therefore provides a good illustration of the advantage of synchrotron radiation over neutrons in this respect.

5.3.3 Small Angle Scattering

It can be seen from Bragg's law (Eq. 5.8) that the angle of coherent elastic scattering depends on the spacing of the scattering objects. In the determination of crystallographic structure, in which the scattering is due to individual atoms, these distances are in the range of typical plane spacings. As the scattering angle decreases, we observe scattering due to increasingly large scattering object spacings. In the low angle region we can therefore observe scattering that originates from fluctuations in the scattering strength over longer length scales. Small angle scattering of both X-rays and neutrons typically probes the range between 1 and 100 nm [95–98], although longer length scales of up to 50 μm can be studied using ultra small angle scattering[13] [98]. In terms of the study of potential hydrogen storage materials, the phenomena that can result in the scattering density fluctuations probed by small angle scattering include the spatial inhomogeneity of the hydrogen concentration in metal hydrides, the lattice strain and dislocation formation that accompanies hydride phase formation, and the porosity of hydrogen adsorbents.

Orimo et al. [99, 100] used Small Angle Neutron Scattering (SANS) to study the spatial inhomogeneity of the deuterium concentration in ball milled Mg_2Ni deuteride and found inhomogeneities that they attributed to the intra- and inter-grain regions. Fultz et al. [101, 102] used in situ SANS to study the difference in the spatial homogeneity of the deuterium concentration in $LaNi_5H_x$ and $LaNi_{4.75}Sn_{0.25}H_x$ intermetallic hydrides. While no difference was observed between the fully deuterated samples, the homogeneity on an approximate length scale of 4 nm to 15 nm was found to be greater for the partially substituted compound at intermediate deuterium concentrations. The authors propose that the smoothing of the mesoscopic hydrogen concentration profile in Sn-substituted compounds may be responsible for their greater long term cycling stability (see Sect. 3.1.2). SANS was also used by Flanagan et al. [103] to study phase separation in a $Pd_{0.85}Ni_{0.15}$ alloy following exposure to hydrogen. A greater scattering intensity was observed in the low angle region for the samples treated under a high hydrogen pressure, which was attributed to the existence of Pd-rich and Pd-poor regions in the materials. SANS has also been used to study the trapping of deuterium by dislocations in Pd [104–106] and the formation of Pd deuteride [107, 108]. Although the latter studies are of a more fundamental nature, it is still worth noting the complementary information that such studies can provide in the practical characterisation of hydrogen storage by hydrides.

Relevant work related to the study of particle morphology and the dispersion of dopants in potential hydrogen storage media has also been presented recently. Pranzas et al. [109, 110] have used both SANS and USANS, as well as Small Angle X-ray Scattering (SAXS) and USAXS [110], to study the particle size of nanocrystalline MgH_2 as a result of ball milling with and without additives. Dobbins et al. [111], meanwhile, used USAXS to study changes in the powder morphology of $NaAlH_4$ as a function of milling time with four different dopants, namely $TiCl_2$, $TiCl_3$, $ZrCl_3$ and VCl_3. Although the results were not directly compared to the hydrogen storage properties of the particular samples studied, it was found that the sample doped with $TiCl_3$, which is the most effective additive

[13] This is known as either Ultra Small Angle X-ray Scattering (USAXS) or Ultra Small Angle Neutron Scattering (USANS).

from a hydrogen storage point of view (see Sect. 2.3.1), had the highest powder particle surface area. In addition, Sartori et al. [112] recently applied SANS to study the dispersion of $Mg(BH_4)_2$ in an activated carbon matrix.

A range of porous materials relevant to hydrogen storage applications have also been studied recently using small angle scattering, including porous silica [113], CDCs [114], zeolites [115] and MOFs [116]. Sheppard et al. [113] used laboratory SAXS to complement hydrogen adsorption measurements performed on the mesoporous silica MCM-41. The SAXS data from this preliminary study allowed the identification of the lower limit of the pore size of the MCM-41 ($d_p \geq 2.3$ nm). Laudisio et al. [114] used laboratory SAXS to complement gas adsorption studies of the PSD of CDCs synthesised under different conditions. Similarly, Du and Wu [115] used synchrotron SAXS to analyse the porosity of zeolites A and X, and ZSM-5. Tsao et al. [116] recently used laboratory SAXS to complement nitrogen adsorption characterisation methods for MOF-5 (IRMOF-1). They examined samples that had been synthesised using different methods, which resulted in significant discrepancies between the BET and Langmuir SSAs (see Sect. 5.2) measured for samples prepared using different synthesis routes. Mesopores of approximately 8 and 12 nm were identified in samples prepared using differing synthesis methods and conditions.[14] Although the X-ray powder diffraction patterns of each sample showed significant differences, the mesoporosity of the samples would not have been detected from the crystallographic information obtained from the (wide angle) powder diffraction measurements. This latter study is particularly pertinent to our later discussion of the importance of sample purity to hydrogen sorption characterisation (see Sect. 6.2.4).

5.4 Spectroscopy

Various types of spectroscopy can be used for the complementary characterisation of hydrogen storage materials. Inelastic Neutron Scattering (INS) has been used extensively to study the dynamics of hydrogen in metals and has been applied recently to the study of the adsorption of molecular hydrogen in microporous adsorbents. Quasi-Elastic Neutron Scattering (QENS) is a closely related technique that can be used to study the bulk and surface diffusion of atomic and molecular hydrogen, respectively. Nuclear Magnetic Resonance (NMR), meanwhile, provides another method for the determination of diffusion coefficients and jump rates for hydrogen in metals. Finally, although it is used widely in materials chemistry, infrared spectroscopy has recently been applied in a variable

[14] Unfortunately, Tsao et al. [116] only measured hydrogen adsorption at ambient temperature for these samples and did not detect a strong dependence on the degree of mesoporosity shown by the low SSA materials. This was due, partly, to the low uptake and large error attributed to the ambient temperature measurements, which were performed at 6.9 MPa. However, it is likely that the hydrogen adsorption at 77 K would vary significantly between the samples.

temperature form to determine the enthalpy and entropy of molecular hydrogen adsorption in microporous adsorbents. In this section we look at each of these techniques in turn.

5.4.1 Inelastic Neutron Scattering

Inelastic Neutron Scattering (INS) experiments involve the determination of the energy transfer between neutrons and the scattering objects. This technique can be used to probe the local environment of atomic hydrogen in matter or to determine the quantized excitation energies of adsorbed molecular hydrogen [117]. The scattering can be coherent or incoherent, but we are primarily concerned with incoherent INS.[15] Some of the advantages of neutrons covered in Sect. 5.3 for diffraction studies of hydrogen storage materials also apply to INS, including their penetrating nature and the strong scattering exhibited by hydrogen relative to other heavier elements. In addition, INS also has further advantages over other spectroscopic techniques, such as IR and Raman spectroscopy, which are both restricted by the symmetry requirements that lead to selection rules. These rules mean that not all of the vibrational modes are visible. In contrast, the symmetry requirements are absent for neutrons, so all vibrational modes are neutron active and can therefore be identified using INS. INS spectroscopy is also sensitive to modes at all wavevectors across the Brillouin zone, again in contrast to IR and Raman Spectroscopy [118]. Quasi-Elastic Neutron Scattering (QENS) is a related technique in which the broadening of the elastic scattering peak is measured to determine information regarding the diffusion rate of hydrogen. A thorough introduction to the use of both INS and QENS for the study of potential hydrogen storage materials was given by Ross [119], while a more detailed treatment of the theory of neutron scattering can be found in the book by Squires [63]. The study of hydrogen in binary hydrides using INS was addressed in some detail by Fukai [120], together with the use of QENS for the investigation of diffusion processes. Other recent reviews have also covered the use of INS to study potential hydrogen storage materials [118, 121–123].

INS spectroscopy has been used to study many ternary hydrides. For example, a number of studies have focussed on the transition metal complex hydrides [124–128]. INS, used in conjunction with IR and Raman spectroscopy, allowed the identification of the vibrational modes (stretching, bending, librational, translational and acoustic) in Mg_2FeH_6 [124], Mg_2CoH_5 [125], a series of Pd-H based complex hydrides [126], Rb_2PtH_6 [127] and $BaReH_9$ [128]. Although these studies focus on the crystal chemistry of these materials, in terms of complementarity, INS clearly provides an excellent method of actually 'seeing' the hydrogen bound in

[15] Coherent INS can be used to study excitations in solids but this is beyond the scope of our discussion.

the bulk of metal hydrides. INS also has an advantage over diffraction in certain circumstances because, unlike the latter, it does not require a sample to exhibit long range crystalline order. For example, Schimmel et al. [129] compared the vibrational spectrum of ball milled MgH_2, both before and after a single hydrogenation cycle, with the spectrum from the unmilled material [130]. Although the spectrum from the ball milled sample indicated the presence of stresses in the material, the spectrum for the hydrogen cycled sample matched that for un-milled bulk magnesium hydride. This showed that the dehydrogenation and hydrogenation process annealed the structure, whilst also demonstrating the important role that INS can play in the study of hydrogen in nanocrystalline materials that exhibit only broad peaks in their powder diffraction patterns (see Sect. 5.3). Fu et al. [131], meanwhile, used INS to identify the presence of molecular aluminium hydride during the hydrogenation of Ti and Ti/Sn doped sodium alanate, and borohydrides have also been studied using the technique [132, 133].

With regard to microporous materials, INS can be used to observe the rotational transitions of adsorbed molecular hydrogen. Mulder et al. [134] determined the energies of these transitions for hydrogen in MOF-5 and were able to identify five different adsorption sites in this material. Brown et al. [135] used INS to study hydrogen adsorption in the metal–organic framework HKUST-1, identifying three main adsorption sites from the observed rotational transition peaks, while Nouar et al. [136] used INS to observe differences in the binding of molecular hydrogen in lithium and magnesium ion-exchanged Zeolite-like Metal–Organic Frameworks (ZMOFs). Although in the latter study the spectra were broadly similar, at low hydrogen loading an additional low energy peak was observed in the Li-exchanged compound, suggesting that there is a stronger hydrogen interaction with this material. As exemplified by the first two examples given above [134, 135], INS spectroscopy is often combined with first principles calculations to aid the assignment of peaks in the neutron energy transfer spectra to particular transitions. Other microporous materials that have been studied by INS include zeolites [137, 138] and nanostructured carbons [119, 139–143]. In the latter case, INS played an important role in confirming that the highest, and most controversial, hydrogen storage capacities reported for carbon nanotubes and nanofibres were erroneous [139].

THF-stabilised clathrate hydrates (see Sect. 2.4.1) have also been studied recently using INS [144–146] due to its ability to observe the behaviour of hydrogen held in the cages. Tait et al. [144] observed the rotational transitions of the molecular hydrogen, but also identified a series of transitions between translational quantum states. Similarly, translational as well as rotational transitions were also observed by Ulivi et al. [145, 146]. Further details regarding TOSCA, the INS spectrometer at the ISIS neutron source used by Ulivi et al. [145, 146] and in a number of the other INS studies mentioned above, have been given by Parker and co-workers [147, 148]. These reports are recommended to interested readers.

In quasi-elastic neutron scattering, the diffusion of hydrogen results in the Lorentzian broadening of the elastic scattering peak, and the overall peak shape can therefore be represented by one or more Lorentzian functions convoluted with the instrumental resolution. The width of the Lorentzian broadening can then be related to the diffusion behaviour by appropriate models, of which the jump diffusion model due to Chudley and Elliot [149] is the best known. QENS has been used to study hydrogen dynamics in a range of materials, including microporous absorbents such as carbons [143], zeolites [150–154], MOFs [155, 156] and microporous amorphous silica [157], and hydrides, such as the alanates [158, 159] and AB_2 [160–164] and AB_5 [165] intermetallics. Although QENS does not provide quantitative information regarding the hydrogen content, it allows direct observation of hydrogen in a particular sample and the determination of hydrogen diffusion coefficients, which can be related to the kinetics of the hydrogen sorption process. It can therefore provide important complementary information for kinetic studies.

5.4.2 Nuclear Magnetic Resonance

Nuclear Magnetic Resonance (NMR) spectroscopy can use any isotope with a nuclear magnetic moment, and can therefore be performed with hydrogen, deuterium and tritium. There are many different ways to implement NMR and it can be used to study both the structure and dynamics of metal hydrides [166]. Recently it has been used for the study of the microstructural composition and evolution of complex hydrides, such as the sodium [167, 168] and lithium [169, 170] alanates and Li amide [171], during or as a result of hydrogenation, dehydrogenation, or ball milling. However, we will concentrate here on the use of proton (^1H) NMR to study hydrogen diffusion and, in particular, the determination of atomic hydrogen jump rates and diffusion coefficients.

In NMR spectroscopy, a Radio Frequency (RF) magnetic field is applied in the presence of an applied static field, while the response is detected by a receiver. The technique can involve the measurement of the position, or frequency, of the resonance line, the strength of the line, the linewidth or one of the relaxation times. The study of hydrogen diffusion usually involves the measurement of relaxation times, although pulsed field gradient methods and the determination of linewidth can also be used [120, 172]. With regard to the former, various relaxation times can be determined, and these include T_1, the spin–lattice (longitudinal) relaxation time, $T_{1\rho}$, the spin–lattice relaxation in the rotating frame, T_{1D}, the dipolar relaxation time, $T_{1D\rho}$, the dipolar relaxation in the rotating frame, and T_2, the spin–spin (transverse) relaxation time. Each of these represents the time that the nuclear spin system takes to return to equilibrium following perturbation by the RF pulse [173–175]. The spin–lattice relaxation times represent the time taken for the lattice to absorb energy from the spin system and the spin–spin relaxation times represent the timescale for the exchange of energy within the spin system. The relaxation times are indirectly related to the diffusion properties of a system and can therefore be related to the hydrogen diffusion process, although this requires the use of a model and a number of assumptions.

Each relaxation time mentioned above applies to a different timescale and so can be used to probe hydrogen diffusion over a range of regimes. According to Richter et al. [172], T_1 probes an approximate diffusion coefficient (D) range of 10^{-10} cm^2 s^{-1} to 10^{-6} cm^2 s^{-1}, $T_{1\rho}$, the range 10^{-13} cm^2 s^{-1} to 10^{-7} cm^2 s^{-1}, T_{1D}, the range 10^{-15} cm^2 s^{-1} to 10^{-11} cm^2 s^{-1}, and T_2 can be used to determine D in the range 10^{-11} cm^2 s^{-1} to 10^{-7} cm^2 s^{-1}. In each case, the diffusion coefficient is calculated by determining the interval between hydrogen jumps, τ_H, using the expression,

$$D = \frac{fl^2}{6\tau_H} \tag{5.12}$$

where l is the jump distance and f is the tracer correlation factor. Complications can arise due to jumps occurring over different length scales, but Pulsed Field Gradient (PFG) methods, which involve the determination of the attenuation of spin echo signals, can be used to directly measure D.[16] By combining relaxation rate measurements to determine τ_H and PFG methods to determine D, the jump length can therefore be determined. For our purposes, the important point is that ^{1}H NMR potentially allows the determination of the hydrogen diffusion rates in metal hydrides, whether inferred from τ_H or measured directly using PFG techniques.

It is worth noting that relaxation times, and hence the measured diffusion parameters, can be significantly affected by the presence of paramagnetic impurities [176] and NMR spectroscopy cannot be performed on ferromagnetic and strongly paramagnetic materials [172]. The determined jump rates can also be affected by the choice of model used to relate them to the relaxation times. An example is the BPP (Bloembergen–Purcell–Pound) model [177], which can result in a significant error in the calculated τ_H, but a number of alternative models have been proposed and applied to metal-hydrogen systems [172, 178].

With regard to practical experimental NMR systems, Baker and Conradi [179] presented apparatus for the performance of in situ NMR studies on hydrides up to hydrogen pressures of 55 atm (5.57 MPa) and temperatures of 1,300 K. In this system, the sample is contained in an alumina ceramic tube. The tube is protected against rupture by the external application of an overpressure of argon. The temperature of the sample is controlled by a serpentine heater wound through ceramic tubes around an outer ceramic tube in the Argon environment at the lower end of the apparatus, with a series of baffles situated above. The gaseous hydrogen is delivered to the sample through the top of the ceramic sample tube.

NMR spectroscopy has been used to study hydrogen diffusion in a range of potential storage materials, including nanostructured H-graphite [180–182], amorphous hydrides [183, 184], AB_2 [161, 185] and AB_5 [186–188] intermetallic hydrides, coarse-grained and ball milled MgH_2 [189], Mg-based ternary and

[16] Note that PFG methods measure relatively rapid diffusion rates in the range 10^{-8} cm^2 s^{-1} to 10^{-4} cm^2 s^{-1}.

quaternary hydrides [190–192], BCC alloys [193], quasicrystalline alloys [178, 194, 195] and Na alanate [196]. The technique has also been used to study the diffusion of molecular hydrogen in microporous materials [151], and to study both H_2 diffusion [197–199] and cage occupancy [200] in clathrate hydrates. Although NMR spectroscopy does not probe hydrogen in materials as directly as either INS or QENS and has some disadvantages, including its inapplicability to ferromagnetic materials, it can clearly provide useful information regarding the motion and state of hydrogen in many different types of solid state storage material.

5.4.3 Infrared Spectroscopy

In Infrared (IR) spectroscopy, infrared radiation is passed through a sample and an absorption spectrum recorded. Depending on the sample, the spectrum will include characteristic IR absorption bands corresponding to the wavenumbers of the infrared active vibrational modes in the sample. IR spectroscopy can be used to determine the energy of gas–solid interactions, providing the adsorptive molecule has at least one IR absorption mode which is altered, or appears, as a result of its interaction with the surface during adsorption. If this is the case, the integrated intensity of the corresponding absorption band should be proportional to the surface coverage, and this relationship is exploited in Variable Temperature Infrared (VTIR) spectroscopy.

Two methods have been used to determine the enthalpy of adsorption of hydrogen by microporous materials using VTIR. The first requires the integrated intensity of the IR absorption band and the equilibrium hydrogen pressure to be recorded at a series of temperatures. The second approach, termed the *pseudo-isobaric method* [201], assumes that the coverage is sufficiently low for the equilibrium pressure to be ignored. In this case, only the integrated intensity of the IR band is required at each of the measurement temperatures.

In the first method, the IR transmission spectra are measured using a Fourier Transform Infrared (FTIR) spectrometer, with the sample mounted in a variable temperature cell that allows the measurement of the equilibrium gas pressure at the measurement temperature [202, 203]. A known amount of hydrogen is dosed to the cell at a low temperature and a transmission spectrum recorded, together with the equilibrium pressure. The cell is sealed and IR spectra are then determined at a series of increasing temperatures whilst the rise in pressure due to the temperature change in the closed system is recorded. The pressure rise partially counteracts the desorption that occurs due to the increasing temperature [204]. To analyse the data, the Langmuir equation (Eq. 3.42) is combined with the van ‘t Hoff relation (Eq. 3.1) to give the following expression for the surface coverage, θ,

$$\theta = \frac{P\exp\left(\frac{\Delta S^0}{R}\right)\exp\left(-\frac{\Delta H^0}{RT}\right)}{1 + P\exp\left(\frac{\Delta S^0}{R}\right)\exp\left(-\frac{\Delta H^0}{RT}\right)} \tag{5.13}$$

where P is the pressure, ΔS^0 and ΔH^0 are the entropy and enthalpy changes relative to standard pressure, R is the universal gas constant and T is the temperature. By assuming the validity of the Beer–Lambert law, which results in the intensity, A, of the characteristic IR absorption band being proportional to the amount adsorbed, the surface coverage can be expressed as the ratio of the measured intensity to the maximum intensity, A_m, so that $\theta = A/A_m$. Eq. 5.13 can be expressed in the following linearised form,

$$\ln\left(\frac{\theta}{P(1-\theta)}\right) = -\frac{\Delta H^0}{RT} + \frac{\Delta S^0}{R} \tag{5.14}$$

so that a plot of the left hand side versus inverse temperature gives a plot of gradient, $-\Delta H^0/R$, and intercept, $\Delta S^0/R$ [205].

In the pseudo-isobaric method, using the ideal gas law (Eq. 6.1) and the assumption of $\theta \ll 1$, the IR absorption band intensity is expressed as,

$$A \approx A_m N_t \left(\frac{RT}{V_c}\right) \exp\left(\frac{\Delta S^0}{R}\right) \exp\left(-\frac{\Delta H^0}{RT}\right) \tag{5.15}$$

where N_t is the total number of moles of gas in the sample cell volume, V_c. This can be written as,

$$A \approx CT \exp\left(-\frac{\Delta H^0}{RT}\right) \tag{5.16}$$

where C is a constant. An approximate value for the enthalpy can then be determined from a plot of $\ln(A/T)$ versus inverse temperature [201].

VTIR spectroscopy has been used to determine the enthalpy of adsorption for hydrogen on a range of microporous adsorbents. Garrone et al. [206] reported enthalpy values in the range -3.5 kJ mol^{-1} to -18 kJ mol^{-1} for a number of zeolites, with the largest of these values being found for (Mg,Na)-Y zeolite [207]. Spoto et al. [208] determined the enthalpy of adsorption for a Hypercrosslinked Polymer (HCP) using the technique, reporting a value of approximately -4 kJ mol^{-1} using the first of the two methods described above in the temperature range 14–100 K. Vitillo et al. [209] determined the enthalpy of adsorption for hydrogen on three metal–organic framework materials, MOF-5, HKUST-1 and CPO-27-Ni, using VTIR to examine the role of exposed metal sites in these materials. They obtained values of -10.1 and -13.5 kJ mol^{-1} for HKUST-1 and CPO-27-Ni, respectively, and either -3.5 or -7.4 kJ mol^{-1} for MOF-5, depending on which peak was selected for the calculation. This method can clearly provide important complementary information on the enthalpy, and hence the strength, of hydrogen adsorption by different host adsorbents.

5.5 Other Techniques

In this section, we will note some further examples of complementary techniques that are commonly used in relation to hydrogen storage materials. Firstly, electrochemical PCI determination is an alternative to the use of gas phase sorption techniques. The development of Ni-MH batteries using $LaNi_5$-based intermetallics as cathode materials has been the most commercially successful practical application of metal hydride technology, and electrochemical characterisation of the electrode material is obviously crucial for this application. The electrochemical charge and discharge process is directly analogous to gas phase hydrogenation and dehydrogenation, and so a direct comparison between data from electrochemical and gas phase techniques can be made. If an electrode is constructed from a hydride-forming alloy or a hydrogen adsorbing material and equilibrium electrochemical measurements are performed in a cell to produce a plot of potential (V) against discharge capacity (mAh g^{-1}), the data can be directly compared to the gas phase equivalent using the Nernst equation [210, 211]. The operating temperature of a cell is limited and the material must be suited to electrochemical charging and discharging, but electrochemical characterisation can nevertheless serve as a useful comparative technique. See, for example, the work of Züttel and co-workers on the adsorption of hydrogen by carbon nanotubes [212–214]. A large number of reports regarding the electrochemical behaviour and characterisation of metal hydrides can be found in the literature and a number of reviews cover the topic [210, 211, 215, 216]. It was recently argued by Bliznakov et al. [217] that electrochemical PCT determination performed over a limited temperature range, in order to determine the thermodynamic properties of a metal hydride, offers a low cost, safe and affordable alternative to gas phase measurements, providing the material in question can be electrochemically charged and discharged.

Electron microscopy, including Transmission Electron Microscopy (TEM) and Scanning Electron Microscopy (SEM), is used widely in hydrogen storage material research although primarily for the microstructural characterisation of materials. Such microstructural characterisation is beyond the scope of our discussion; however, it is worth noting the use of the technique to examine the effects of hydrogen cycling, which together with the analysis of anisotropic peak broadening in X-ray and neutron powder diffraction patterns and the use of Positron Annihilation Spectrocopy (PAS) [218, 219], can greatly aid our understanding of the effects of the activation process (see Sect. 3.1.4) and long term hydrogen cycling (see Sect. 3.1.2). Studies of the effects of hydrogen cycling on intermetallics include the work of Kim et al. [220, 221], who used TEM to examine the microstructural effects of the hydrogen absorption and desorption process on $LaNi_5$. In addition to $LaNi_5$, Decamps et al. [222] studied partially substituted AB_5 materials, also using TEM, while Inui et al. [223] reported TEM for both cycled $LaNi_5$ and FeTi. Recently, however, in situ SEM and TEM have been used to study the dehydrogenation of a number of materials, including AlH_3 [224–226], NbF_5-doped MgH_2 [227], and sodium and lithium alanates [228].

There are a number of other techniques that have been used for the study of hydrogen diffusion in hydrides, a summary of which is provided by Fukai [120] with regard to binary hydrides. These include gas phase permeation measurements, electrochemical methods, the measurement of resistivity, and mechanical relaxation measurements using the Gorsky effect, the Snoek effect and Zener relaxation. The use of the Gorsky and Snoek effects for the study of hydrogen diffusion was also covered briefly by Richter et al. [172]. Anelastic relaxation spectroscopy has been used recently to analyse the dynamics of the hydrogenation and dehydrogenation process in sodium alanates [229, 230]. The application of this technique to alanates, as well as lithium nitrides and ammonia borane, was reviewed recently by Palumbo et al. [231].

5.6 Summary

In this chapter we have covered a number of complementary techniques that can be used to study hydrogen storage materials. We have seen how thermal analysis and calorimetry can be used to determine thermodynamic parameters that can be compared to the results obtained using laboratory gas sorption measurements. We have also covered the gas adsorption methods that allow the determination of the surface area, pore volume and pore size distribution of microporous materials. The microstructural evolution of hydride compounds can be followed using in situ neutron and X-ray diffraction, and these techniques can be used to directly observe the hydrogen sorption processes that are measured indirectly using the laboratory gas sorption measurement techniques. Small angle scattering of X-rays and neutrons can determine fluctuations in the density of hydrogen and deuterium over larger length scales than standard powder diffraction, as well as probing other microstructure-related spatial inhomogeneities. We have also covered various spectroscopic techniques that can provide valuable complementary information regarding the dynamics of both atomic and molecular hydrogen in solids, and to directly probe hydrogen in matter, as well as to determine thermodynamic information such as the enthalpy of molecular hydrogen adsorption in microporous adsorbents. Finally, we have briefly covered some of the other complementary techniques that do not fit readily into the above categories, including the electrochemical determination of PCIs, electron microscopy, and some kinetic measurement methods. The use of one or more of these methods should be considered during the characterisation of hydrogen storage materials in order to support the results obtained from either equilibrium or kinetic gas phase hydrogen sorption measurements.

References

1. Brown ME (2001) Introduction to thermal analysis: techniques and applications, 2nd edn edn. Kluwer, Dordrecht
2. Gabbot P (2008) A practical introduction to differential scanning calorimetry. In: Gabbott P (ed) Principles and applications of thermal analysis. Blackwell, Oxford
3. Rouquerol J, Wadsö I, Lever T, Haines P (2008) Developments in nomenclature. In: Brown ME, Gallagher PK (eds) Handbook of thermal analysis and calorimetry: recent advances, techniques and applications, vol 5. Elsevier, Amsterdam
4. Fernández JF, Cuevas F, Sánchez C (2000) Simultaneous differential scanning calorimetry and thermal desorption spectroscopy measurements for the study of the decomposition of metal hydrides. J Alloy Compd 298:244–253
5. Wenzl H, Lebsanft E (1980) Phase diagram and thermodynamic parameters of the quasibinary interstitial alloy $Fe_{0.5}Ti_{0.5}H_x$ in equilibrium with hydrogen gas. J Phys F: Met Phys 10:2147–2156
6. Luo W, Clewley JD, Flanagan TB, Oates WA (1992) Thermodynamic characterization of the Zr-Mn-H system part 1. Reaction of H_2 with single-phase $ZrMn_{2+x}$ C-14 Laves phase alloy. J Alloy Compd 185:321–338
7. Bohmhammel K, Christ B, Wolf G (1996) Isobaric and isothermal DSC measurements of metal-hydrogen systems. Thermochim Acta 271:67–73
8. Bowerman BS, Wulff CA, Biehl GE, Flanagan TB (1980) Calorimetry within hysteresis loops: application to $LaNi_5$-H. J Less-Common Met 73:1–13
9. Murray JJ, Post ML, Taylor JB (1980) Differential heat flow calorimetry of the hydrides of intermetallic compounds. J Less-Common Met 73:33–40
10. Murray JJ, Post ML, Taylor JB (1983) The thermodynamics of the system $CaNi_5$–H_2 using differential heat conduction calorimetry. J Less-Common Met 90:65–73
11. Hubbard WN, Rawlins PL, Connick PA, Stedwell RE, O'Hare PAG (1983) The standard enthalpy of formation of $LaNi_5$. The enthalpies of hydriding of $LaNi_{5-x}Al_x$. J Chem Thermodyn 15:785–798
12. Luo S, Luo W, Clewley JD, Flanagan TB, Bowman RC (1995) Thermodynamic and degradation studies of $LaNi_{4.8}Sn_{0.2}$-H using isotherms and calorimetry. J Alloy Compd 231:473–478
13. Spit FHM, Drijver JW, Radelaar S (1980) Hydrogen sorption by the metallic glass $Ni_{64}Zr_{36}$ and by related crystalline compounds. Scr Metall 14:1071–1076
14. Zaluska A, Zaluski L, Ström-Olsen JO (2001) Structure, catalysis and atomic reactions on the nano-scale: a systematic approach to metal hydrides for hydrogen storage. Appl Phys A 72:157–165
15. Rongeat C, Llamas-Jansa I, Doppiu S, Deledda S, Borgschulte A, Schultz L, Gutfleisch O (2007) Determination of the heat of hydride formation/decomposition by high-pressure differential scanning calorimetry (HP-DSC). J Phys Chem B 111:13301–13306
16. Dilts JA, Ashby EC (1972) A study of the thermal decomposition of complex metal hydrides. Inorg Chem 11(6):1230–1236
17. Mamatha M, Bogdanović B, Felderhoff M, Pommerin A, Schmidt W, Schüth F, Weidenthaler C (2006) Mechanochemical preparation and investigation of properties of magnesium, calcium and lithium-magnesium alanates. J Alloy Compd 407:78–86
18. Wang F, Liu Y, Gao M, Luo K, Pan H, Wang Q (2009) Formation reactions and the thermodynamics and kinetics of dehydrogenation reaction of mixed alanate Na_2LiAlH_6. J Phys Chem C 113:7978–7984
19. Hanada N, Chłopek K, Frommen C, Lohstroh W, Fichtner M (2008) Thermal decomposition of $Mg(BH_4)_2$ under He flow and H_2 pressure. J Mater Chem 18:2611–2614
20. Soloveichik GL, Gao Y, Rijssenbeek J, Andrus M, Kniajanski S, Bowman RC Jr, Hwang S-J, Zhao J-C (2009) Magnesium borohydride as a hydrogen storage material:

properties and dehydrogenation pathway of unsolvated $Mg(BH_4)_2$. Int J Hydrogen Energy 34:916–928
21. Yan Y, Li H-W, Sato T, Umeda N, Miwa K, Towata S, Orimo S (2009) Dehydriding and rehydriding properties of yttrium borohydride $Y(BH_4)_3$ prepared by liquid-phase synthesis. Int J Hydrogen Energy 34:5732–5736
22. Zakotnik M, Prosperi D, Williams AJ (2009) Kinetic studies of hydrogen desorption in SmCo 2/17 sintered magnets. Thermochim Acta 486:41–45
23. Rouquerol F, Rouquerol J, Sing K (1999) Adsorption by powders and porous solids: principles, methodology and applications. Academic Press, London
24. Lowell S, Shields JE, Thomas MA, Thommes M (2004) Characterization of porous solids and powders: surface area, pore size and density. Springer, Dordrecht
25. Brunauer S (1943) The adsorption of gases and vapors vol 1—physical adsorption. Princeton University Press, Princeton
26. Do DD (1998) Adsorption analysis: equilibria and kinetics. Imperial College Press, London
27. Rouquerol J, Llewellyn P, Rouquerol F (2007) Is the BET equation applicable to microporous adsorbents? Stud Surf Sci Catal 160:49–56
28. Moellmer J, Celer EB, Luebke R, Cairns AJ, Staudt R, Eddaoudi M, Thommes M (2010) Insights on adsorption characterization of metal-organic frameworks: a benchmark study on the novel *soc*-MOF. Microporous Mesoporous Mater 129:345–353
29. McClellan AL, Harnsberger HF (1967) Cross-sectional areas of molecules adsorbed on solid surfaces. J Colloid Interface Sci 23:577–599
30. Gelb LD, Gubbins KE, Radhakrishnan R, Sliwinska-Bartkowiak M (1999) Phase separation in confined systems. Rep Prog Phys 62:1573–1659
31. Lastoskie C, Gubbins KE, Quirke N (1993) Pore size distribution analysis of microporous carbons: a density functional theory approach. J Phys Chem 97:4786–4796
32. Tarazona P, Marconi UMB, Evans R (1987) Phase equilibria of fluid interfaces and confined fluids. Mol Phys 60(3):573–595
33. House WA (1978) Adsorption on a random configuration of adsorptive hetereogeneities. J Colloid Interface Sci 67(1):166–180
34. Merz PH (1980) Determination of adsorption energy distribution by regularization and a characterization of certain adsorption isotherms. J Comput Phys 38:64–85
35. Szombathely MV, Bräuer P, Jaroniec M (1992) The solution of adsorption integral equations by means of the regularization method. J Comput Chem 13(1):17–32
36. Jagiełło J (1994) Stable numerical solution of the adsorption integral equation using splines. Langmuir 10:2778–2785
37. Jagiello J, Thommes M (2004) Comparison of DFT characterization methods based on N_2, Ar, CO_2, and H_2 adsorption applied to carbons with various pore size distributions. Carbon 42:1227–1232
38. Jagiello J, Ansón A, Martínez MT (2006) DFT-based prediction of high-pressure H_2 adsorption on porous carbons at ambient temperatures from low-pressure adsorption data measured at 77 K. J Phys Chem B 110:4531–4534
39. Jagiello J, Ania CO, Parra JB, Jagiello L, Pis JJ (2007) Using DFT analysis of adsorption data of multiple gases including H_2 for the comprehensive characterization of microporous carbons. Carbon 45:1066–1071
40. Jagiello J, Betz W (2008) Characterization of pore structure of carbon molecular sieves using DFT analysis of Ar and H_2 adsorption data. Microporous Mesoporous Mater 108:117–122
41. Thomas KM (2009) Adsorption and desorption of hydrogen on metal-organic framework materials for storage applications: comparison with other nanoporous materials. Dalton Trans 9:1487–1505
42. Silvestre-Albero J, Sepúlveda-Escribano A, Rodríguez-Reinoso F, Kouvelos V, Pilatos G, Kanellopoulos NK, Krutyeva M, Grinberg F, Kaerger J, Spjelkavik AI, Stöcker M, Ferreira A, Brouwer S, Kapteijn F, Weitkamp J, Sklari SD, Zaspalis VT, Jones DJ, de Menorval LC, Lindheimer M, Caffarelli P, Borsella E, Tomlinson AAG, Linders MJG, Tempelman JL, Bal

EA (2009) Characterisation measurements of common reference nanoporous materials by gas adsorption (Round Robin tests). In: Kaskel S, Llewellyn P, Rodríguez-Reinoso F, Seaton NA (eds) Characterisation of porous solids VIII: proceedings of the 8th international symposium on the characterisation of porous solids. RSC Publishing, Cambridge
43. Sing KSW, Everett DH, Haul RAW, Moscou L, Pierotti RA, Rouquérol J, Siemieniewska T (1985) Reporting physisorption data for gas/solid systems with special reference to the determination of surface area and porosity. Pure Appl Chem 57(4):603–619
44. Texier-Mandoki N, Dentzer J, Piquero T, Saadallah S, David P, Vix-Guterl C (2004) Hydrogen storage in activated carbon materials: role of the nanoporous texture. Carbon 42:2744–2747
45. Gadiou R, Saadallah S-E, Piquero T, David P, Parmentier J, Vix-Guterl C (2005) The influence of textural properties on the adsorption of hydrogen on ordered nanostructured carbons. Microporous Mesoporous Mater 79:121–128
46. Jordá-Beneyto M, Suárez-García F, Lozano-Castelló D, Cazorla-Amorós D, Linares-Solano A (2007) Hydrogen storage on chemically activated carbons and carbon nanomaterials at high pressures. Carbon 45:293–303
47. Jordá-Beneyto M, Lozano-Castelló D, Suárez-García F, Cazorla-Amorós D, Linares-Solano Á (2008) Advanced activated carbon monoliths and activated carbons for hydrogen storage. Microporous Mesoporous Mater 112:235–242
48. Zubizarreta L, Gomez EI, Arenillas A, Ania CO, Parra JB, Pis JJ (2008) H_2 storage in carbon materials. Adsorption 14:557–566
49. Gogotsi Y, Dash RK, Yushin G, Yildirim T, Laudisio G, Fischer JE (2005) Tailoring of nanoscale porosity in carbide-derived carbons for hydrogen storage. J Am Chem Soc 127:16006–16007
50. Yushin G, Dash R, Jagiello J, Fischer JE, Gogotsi Y (2006) Carbide-derived carbons: effect of pore size on hydrogen uptake and heat of adsorption. Adv Funct Mater 16:2288–2293
51. Walton KS, Snurr RQ (2007) Applicability of the BET method for determining surface areas of microporous metal-organic frameworks. J Am Chem Soc 129:8552–8556
52. Düren T, Millange F, Férey G, Walton KS, Snurr RQ (2007) Calculating geometric surface areas as a characterization tool for metal-organic frameworks. J Phys Chem C 111(42): 15350–15356
53. Frost H, Düren T, Snurr RQ (2006) Effects of surface area, free volume, and heat of adsorption on hydrogen uptake in metal-organic frameworks. J Phys Chem B 110:9565–9570
54. Hauback BC (2008) Structural characterization of hydride materials. In: Walker G (ed) Solid-state hydrogen storage: materials and chemistry. Woodhead Publishing, Cambridge
55. Yvon K (2003) Hydrogen in novel solid-state metal hydrides. Z Kristallogr 218:108–116
56. Bailey IF (2003) A review of sample environments in neutron scattering. Z Kristallogr 218:84–95
57. Rietveld HM (1969) A profile refinement method for nuclear and magnetic structures. J Appl Crystallogr 2:65–71
58. Young RA (ed) (1993) The Rietveld method. Oxford University Press, Oxford
59. Larson AC, Von Dreele RB (2004) General Structure Analysis System (GSAS). Los Alamos National Laboratory Report LAUR 86-748
60. Rodriguez-Carvajal J (2001) Recent developments of the program FULLPROF. IUCr Commission on Powder Diffraction Newsletter 26:12–19
61. Choi YN, Oh HS, Em VT, Somenkov VA, Lee C-H, Park SD (2002) Measurement of very small hydrogen content in zirconium alloys by measuring thermal neutron incoherent scattering. Appl Phys A 74:S1710–S1712
62. Perego RC, Blaauw M (2005) Incoherent neutron-scattering determination of hydrogen content: theory and modeling. J Appl Phys 97:123533
63. Squires GL (1978) Introduction to the theory of thermal neutron scattering. Cambridge University Press, Cambridge

64. Gray EM, Bailey IF (2008) Embrittlement of titanium-zirconium 'null-matrix' alloy by deuterium. J Neutron Res 16(3–4):127–132
65. Bull DJ, Weidner E, Shabalin IL, Telling MTF, Jewell CM, Gregory DH, Ross DK (2010) Pressure-dependent deuterium reaction pathways in the Li-N-D system. Phys Chem Chem Phys 12:2089–2097
66. Yildirim T, Hartman MR (2005) Direct observation of hydrogen adsorption sites and nanocage formation in metal-organic frameworks. Phys Rev Lett 95:215504
67. Dincă M, Dailly A, Liu Y, Brown CM, Neumann DA, Long JR (2006) Hydrogen storage in a microporous metal-organic framework with exposed Mn^{2+} coordination sites. J Am Chem Soc 128:16876–16883
68. Peterson VK, Liu Y, Brown CM, Kepert CJ (2006) Neutron powder diffraction study of D_2 sorption in Cu_3(1, 3, 5-benzenetricarboxylate$)_2$. J Am Chem Soc 128:15578–15579
69. Liu Y, Kabbour H, Brown CM, Neumann DA, Ahn CC (2008) Increasing the density of adsorbed hydrogen with coordinatively unsaturated metal centers in metal-organic frameworks. Langmuir 24:4772–4777
70. Brown CM, Liu Y, Neumann DA (2008) Neutron powder diffraction of metal-organic frameworks for hydrogen storage. Pramana J Phys 71(4):755–760
71. Lokshin KA, Zhao Y, He D, Mao WL, Mao H-K, Hemley RJ, Lobanov MV, Greenblatt M (2004) Structure and dynamics of hydrogen molecules in the novel clathrate hydrate by high pressure neutron diffraction. Phys Rev Lett 93(12):125503
72. Gross KJ, Guthrie S, Takara S, Thomas G (2000) In situ X-ray diffraction study of the decomposition of $NaAlH_4$. J Alloy Compd 297:270–281
73. Jensen CM, Gross KJ (2001) Development of catalytically enhanced sodium aluminium hydride as a hydrogen-storage material. Appl Phys A 72:213–219
74. Baldé CP, Hereijgers BPC, Bitter JH, de Jong KP (2008) Sodium alanate nanoparticles—linking size to hydrogen storage properties. J Am Chem Soc 130:6761–6765
75. Huot J, Pelletier JF, Liang G, Sutton M, Schulz R (2002) Structure of nanocomposite metal hydrides. J Alloy Compd 330–332:727–731
76. Jensen TR, Andreasen A, Vegge T, Andreasen JW, Ståhl K, Pedersen AS, Nielsen MM, Molenbroek AM, Besenbacher F (2006) Dehydrogenation kinetics of pure and nickel-doped magnesium hydride investigated by in situ time-resolved powder X-ray diffraction. Int J Hydrogen Energy 31:2052–2062
77. Sun D, Gingl F, Nakamura Y, Enoki H, Bououdina M, Akiba E (2002) In situ X-ray diffraction study of hydrogen-induced phase decomposition in $LaMg_{12}$ and La_2Mg_{17}. J Alloy Compd 333:103–108
78. Narehood DG, Kishore S, Goto H, Adair JH, Nelson JA, Gutiérrez HR, Eklund PC (2009) X-ray diffraction and H-storage in ultra-small palladium particles. Int J Hydrogen Energy 34:952–960
79. Phan T-H, Schaak RE (2009) Polyol synthesis of palladium hydride: bulk powders *vs.* nanocrystals. Chem Commun 3026–3028
80. Mosegaard L, Møller B, Jørgensen J-E, Filinchuk Y, Cerenius Y, Hanson JC, Dimasi E, Besenbacher F, Jensen TR (2008) Reactivity of $LiBH_4$: in situ synchrotron radiation X-ray diffraction study. J Phys Chem C 112(4):1299–1303
81. Riktor MD, Sørby MH, Chłopek K, Fichtner M, Buchter F, Züttel A, Hauback BC (2007) *In situ* synchrotron diffraction studies of phase transitions and thermal decomposition of $Mg(BH_4)_2$ and $Ca(BH_4)_2$. J Mater Chem 17:4939–4942
82. Brinks HW, Hauback BC, Norby P, Fjellvåg H (2003) The decomposition of $LiAlD_4$ studied by in situ X-ray and neutron diffraction. J Alloy Compd 351:222–227
83. Blanchard D, Brinks HW, Hauback BC, Norby P (2004) Desorption of $LiAlH_4$ with Ti- and V-based additives. Mater Sci Eng B 108:54–59
84. Fossdal A, Brinks HW, Fichtner M, Hauback BC (2005) Thermal decomposition of $Mg(AlH_4)_2$ studied by in situ synchrotron X-ray diffraction. J Alloy Compd 404–406:752–756

85. Maehlen JP, Yartys VA, Denys RV, Fichtner M, Frommen C, Bulychev BM, Pattison P, Emerich H, Filinchuk YE, Chernyshov D (2007) Thermal decomposition of AlH_3 studied by *in situ* synchrotron X-ray diffraction and thermal desorption spectroscopy. J Alloy Compd 446–447:280–289
86. Huot J, Pelletier JF, Lurio LB, Sutton M, Schulz R (2003) Investigation of dehydrogenation mechanism of MgH_2-Nb nanocomposites. J Alloy Compd 348:319–324
87. Yavari AR, de Castro JFR, Vaughan G, Heunen G (2003) Structural evolution and metastable phase detection in MgH2–5%NbH nanocomposite during in situ H-desorption in a synchrotron beam. J Alloy Compd 353:246–251
88. Denys RV, Riabov AB, Maehlen JP, Lototsky MV, Solberg JK, Yartys VA (2009) *In situ* synchrotron X-ray diffraction studies of hydrogen desorption and absorption properties of Mg and Mg-Mm-Ni after reactive ball milling in hydrogen. Acta Mater 57:3989–4000
89. Bösenberg U, Doppiu S, Mosegaard L, Barkhordarian G, Eigen N, Borgschulte A, Jensen TR, Cerenius Y, Gutfleisch O, Klassen T, Dornheim M, Bormann R (2007) Hydrogen sorption properties of MgH_2-$LiBH_4$ composites. Acta Mater 55:3951–3958
90. Nakamura Y, Hino S, Ichikawa T, Fujii H, Brinks HW, Hauback BC (2008) Dehydrogenation reaction of Li-Mg-N-H systems studied by in situ synchrotron powder X-ray diffraction and powder neutron diffraction. J Alloy Compd 457:362–367
91. Joubert J-M, Černý R, Latroche M, Percheron-Guégan A, Schmitt B (2006) Hydrogenation of $LaNi_5$ studied by in situ synchrotron powder diffraction. Acta Mater 54:713–719
92. Stange M, Maehlen JP, Yartys VA, Norby P, van Beek W, Emerich H (2005) In situ SR-XRD studies of hydrogen absorption-desorption in $LaNi_{4.7}Sn_{0.3}$. J Alloy Compd 404–406:604–608
93. Gray EM, Cookson DJ, Blach TP (2006) X-ray diffraction cell for studying solid-gas reactions under gas pressures to 100 bar. J Appl Crystallogr 39:850–855
94. Černý R, Joubert J-M, Latroche M, Percheron-Guégan A, Yvon K (2000) Anisotropic diffraction peak broadening and dislocation substructure in hydrogen-cycled $LaNi_5$ and substitutional derivatives. J Appl Crystallogr 33:997–1005
95. Fratzl P (2003) Small-angle scattering in materials science—a short review of applications in alloys, ceramics and composite materials. J Appl Crystallogr 36:397–404
96. Simon J-P (2007) Contribution of synchrotron radiation to small-angle X-ray scattering studies in hard condensed matter. J Appl Crystallogr 40:S1–S9
97. Melnichenko YB, Wignall GD (2007) Small-angle neutron scattering in materials science: recent practical applications. J Appl Phys 102:021101
98. Sharp MA, Pranzas PK, Schreyer A (2009) Going ultra: how we can increase the length scales studied in small-angle neutron scattering. Adv Eng Mater 11(6):441–445
99. Orimo S, Seto H, Ikeda K, Nagao M, Fujii H (1996) Small angle neutron scattering measurements of a nanostructured Mg_2Ni-D system. Physica B 226:370–374
100. Orimo S, Fujii H (2001) Materials science of Mg-Ni-based new hydrides. Appl Phys A 72:167–186
101. Fultz B, Witham CK, Udovic TJ (2002) Distributions of hydrogen and strains in $LaNi_5$ and $LaNi_{4.75}Sn_{0.25}$. J Alloy Compd 335:165–175
102. Bowman RC, Fultz B (2002) Metallic hydrides 1: hydrogen storage and other gas-phase applications. MRS Bull 27(9):688–693
103. Flanagan TB, Noh H, Clewley JD, Barker JG (1998) Evidence from SANS and H_2 solubilities for H-enhanced metal atom diffusion in Pd-Ni alloys. Scr Mater 39(11):1607–1611
104. Heuser BJ, King JS (1997) SANS measurements of deuterium-dislocation trapping in deformed single crystal Pd. J Alloy Compd 261:225–230
105. Ross DK, Stefanopoulos K, Kemali M (1999) The use of small angle neutron scattering in the study of hydrogen trapping at defects in metals. J Alloy Compd 293–295:346–350
106. Maxelon M, Pundt A, Pyckhout-Hintzen W, Barker J, Kirchheim R (2001) Interaction of hydrogen and deuterium with dislocations in palladium as observed by small angle neutron scattering. Acta Mater 49:2625–2634

107. Heuser BJ, King JS, Chen WC (1999) SANS measurements of deuteride (hydride) formation in single crystal Pd. J Alloy Compd 292:134–147
108. Chen WC, Heuser BJ, King JS (2000) Small-angle neutron scattering investigation of deuteride (hydride) precipitation and decomposition in single-crystal Pd. J Appl Crystallogr 33:442–446
109. Pranzas PK, Dornheim M, Bellmann D, Aguey-Zinsou K-F, Klassen T, Schreyer A (2006) SANS/USANS investigations of nanocrystalline MgH_2 for reversible storage of hydrogen. Physica B 385–386:630–632
110. Pranzas PK, Dornheim M, Bösenberg U, Fernandez JRA, Goerigk G, Roth SV, Gehrke R, Schreyer A (2007) Small-angle scattering investigations of magnesium hydride used as a hydrogen storage material. J Appl Crystallogr 40:S383–S387
111. Dobbins TA, Bruster EL, Oteri EU, Ilavsky J (2007) Ultrasmall-angle X-ray scattering (USAXS) studies of morphological trends in high energy milled $NaAlH_4$ powders. J Alloy Compd 446–447:248–254
112. Sartori S, Knudsen KD, Zhao-Karger Z, Bardaij EG, Fichtner M, Hauback BC (2009) Small-angle scattering investigations of Mg-borohydride infiltrated in activated carbon. Nanotechnology 20:505702
113. Sheppard DA, Maitland CF, Buckley CE (2005) Preliminary results of hydrogen adsorption and SAXS modelling of mesoporous silica: MCM-41. J Alloy Compd 404–406:405–408
114. Laudisio G, Dash RK, Singer JP, Yushin G, Gogotsi Y, Fischer JE (2006) Carbide-derived carbons: a comparative study of porosity based on small-angle scattering and adsorption isotherms. Langmuir 22:8945–8950
115. Du X, Wu E (2007) Porosity of microporous zeolites A, X and ZSM-5 studied by small angle X-ray scattering and nitrogen adsorption. J Phys Chem Solids 68:1692–1699
116. Tsao C-S, Yu M-S, Chung T-Y, Wu H-C, Wang C-Y, Chang K-S, Chen H-L (2007) Characterization of pore structure in metal-organic framework by small-angle X-ray scattering. J Am Chem Soc 129:15997–16004
117. Fultz B (2006) Materials science applications of inelastic neutron scattering. JOM 58(3):58–63
118. Parker SF (2010) Spectroscopy and bonding in ternary metal hydride complexes—potential hydrogen storage media. Coord Chem Rev 254:215–234
119. Ross DK (2008) Neutron scattering studies for analysing solid-state hydrogen storage. In: Walker G (ed) Solid-state hydrogen storage: materials and chemistry. Woodhead Publishing, Cambridge
120. Fukai Y (2005) The metal-hydrogen system. Basic bulk properties, 2nd edn. Springer, Berlin
121. Neumann D (2006) Neutron scattering and hydrogenous materials. Mater Today 9(1–2):34–41
122. Ramirez-Cuesta AJ, Jones MO, David WIF (2009) Neutron scattering and hydrogen storage. Mater Today 12(11):54–61
123. Kearley GJ, Johnson MR (2010) Vibrational spectroscopy with neutrons—where are we now? Vib Spectrosc 53:54–59
124. Parker SF, Williams KPJ, Bortz M, Yvon K (1997) Inelastic neutron scattering, infrared, and Raman spectroscopic studies of Mg_2FeH_6 and Mg_2FeD_6. Inorg Chem 36:5218–5221
125. Parker SF, Jayasooriya UA, Sprunt JC, Bortz M, Yvon K (1998) Inelastic neutron scattering, IR and Raman spectroscopic studies of Mg_2CoH_5 and Mg_2CoD_5. J Chem Soc Faraday Trans 94(17):2595–2599
126. Olofsson-Mårtensson M, Häussermann U, Tomkinson J, Noréus D (2000) Stabilization of electron-dense palladium–hydrido complexes in solid-state hydrides. J Am Chem Soc 122:6960–6970
127. Parker SF, Bennington SM, Ramirez-Cuesta AJ, Auffermann G, Bronger W, Herman H, Williams KPJ, Smith T (2003) Inelastic neutron scattering and Raman spectroscopies and periodic DFT studies of Rb_2PtH_6 and Rb_2PtD_6. J Am Chem Soc 125:11656–11661
128. Parker SF, Refson K, Williams KPJ, Braden DA, Hudson BS, Yvon K (2006) Spectroscopic and ab initio characterization of the $[ReH_9]^{2-}$ ion. Inorg Chem 45(26):10951–10957

129. Schimmel HG, Johnson MR, Kearley GJ, Ramirez-Cuesta AJ, Huot J, Mulder FM (2005) Structural information on ball milled magnesium hydride from vibrational spectroscopy and ab initio calculations. J Alloy Compd 393:1–4
130. Schimmel HG, Johnson MR, Kearley GJ, Ramirez-Cuesta AJ, Huot J, Mulder FM (2004) The vibrational spectrum of magnesium hydride from inelastic neutron scattering and density functional theory. Mater Sci Eng B 108:38–41
131. Fu QJ, Ramirez-Cuesta AJ, Tsang SC (2006) Molecular aluminium hydrides identified by inelastic neutron scattering during H_2 regeneration of catalyst-doped $NaAlH_4$. J Phys Chem B 110:711–715
132. Allis DG, Hudson BS (2004) Inelastic neutron scattering spectra of $NaBH_4$ and KBH_4: reproduction of anion mode shifts via periodic DFT. Chem Phys Lett 385:166–172
133. Hartman MR, Rush JJ, Udovic TJ, Bowman RC, Hwang S-J (2007) Structure and vibrational dynamics of isotopically labeled lithium borohydride using neutron diffraction and spectroscopy. J Solid State Chem 180:1298–1305
134. Mulder FM, Dingemans TJ, Schimmel HG, Ramirez-Cuesta AJ, Kearley GJ (2008) Hydrogen adsorption strength and sites in the metal organic framework MOF5: comparing experiment and model calculations. Chem Phys 351(1–3):72–76
135. Brown CM, Liu Y, Yildirim T, Peterson VK, Kepert CJ (2009) Hydrogen adsorption in HKUST-1: a combined inelastic neutron scattering and first-principles study. Nano technology 20:204025
136. Nouar F, Eckert J, Eubank JF, Forster P, Eddaoudi M (2009) Zeolite-*like* metal-organic frameworks (ZMOFs) as hydrogen storage platform: lithium and magnesium ion-exchange and H_2-(*rho*-ZMOF) interaction studies. J Am Chem Soc 131:2864–2870
137. Ramirez-Cuesta AJ, Mitchell PCH (2007) Hydrogen adsorption in a copper ZSM5 zeolite: an inelastic neutron scattering study. Catal Today 120:368–373
138. Ramirez-Cuesta AJ, Mitchell PCH, Ross DK, Georgiev PA, Anderson PA, Langmi HW, Book D (2007) Dihydrogen in cation-substituted zeolites X—an inelastic neutron scattering study. J Mater Chem 17:2533–2539
139. Schimmel HG, Kearley GJ, Nijkamp MG, Visser CT, de Jong KP, Mulder FM (2003) Hydrogen adsorption in carbon nanostructures: comparison of nanotubes, fibers, and coals. Chem Eur J 9:4764–4770
140. Georgiev PA, Ross DK, De Monte A, Montaretto-Marullo U, Edwards RAH, Ramirez-Cuesta AJ, Colognesi D (2004) Hydrogen site occupancies in single-walled carbon nanotubes studied by inelastic neutron scattering. J Phys Condens Matter 16:L73–L78
141. Georgiev PA, Ross DK, De Monte A, Montaretto-Marullo U, Edwards RAH, Ramirez-Cuesta AJ, Adams MA, Colognesi D (2005) In situ inelastic neutron scattering studies of the rotational and translational dynamics of molecular hydrogen adsorbed in single-wall carbon nanotubes (SWNTs). Carbon 43:895–906
142. Georgiev PA, Ross DK, Albers P, Ramirez-Cuesta AJ (2006) The rotational and translational dynamics of molecular hydrogen physisorbed in activated carbon: a direct probe of microporosity and hydrogen storage performance. Carbon 44:2724–2738
143. Fernandez-Alonso F, Bermejo FJ, Cabrillo C, Loutfy RO, Leon V, Saboungi ML (2007) Nature of the bound states of molecular hydrogen in carbon nanohorns. Phys Rev Lett 98:215503
144. Tait KT, Trouw F, Zhao Y, Brown CM, Downs RT (2007) Inelastic neutron scattering study of hydrogen in d_8-THF/D_2O ice clathrate. J Chem Phys 127:134505
145. Ulivi L, Celli M, Giannasi A, Ramirez-Cuesta AJ, Bull DJ, Zoppi M (2007) Quantum rattling of molecular hydrogen in clathrate hydrate nanocavities. Phys Rev B 76:161401
146. Ulivi L, Celli M, Giannasi A, Ramirez-Cuesta AJ, Zoppi M (2008) Inelastic neutron scattering from hydrogen clathrate hydrates. J Phys Condens Matter 20:104242
147. Parker SF (2002) The design and performance of the indirect geometry spectrometer TOSCA. J Neutron Res 10(3–4):173–177

148. Colognesi D, Celli M, Cilloco F, Newport RJ, Parker SF, Rossi-Albertini V, Sacchetti F, Tomkinson J, Zoppi M (2002) TOSCA neutron spectrometer: the final configuration. Appl Phys A 74:S64–S66
149. Chudley CT, Elliott RJ (1961) Neutron scattering from a liquid on a jump diffusion model. Proc Phys Soc 77(2):353–361
150. Kahn R, Cohen De Lara E, Viennet E (1989) Diffusivity of the hydrogen molecule sorbed in NaA zeolite by a neutron scattering experiment. J Chem Phys 91(8):5097–5102
151. Bär N-K, Ernst H, Jobic H, Kärger J (1999) Combined quasi-elastic neutron scattering and NMR study of hydrogen diffusion in zeolites. Magn Reson Chem 37:S79–S83
152. Jobic H, Kärger J, Bée M (1999) Simultaneous measurement of self- and transport diffusivities in zeolites. Phys Rev Lett 82(21):4260–4263
153. Kumar AVA, Jobic H, Bhatia SK (2006) Quantum effects on adsorption and diffusion of hydrogen and deuterium in microporous materials. J Phys Chem B 110:16666–16671
154. Pantatosaki E, Papadopoulos GK, Jobic H, Theodorou DN (2008) Combined atomistic simulation and quasielastic neutron scattering study of the low-temperature dynamics of hydrogen and deuterium confined in NaX zeolite. J Phys Chem B 112:11708–11715
155. Salles F, Jobic H, Maurin G, Koza MM, Llewellyn PL, Devic T, Serre C, Ferey G (2008) Experimental evidence supported by simulations of a very high H_2 diffusion in metal organic framework materials. Phys Rev Lett 100:245901
156. Salles F, Kolokolov DI, Jobic H, Maurin G, Llewellyn PL, Devic T, Serre C, Ferey G (2009) Adsorption and diffusion of H_2 in the MOF type systems MIL-47(V) and MIL-53(Cr): a combination of microcalorimetry and QENS experiments with molecular simulations. J Phys Chem C 113:7802–7812
157. Benes NE, Jobic H, Réat V, Bouwmeester HJM, Verweij H (2003) Mobility of hydrogen in microporous silica studied with quasi-elastic neutron scattering. Sep Purif Technol 32:9–15
158. Voss J, Shi Q, Jacobsen HS, Zamponi M, Lefmann K, Vegge T (2007) Hydrogen dynamics in Na_3AlH_6: a combined density functional theory and quasielastic neutron scattering study. J Phys Chem B 111:3886–3892
159. Shi Q, Voss J, Jacobsen HS, Lefmann K, Zamponi M, Vegge T (2007) Point defect dynamics in sodium aluminium hydrides–a combined quasielastic neutron scattering and density functional theory study. J Alloy Compd 446–447:469–473
160. Bull DJ, Broom DP, Ross DK (2003) Monte Carlo simulation of quasielastic neutron scattering from localised and long-range hydrogen motion in C15 Laves phase intermetallic compounds. Chem Phys 292:153–160
161. Skripov AV (2005) Hydrogen jump motion in Laves-phase hydrides: two frequency scales. J Alloy Compd 404–406:224–229
162. Skripov AV, Gonzalez MA, Hempelmann R (2006) Evidence for a two-site localized hydrogen motion in C15-type YMn_2H_x. J Phys Condens Matter 18:7249–7256
163. Skripov AV, Udovic TJ, Rush JJ (2007) Hydrogen jump diffusion in C14-type $ZrMn_2H_3$: quasielastic neutron scattering study. Phys Rev B 76:104305
164. Skripov AV, Gonzalez MA, Hempelmann R (2008) Localized hydrogen motion in C15-type $TaV_2H_{0.65}$: temperature dependence of the H jump rate. J Phys Condens Matter 20:085213
165. Richter D, Hempelmann R, Vinhas LA (1982) Hydrogen diffusion in $LaNi_5H_6$ studied by quasi-elastic neutron scattering. J Less-Common Met 88:353–360
166. Cotts RM (1978) Nuclear magnetic resonance on metal-hydrogen systems. In: Alefeld G, Völkl J (eds) Topics in applied physics vol. 28: hydrogen in metals I. Basic properties. Springer, Berlin
167. Bogdanović B, Felderhoff M, Germann M, Härtel M, Pommerin A, Schüth F, Weidenthaler C, Zibrowius B (2003) Investigation of hydrogen discharging and recharging processes of Ti-doped $NaAlH_4$ by X-ray diffraction analysis (XRD) and solid-state NMR spectroscopy. J Alloy Compd 350:246–255
168. Herberg JL, Maxwell RS, Majzoub EH (2006) ^{27}Al and ^{1}H MAS NMR and ^{27}Al multiple quantum studies of Ti-doped $NaAlH_4$. J Alloy Compd 417:39–44

169. Balema VP, Wiench JW, Dennis KW, Pruski M, Pecharsky VK (2001) Titanium catalyzed solid-state transformations in $LiAlH_4$ during high-energy ball-milling. J Alloy Compd 329:108–114
170. Wiench JW, Balema VP, Pecharsky VK, Pruski M (2004) Solid-state ^{27}Al NMR investigation of thermal decomposition of $LiAlH_4$. J Solid State Chem 177:648–653
171. Hu JZ, Kwak JH, Yang Z, Osborn W, Markmaitree T, Shaw LL (2008) Probing the reaction pathway of dehydrogenation of the $LiHH_2$ + LiH mixture using *in situ* ^{1}H NMR spectroscopy. J Power Sources 181:116–119
172. Richter D, Hempelmann R, Bowman RC (1992) Dynamics of hydrogen in intermetallic hydrides. In: Schlapbach L (ed) Topics in applied physics vol. 67: hydrogen in intermetallic compounds II. Surface and dynamic properties. Springer-verlag, Berlin
173. Ailion DC, Ohlsen WD (1983) Magnetic resonance methods for studying defect structure in solids. In: Mundy JN (ed) Solid state: nuclear methods. Methods of experimental physics, vol 21. Academic Press, Orlando
174. Ailion DC (1983) Nuclear magnetic resonance relaxation time methods for studying atomic and molecular motions in solids. In: Mundy JN (ed) Solid state: nuclear methods. Methods of experimental physics, vol 21. Academic Press, Orlando
175. Gerstein BC, Dybowski CR (1985) Transient techniques in NMR of solids: an introduction to theory and practice. Academic Press, Orlando
176. Phua T-T, Beaudry BJ, Peterson DT, Torgeson DR, Barnes RG, Belhoul M, Styles GA, Seymour EFW (1983) Paramagnetic impurity effects in NMR determinations of hydrogen diffusion and electronic structure in metal hydrides. Gd^{3+} in YH_2 and $LaH_{2.25}$. Phys Rev B 28(11):6227–6250
177. Bloembergen N, Purcell EM, Pound RV (1948) Relaxation effects in nuclear magnetic resonance absorption. Phys Rev 73(7):679–712
178. McDowell AF, Adolphi NL, Sholl CA (2001) Site and barrier energy distributions that govern the rate of hydrogen motion in quasicrystalline $Ti_{45}Zr_{38}Ni_{17}H_x$. J Phys Condens Matter 13:9799–9812
179. Baker DB, Conradi MS (2005) Apparatus for high temperatures and intermediate pressures, for *in situ* nuclear magnetic resonance of hydrogen storage systems. Rev Sci Instrum 76:073906
180. Majer G, Stanik E, Orimo S (2003) NMR studies of hydrogen motion in nanostructured hydrogen-graphite systems. J Alloy Compd 356–357:617–621
181. Majer G, Eberle U, Kimmerle F, Stanik E, Orimo S (2003) Hydrogen diffusion in metallic and nanostructured materials. Physica B 328:81–89
182. Stanik E, Majer G, Orimo S, Ichikawa T, Fujii H (2005) Nuclear-magnetic-resonance measurements of the hydrogen dynamics in nanocrystalline graphite. J Appl Phys 98:044302
183. Bowman RC, Maeland AJ (1981) NMR studies of diffusion in the metallic glass $TiCuH_x$. Phys Rev B 24(5):2328–2333
184. Bowman RC, Attalla A, Maeland AJ, Johnson WL (1983) Hydrogen mobility in crystalline and amorphous Zr_2PdH_x. Solid State Commun 47(10):779–782
185. Buzlukov AL, Skripov AV (2004) Nuclear magnetic resonance study of hydrogen motion in C15-type TaV_2H_x ($x \leq 0.18$). J Alloy Compd 366:61–66
186. Bowman RC, Gruen DM, Mendelsohn MH (1979) NMR studies of hydrogen diffusion in β-$LaNi_{5-y}Al_y$ hydrides. Solid State Commun 32:501–506
187. Bowman RC, Craft BD, Attalla A, Mendelsohn MH, Gruen DM (1980) Role of aluminum substitution on hydrogen diffusion in β-$LaNi_{5-y}Al_yH_x$. J Less-Common Met 73:227–232
188. Mendenhall MP, Bowman RC, Ivancic TM, Conradi MS (2007) Rate of hydrogen motion in Ni-substituted $LaNi_5H_x$ from NMR. J Alloy Compd 446–447:495–498
189. Corey RL, Ivancic TM, Shane DT, Carl EA, Bowman RC, Bellosta von Colbe JM, Dornheim M, Bormann R, Huot J, Zidan R, Stowe AC, Conradi MS (2008) Hydrogen motion in magnesium hydride by NMR. J Phys Chem C 112:19784–19790

190. Conradi MS, Mendenhall MP, Ivancic TM, Carl EA, Browning CD, Notten PHL, Kalisvaart WP, Magusin PCMM, Bowman RC, Hwang S-J, Adolphi NL (2007) NMR to determine rates of motion and structures in metal-hydrides. J Alloy Compd 446–447:499–503
191. Magusin PCMM, Kalisvaart WP, Notten PHL, van Santen RA (2008) Hydrogen sites and dynamics in light-weight hydrogen-storage material magnesium-scandium hydride investigated with ^{1}H and ^{2}H NMR. Chem Phys Lett 456:55–58
192. Shane DT, Corey RL, Bowman RC, Zidan R, Stowe AC, Hwang S-J, Kim C, Conradi MS (2009) NMR studies of the hydrogen storage compound $NaMgH_3$. J Phys Chem C 113:18414–18419
193. Grinberg F, Majer G, Skripov AV (2006) Pulsed-field-gradient NMR study of hydrogen diffusivity in random b.c.c. alloys V_yTa_{1-y}. J Alloy Compd 425:24–27
194. Shastri A, Majzoub EH, Borsa F, Gibbons PC, Kelton KF (1998) ^{1}H NMR study of hydrogen in quasicrystalline $Ti_{0.45-x}V_xZr_{0.38}Ni_{0.17}$. Phys Rev B 57(9):5148–5153
195. Shastri A, Majzoub EH, Borsa F, Gibbons PC, Kelton KF (1999) Erratum: ^{1}H NMR study of hydrogen in quasicrystalline $Ti_{0.45-x}V_xZr_{0.38}Ni_{0.17}$ [Phys. Rev. B 57 5148 (1998)]. Phys Rev B 59(21):14108
196. Majer G, Stanik E, Valiente Banuet LE, Grinberg F, Kircher O, Fichtner M (2005) Effects of catalysts on the dehydriding of alanates monitored by proton NMR. J Alloy Compd 404–406:738–742
197. Senadheera L, Conradi MS (2007) Rotation and diffusion of H_2 in hydrogen-ice clathrate by ^{1}H NMR. J Phys Chem B 111:12097–12102
198. Senadheera L, Conradi MS (2008) Hydrogen nuclear spin relaxation in hydrogen-ice clathrate. J Phys Chem A 112:8303–8309
199. Senadheera L, Conradi MS (2008) Hydrogen NMR of H_2-TDF-D_2O clathrate. J Phys Chem B 112:13695–13700
200. Strobel TA, Taylor CJ, Hester KC, Dec SF, Koh CA, Miller KT, Sloan ED (2006) Molecular hydrogen storage in binary THF-H_2 clathrate hydrates. J Phys Chem B 110:17121–17125
201. Garrone E, Rodríguez Delgado M, Bonelli B, Otero Areán C (2005) Ferreting out gas adsorption heats: the pseudo-isobaric method. Phys Chem Chem Phys 7:3519–3522
202. Otero Areán C, Manoilova OV, Tsyganenko AA, Turnes Palomino G, Peñarroya Mentruit M, Geobaldo F, Garrone E (2001) Thermodynamics of hydrogen bonding between CO and the supercage brønsted acid sites of the H-Y zeolite—studies from variable temperature IR spectrometry. Eur J Inorg Chem 2001(7):1739–1743
203. Tsyganenko AA, Storozhev PY, Otero Areán C (2004) IR-spectroscopic study of the binding isomerism of adsorbed molecules. Kinet Catal 45(4):562–573
204. Garrone E, Otero Areán C (2005) Variable temperature infrared spectroscopy: a convenient tool for studying the thermodynamics of weak solid-gas interactions. Chem Soc Rev 34:846–857
205. Otero Areán C, Manoilova OV, Turnes Palomino G, Rodríguez Delgado M, Tsyganenko AA, Bonelli B, Garrone E (2002) Variable-temperature infrared spectroscopy: an access to adsorption thermodynamics of weakly interacting systems. Phys Chem Chem Phys 4:5713–5715
206. Garrone E, Bonelli B, Otero Areán C (2008) Enthalpy-entropy correlation for hydrogen adsorption on zeolites. Chem Phys Lett 456:68–70
207. Otero Areán C, Turnes Palomino G, Llop Carayol MR (2007) Variable temperature FT-IR studies on hydrogen adsorption on the zeolite (Mg, Na)-Y. Appl Surf Sci 253:5701–5704
208. Spoto G, Vitillo JG, Cocina D, Damin A, Bonino F, Zecchina A (2007) FTIR spectroscopy and thermodynamics of hydrogen adsorbed in a cross-linked polymer. Phys Chem Chem Phys 9:4992–4999
209. Vitillo JG, Regli L, Chavan S, Ricchiardi G, Spoto G, Dietzel PDC, Bordiga S, Zecchina A (2008) Role of exposed metal sites in hydrogen storage in MOFs. J Am Chem Soc 130:8386–8396

210. Kleperis J, Wójcik G, Czerwinski A, Skowronski J, Kopczyk M, Beltowska-Brzezinska M (2001) Electrochemical behaviour of metal hydrides. J Solid State Electrochem 5:229–249
211. Iwakura C, Inoue H, Nohara S (2001) Hydrogen-metal systems: electrochemical reactions (fundamentals and applications). In: Buschow KHJ, Cahn RW, Flemings MC, Ilschner MC, Kramer EJ, Mahajan S, Veyssière P (eds) Encyclopedia of materials: science and technology. Elsevier, Amsterdam
212. Züttel A, Sudan P, Mauron P, Kiyobayashi T, Emmenegger C, Schlapbach L (2002) Hydrogen storage in carbon nanostructures. Int J Hydrogen Energy 27:203–212
213. Züttel A, Nützenadel C, Sudan P, Mauron P, Emmenegger C, Rentsch S, Schlapbach L, Weidenkaff A, Kiyobayashi T (2002) Hydrogen sorption by carbon nanotubes and other carbon nanostructures. J Alloy Compd 330–332:676–682
214. Ansón A, Benham M, Jagiello J, Callejas MA, Benito AM, Maser WK, Züttel A, Sudan P, Martínez MT (2004) Hydrogen adsorption on a single-walled carbon nanotube material: a comparative study of three different adsorption techniques. Nanotechnology 15:1503–1508
215. Bittner HF, Badcock CC (1983) Electrochemical utilization of metal hydrides. J Electrochem Soc 130(5):193C–198C
216. Feng F, Geng M, Northwood DO (2001) Electrochemical behaviour of intermetallic-based metal hydrides used in Ni/metal hydride (MH) batteries: a review. Int J Hydrogen Energy 26:725–734
217. Bliznakov S, Lefterova E, Dimitrov N (2008) Electrochemical PCT isotherm study of hydrogen absorption/desorption in AB_5 type intermetallic compounds. Int J Hydrogen Energy 33:5789–5794
218. Shirai Y, Araki H, Mori T, Nakamura W, Sakaki K (2002) Positron annihilation study of lattice defects induced by hydrogen absorption in some hydrogen storage materials. J Alloy Compd 330–332:125–131
219. Sakaki K, Akiba E, Mizuno M, Araki H, Shirai Y (2009) The effect of substitutional elements (Al, Co) in $LaNi_{4.5}M_{0.5}$ on the lattice defect formation in the initial hydrogenation and dehydrogenation. J Alloy Compd 473(1–2):87–93
220. Kim G-H, Chun C-H, Lee S-G, Lee J-Y (1994) TEM study on the nucleation and growth of hydride in $LaNi_5$ alloy. Acta Metall Mater 42(9):3157–3161
221. Kim G-H, Lee S-G, Lee K-Y, Chun C-H, Lee J-Y (1995) Observation of the defects induced by hydrogen absorption and desorption in $LaNi_5$. Acta Metall Mater 43(6):2233–2240
222. Décamps B, Joubert J-M, Cerny R, Percheron-Guégan A (2005) TEM study of the dislocations generated by hydrogen absorption/desorption in $LaNi_5$ and derivatives. J Alloy Compd 404–406:570–575
223. Inui H, Yamamoto T, Hirota M, Yamaguchi M (2002) Lattice defects introduced during hydrogen absorption-desorption cycles and their effects on *P-C* characteristics in some intermetallic compounds. J Alloy Compd 330–332:117–124
224. Beattie SD, Humphries T, Weaver L, McGrady GS (2008) Temporal and spatial imaging of hydrogen storage materials: watching solvent and hydrogen desorption from aluminium hydride by transmission electron microscopy. Chem Commun 4448–4450
225. Ikeda K, Muto S, Tatsumi K, Menjo M, Kato S, Bielmann M, Züttel A, Jensen CM, Orimo S (2009) Dehydriding reaction of AlH_3: *in situ* microscopic observations combined with thermal and surface analyses. Nanotechnology 20:204004
226. Muto S, Tatsumi K, Ikeda K, Orimo S (2009) Dehydriding process of α-AlH_3 observed by transmission electron microscopy and electron energy-loss spectroscopy. J Appl Phys 105:123514
227. Kim JW, Ahn J-P, Kim DH, Chung H-S, Shim J-H, Cho YW, Oh KH (2010) In situ transmission electron microscopy study on microstructural changes in NbF_5-doped MgH_2 during dehydrogenation. Scr Mater 62:701–704
228. Beattie SD, McGrady GS (2009) Hydrogen desorption studies of $NaAlH_4$ and $LiAlH_4$ by in situ heating in an ESEM. Int J Hydrogen Energy 34:9151–9156

229. Palumbo O, Cantelli R, Paolone A, Jensen CM, Srinivasan SS (2005) Point defect dynamics and evolution of chemical reactions in alanates by anelastic spectroscopy. J Phys Chem B 109:1168–1173
230. Palumbo O, Paolone A, Cantelli R, Jensen CM, Sulic M (2006) Fast H-vacancy dynamics during alanate decomposition by anelastic spectroscopy. Proposition of a model for Ti-enhanced hydrogen transport. J Phys Chem B 110:9105–9111
231. Palumbo O, Paolone A, Rispoli P, Cantelli R (2009) Novel materials for solid-state hydrogen storage: anelastic spectroscopy studies. Mater Sci Eng A 521–522:134–138

Chapter 6
Experimental Considerations

A careful assessment of the potential sources of error is a crucial precursor to the performance of an accurate hydrogen sorption measurement and so we address some of these issues in this chapter.[1] Firstly, we will cover some of the properties of hydrogen that are relevant to the accurate measurement of hydrogen uptake by materials. We begin with the compressibility, a property that must be described with sufficient accuracy for elevated pressure measurements, and we present some equations of state that can describe the real gas behaviour of hydrogen with varying degrees of accuracy. We then describe the Joule–Thomson effect and discuss the thermal conductivity of hydrogen, both of which can affect the necessary isothermal conditions and the accurate determination of the sample temperature. The transition between the continuum (viscous flow) and free molecule regimes, in terms of the mean free path of hydrogen and the system length scale, is then discussed and we also describe thermal transpiration effects, which can affect low pressure measurements. We conclude the first section by discussing the importance of gas purity. In the second section, the properties of the storage material itself are covered. These include the interrelated properties of sample volume, density and mass, the sensitivity of the sample to air and moisture, and the likelihood of the gettering of impurities from the gas phase hydrogen. The third section addresses general instrumentation issues. We first discuss the compatibility of instrumentation for both vacuum and high pressure hydrogen operation, in terms of leakage and permeation. We then look at the importance of temperature homogeneity, stability and control, and accurate temperature and pressure measurement. Aspects of the experimental methodology are then covered, including the conditions under which the sample is degassed prior to a measurement and the monitoring of equilibration times at each isotherm point.

In the last three sections, we discuss error sources that apply to the volumetric, gravimetric and thermal desorption techniques, respectively. The volumetric measurement issues include the existence of thermal gradients, the sample

[1] The content of this chapter is based on previous discussions of the topic [1–3].

D. P. Broom, *Hydrogen Storage Materials*, Green Energy and Technology,
DOI: 10.1007/978-0-85729-221-6_6,

size-to-system volume ratio, the dead volume calibration or corrections, the accumulative errors inherent in multipoint volumetric or manometric isotherm determination, and hydrogen leakage. The gravimetric measurement issues considered here include sample size, buoyancy effect corrections and the disturbance of the balance. The issues relating to the thermal desorption techniques include the sample size, the importance of the temperature ramp rate and, where applicable, the signal calibration method.

6.1 Properties of Gas Phase Hydrogen

In this first section we cover some of the properties of gaseous hydrogen that are relevant to the performance of accurate hydrogen sorption measurements. Most of these properties are relevant to other error sources discussed later in this chapter, and we will therefore expand upon their influence later, where appropriate. The properties addressed are the compressibility of hydrogen, the Joule–Thomson effect, thermal conductivity, the transition of the gas between the continuum and free molecule regimes as a function of pressure, thermal transpiration (thermo-molecular flow) effects and gas purity.[2] Some of these can cause errors directly, such as insufficient gas purity, whereas others, such as the thermal conductivity, should be considered in relation to other error sources.

Molecular hydrogen exists in two forms, known as *ortho-* and *para-hydrogen*, which differ in the orientation of the nuclear spins of the two component atoms. *Normal hydrogen* is the equilibrium mixture of the two spin isomers, which varies from nearly 100% para-hydrogen below the boiling point to 25% para-hydrogen and 75% ortho-hydrogen at temperatures above 200 K [4]. We refer here only to normal hydrogen, the natural mixture for any given temperature, because in the context of sorption measurement accuracy the differences in the properties of the two forms of hydrogen are not sufficiently significant to be of concern.

6.1.1 Compressibility

As for any real gas, the pressure-density relationship for hydrogen deviates to a certain extent from the linearity expressed by the ideal gas law,

$$PV = nRT \tag{6.1}$$

---

[2] An extensive compilation of thermophysical data for hydrogen was compiled by McCarty [4], which includes many more properties than we cover here. The report also includes an extensive reference list of data published prior to 1975. A later report [5] updating and correcting some of the content of the earlier report, but also including other useful hydrogen property information, is also recommended.

where P is the pressure, V is the volume, n is the number of moles, R is the universal gas constant and T is the temperature. The relationship must therefore be represented by an *Equation of State* (EOS), which can produce either the density of the hydrogen directly from the pressure and temperature, or the compressibility factor, Z, for use in the real gas law,

$$PV = nZRT. \tag{6.2}$$

The most prominent way in which this can affect the result of a hydrogen sorption measurement is during the volumetric determination of hydrogen sorption at above ambient pressure. In this case, any error introduced to the calculation of the sorbed quantity accumulates through the measurement and an inaccurate representation of the compressibility can contribute significantly to this error accumulation (see Sect. 6.5.4). However, the description of the compressibility of hydrogen is also important for gravimetric measurement in which an accurate representation of the pressure-density relationship of hydrogen is required in order to apply the buoyancy effect corrections (see Sect. 4.2.1.1 and Sect. 6.6.2).

Numerous equations of state are available that can be used to calculate the density of pure fluids as a function of pressure and temperature. The accuracy of each equation differs and this tends to depend, to a certain extent, on the mathematical complexity of the expression; however, different EOS can also apply to different temperature and pressure regimes with varying degrees of accuracy.[3] The expressions of practical interest can generally be separated into either cubic or multiparameter equations of state. The former, with the exception of the van der Waals equation, describe the pressure-density relationship in terms of the critical temperature, critical pressure and an acentric factor; the former two are well known gas properties and the latter, a term introduced by Pitzer and co-workers [6, 7], expresses the degree of non-sphericity of any given molecule.[4] The multiparameter equations of state, on the other hand, describe the pressure-density relationship from empirical fits to experimental data using the number of empirical parameters required to provide sufficient accuracy [8]. The choice of the EOS for hydrogen sorption measurement has been addressed by a number of authors and, in some cases, different equations have been compared [9–11]; however, a comprehensive survey has not yet been published. Nasrifar [12] recently compared 11 different EOS for hydrogen, for the prediction of various thermodynamic properties, although the report focuses only on cubic expressions.

---

[3] For example, an EOS that may be sufficiently accurate at ambient temperature may not apply at 77 K.

[4] It is perhaps notable in the context of the current discussion that Pitzer [6] specifically excluded hydrogen and helium from his discussion because of the importance of quantum effects for these two fluids. However, our interest is in EOS for the higher temperature supercritical region for which quantum effects are less significant. It is also worth noting that the aim of some EOS is to describe behaviour accurately in the region of the critical point, whereas our interest lies in the accuracy of an expression well above this regime in terms of temperature.

6.1.1.1 Van der Waals

The first EOS we will present is the well known van der Waals (vdW) equation, which, in terms of molar density, ρ_m, is given by,

$$P = \frac{\rho_m RT}{1 - \rho_m b} - a\rho_m^2 \tag{6.3}$$

where a is the attraction parameter and b is the repulsion parameter or effective molecular volume, also known as the *van der Waals co-volume* [13]. In terms of the molar volume, v_m, it is given by,

$$P = \frac{RT}{v_m - b} - \frac{a}{v_m^2}. \tag{6.4}$$

In terms of the compressibility, Z, the vdW EOS becomes,

$$Z^3 - (1 + B)\,Z^2 + AZ - AB = 0 \tag{6.5}$$

where A and B are given by,

$$A = \frac{aP}{(RT)^2} \tag{6.6}$$

and,

$$B = \frac{bP}{RT}. \tag{6.7}$$

The van der Waals EOS has been used to represent the compressibility in previous hydrogen storage studies using the volumetric technique (see, for example, Tibbetts et al. [14]). However, care must be taken with the use of this expression because it is not sufficiently accurate in some important temperature and pressure regimes, as shown by Zhou et al. [11] in their work on low temperature hydrogen and methane adsorption by MOF-5 and ZIF-8. The use of the van der Waals EOS at 77 K up to a pressure of 55 bar (5.5 MPa), in this case, resulted in a greater error[5] than the use of the ideal gas law.

6.1.1.2 Soave–Redlich–Kwong

A modified version of an earlier EOS, due to Redlich and Kwong [15], was presented by Soave [16] and is commonly known as the Soave–Redlich–Kwong (SRK) EOS. In terms of the molar density, the SRK EOS expresses the pressure as,

[5] In this case, the measurement of "false hydrogen adsorption" in an empty sample cell.

$$P = \frac{\rho_m RT}{1 - \rho_m b} - \frac{a(T)\rho_m^2}{1 + \rho_m b} \tag{6.8}$$

and in terms of molar volume, v_m, it is given by,

$$P = \frac{RT}{v_m - b} - \frac{a(T)}{v_m(v_m + b)}. \tag{6.9}$$

For the SRK EOS, the parameters a and b are given by,

$$a = \frac{0.42747 R^2 T_c^2 \alpha}{P_c} \tag{6.10}$$

and,

$$b = \frac{0.08664 R T_c}{P_c} \tag{6.11}$$

where T_c is the critical temperature, P_c is the critical pressure and α is defined using the following pair of expressions,

$$\sqrt{\alpha} = 1 + S\left(1 - \sqrt{\frac{T}{T_c}}\right) \tag{6.12}$$

and,

$$S = 0.480 + 1.574\omega - 0.176\omega^2 \tag{6.13}$$

where ω is the acentric factor. For hydrogen, $\omega = -0.216$, $T_c = 33.2$ K and $P_c = 1.28$ MPa [9].[6]

In terms of the compressibility, the SRK EOS becomes,

$$Z^3 - Z^2 + \left(A - B - B^2\right)Z - AB = 0 \tag{6.14}$$

where A and B are given by Eqs. 6.6 and 6.7.

The SRK expression was one of the EOS considered by Zhou and Zhou [9] and Zhang et al. [10]. Although its use for hydrogen sorption does not seem particularly widespread, it has been used in some of the recent work on low temperature hydrogen adsorption by MOFs [18, 19].

6.1.1.3 Peng–Robinson

The Peng–Robinson EOS [20] is another cubic equation, which expresses the pressure as,

---

[6] Values of $\omega = -0.215$, $T_c = 33.19$ K and $P_c = 1.32$ MPa are given by Perry and Green [17].

$$P = \frac{RT}{v_m - b} - \frac{a(T)}{v_m(v_m + b) + b(v_m - b)} \tag{6.15}$$

where the parameters a and b are given by,

$$a = \frac{0.45724R^2T_c^2\alpha}{P_c} \tag{6.16}$$

and,

$$b = \frac{0.07780RT_c}{P_c}. \tag{6.17}$$

For the Peng–Robinson EOS, α is also expressed using Eq. 6.12, but the parameter S takes the following form,

$$S = 0.37464 + 1.54226\omega - 0.26992\omega^2. \tag{6.18}$$

In terms of the compressibility, the Peng–Robinson EOS becomes,

$$Z^3 - (1 - B)Z^2 + (A - 2B - 3B^2)Z - (AB - B^2 - B^3) = 0 \tag{6.19}$$

where A and B are given by Eqs. 6.6 and 6.7.

The Peng–Robinson EOS has been used in both experimental studies on hydrogen storage [21, 22] and to represent the compressibility of hydrogen for simulation purposes [23], although it was not selected by Zhou and Zhou [9], Zhang et al. [10] or Zhou et al. [11] in their comparisons of different EOS.

6.1.1.4 Hemmes et al.

A hydrogen-specific EOS was developed by Hemmes et al. [24] and is commonly used in metal hydride studies. It is valid for temperatures in the range 100–1,000 K and for pressures up to 100 GPa. It is a modified van der Waals EOS of the following form,

$$\left(P + \frac{a(P)}{V^{\alpha(T)}}\right)(V - b(P)) = RT \tag{6.20}$$

where, for P > 0.1 MPa,

$$a(P) = \exp[a_1 + a_2 \ln(P) + \exp(a_3 + a_4 \ln(P))]. \tag{6.21}$$

For P ≥ 10.0 MPa, the parameter b is given by,

$$b(P) = \sum_{i=0}^{8} b_i \ln(P)^i. \tag{6.22}$$

Table 6.1 Hemmes et al. [24] EOS parameters for hydrogen

Parameter	Value
α_0	2.9315
α_1	-1.531×10^{-3}
α_2	4.154×10^{-6}
b_0	20.285
b_1	−7.44171
b_2	7.318565
b_3	−3.463717
b_4	0.87372903
a_1	19.599
a_2	−0.8946
a_3	−18.608
a_4	2.6013
b_5	−0.12385414
b_6	9.8570583×10^{-3}
b_7	$-4.1153723 \times 10^{-4}$
b_8	7.02499×10^{-6}

For $P < 10.0$ MPa, *b* equals the value at 10.0 MPa. Meanwhile, at $T \leq 300$ K, the exponent $\alpha(T)$ is given by,

$$\alpha(T) = \alpha_0 + \alpha_1 T + \alpha_2 T^2. \tag{6.23}$$

For $T > 300$ K, $\alpha(T)$ equals the value at 300 K. The parameters for hydrogen are given in Table 6.1.

McLennan and Gray [25] used the same EOS to represent the pressure–density relationship for deuterium. The parameter set they presented applies to pressures up to at least 100 MPa in the temperature range from 123 K to greater than 623 K. The authors demonstrate the importance of using an accurate equation of state by showing the significant differences exhibited by pressure-composition isotherms for the Pd-D system, at 523 K up to a pressure of 5.4 MPa, calculated using different EOS. The parameters for deuterium presented by McLennan and Gray [25] are given in Table 6.2.

6.1.1.5 Beattie–Bridgeman

The Beattie–Bridgeman EOS [26] expresses the pressure in terms of five parameters, A_0, a, B_0, b and c,

$$P = \frac{RT}{v_m^2}\left(1 - \frac{c}{v_m T^3}\right)\left[v_m + B_0\left(1 - \frac{b}{v_m}\right)\right] - \frac{A_0}{v_m^2}\left(1 - \frac{a}{v_m}\right). \tag{6.24}$$

It can also be written in the following virial form,

Table 6.2 McLennan and Gray [25] EOS parameters for deuterium

Parameter	Value
α_0	2.75133
α_1	-6.21145×10^{-5}
α_2	4.37677×10^{-7}
b_0	20.285
b_1	−7.44171
b_2	7.318565
b_3	−3.463717
b_4	0.87372903
a_1	18.76040
a_2	−0.74606
a_3	−8.69181
a_4	-1.03114×10^{-3}
b_5	−0.12385414
b_6	9.8570583×10^{-3}
b_7	$-4.1153723 \times 10^{-4}$
b_8	7.02499×10^{-6}

$$Pv_m = RT + \frac{\beta}{v_m} + \frac{\gamma}{v_m^2} + \frac{\delta}{v_m^3} \tag{6.25}$$

where,

$$\beta = RTB_0 - A_0 - \frac{Rc}{T^2} \tag{6.26}$$

$$\gamma = -RTB_0 b + A_0 a - \frac{RB_0 c}{T^2} \tag{6.27}$$

and,

$$\delta = \frac{RB_0 bc}{T^2}. \tag{6.28}$$

To express the pressure as a function of the molar density, this becomes,

$$P = RT\rho_m + \left(RTB_0 - A_0 - \frac{Rc}{T^2}\right)\rho_m^2 + \left(-RTB_0 b + A_0 a - \frac{RB_0 c}{T^2}\right)\rho_m^3 + \left(\frac{RB_0 bc}{T^2}\right)\rho_m^4. \tag{6.29}$$

The compressibility can then be written as,

$$Z = 1 + \left(B_0 - \frac{A_0}{RT} - \frac{c}{T^3}\right)\rho_m + \left(-B_0 b + \frac{A_0 a}{RT} - \frac{B_0 c}{T^3}\right)\rho_m^2 + \left(\frac{B_0 bc}{T^3}\right)\rho_m^3. \tag{6.30}$$

The parameters for hydrogen are $A_0 = 0.02$ Pa m^3 mol^{-2}, $a = -5.06 \times 10^6$ m^3 mol^{-1}, $B_0 = 20.96 \times 10^6$ m^3 mol^{-1}, $b = -43.59 \times 10^6$ m^3 mol^{-1}, and $c = 5.04$ m^3 K^3 mol^{-1} [26, 27].

In a recent study of the thermodynamics of MgH_2 nanoparticles, Paskevicius et al. [28] note the importance of expressing the equilibrium properties of the system in terms of the fugacity rather than pressure in order to represent the non-ideality of the hydrogen.[7] The conversion to fugacity uses an EOS and they note that the literature values for bulk MgH_2, determined by Stampfer et al. [29] using the Beattie–Bridgeman EOS, exhibit a deviation of approximately 2% from the values obtained using either the Hemmes et al. EOS (Sect. 6.1.1.4) or the US National Institute of Standards and Technology (NIST) thermophysical properties database [30]. This suggests that the Beattie–Bridgeman EOS is not sufficiently accurate for the high temperature region (573–633 K) considered in this study.

6.1.1.6 Benedict–Webb–Rubin

The Benedict–Webb–Rubin (BWR) EOS [31] expresses the pressure in terms of eight empirical parameters,

$$P = RT\rho_m + \left(B_0RT - A_0 - \frac{C_0}{T^2}\right)\rho_m^2 + (bRT - a)\,\rho_m^3 + a\alpha\rho_m^6 + \frac{c}{T^2}\left(1 + \gamma\rho_m^2\right)\exp\left(-\gamma\rho_m^2\right)\rho_m^3. \tag{6.31}$$

The compressibility can therefore be expressed as,

$$Z = 1 + \left(B_0 - \frac{A_0}{RT} - \frac{C_0}{RT^3}\right)\rho_m + \left(b - \frac{a}{RT}\right)\rho_m^2 + \frac{a\alpha}{RT}\rho_m^5 + \frac{c}{RT^3}\left(1 + \gamma\rho_m^2\right)\exp\left(-\gamma\rho_m^2\right)\rho_m^2. \tag{6.32}$$

Zhou and Zhou [9] compared this expression with the SRK EOS, using the following values for the eight parameters: $A_0 = 9.7319 \times 10^{-2}$ L^2 mol^{-2} atm, $a = -9.2211 \times 10^{-3}$ L^3 mol^{-3} atm, $B_0 = 1.8041 \times 10^{-2}$ L mol^{-1}, $b = 1.7976 \times 10^{-4}$ L^2 mol^{-2}, $C_0 = 3.8914 \times 10^2$ L^2 mol^{-2} K atm, $c = -2.4613 \times 10^2$ L^3 mol^{-3} K^2 atm, $\alpha = -3.4215 \times 10^{-6}$ L^3 mol^{-3}, and $\gamma = 1.89 \times 10^{-3}$ L^2 mol^{-2}. These values are valid for the temperature range 273–423 K, and $\rho/\rho_c < 2.5$. In this limited range of applicability they found that the BWR EOS showed greater precision than the SRK EOS.

[7] In Chap. 3, we considered the properties in terms of only the pressure, an approach that is commonly taken in the literature.

6.1.1.7 Modified Benedict–Webb–Rubin

Many modified forms of the BWR EOS have been proposed. A well known version was formulated by Starling [32], but we shall cover only the 32-term modified BWR equation used in the NIST REFPROP database before the inclusion of the Leachman and co-workers [33, 34] EOS (see next section). The 32-term modified BWR EOS expresses the pressure as,

$$
\begin{aligned}
P = {} & \rho_m RT + \rho_m^2\left(G_1 T + G_2 T^{1/2} + G_3 + \frac{G_4}{T} + \frac{G_5}{T^2}\right) \\
& + \rho_m^3\left(G_6 T + G_7 + \frac{G_8}{T} + \frac{G_9}{T^2}\right) + \rho_m^4\left(G_{10} T + G_{11} + \frac{G_{12}}{T}\right) \\
& + \rho_m^5 G_{13} + \rho_m^6\left(\frac{G_{14}}{T} + \frac{G_{15}}{T^2}\right) + \rho_m^7\left(\frac{G_{16}}{T}\right) + \rho_m^8\left(\frac{G_{17}}{T} + \frac{G_{18}}{T^2}\right) \\
& + \rho_m^9\left(\frac{G_{19}}{T^2}\right) + \rho_m^3\left(\frac{G_{20}}{T^2} + \frac{G_{21}}{T^3}\right)\exp\left(\gamma\rho_m^2\right) \\
& + \rho_m^5\left(\frac{G_{22}}{T^2} + \frac{G_{23}}{T^4}\right)\exp\left(\gamma\rho_m^2\right) + \rho_m^7\left(\frac{G_{24}}{T^2} + \frac{G_{25}}{T^3}\right)\exp\left(\gamma\rho_m^2\right) \\
& + \rho_m^9\left(\frac{G_{26}}{T^2} + \frac{G_{27}}{T^4}\right)\exp\left(\gamma\rho_m^2\right) + \rho_m^{11}\left(\frac{G_{28}}{T^2} + \frac{G_{29}}{T^3}\right)\exp\left(\gamma\rho_m^2\right) \\
& + \rho_m^{13}\left(\frac{G_{30}}{T^2} + \frac{G_{31}}{T^3} + \frac{G_{32}}{T^4}\right)\exp\left(\gamma\rho_m^2\right)
\end{aligned}
\tag{6.33}
$$

where γ is $1/\rho_c^2$. The values of the parameters G_1 to G_{32} are given in Table 6.3. ρ_c for normal hydrogen for this expression is 14.94 mol L^{-1}. P is the pressure in MPa and ρ_m is the molar density in mol L^{-1}. The compressibility is given by dividing the right hand side of Eq. 6.33 by $\rho_m RT$.

6.1.1.8 Leachman et al.

This recently developed expression is a multiparameter EOS that is implemented in the current version of the NIST REFPROP thermophysical properties database [30]. The modified BWR and some other multiparameter equations of state are explicit in pressure, so that density and temperature are used to calculate a pressure. The Leachman et al. EOS, however, is explicit in the Helmholtz free energy [33, 34]. This means that the derivation of other thermophysical properties from the EOS is more convenient, although this is not an important factor for our requirements.

The Helmholtz free energy, a, is expressed as,

$$\frac{a(T,\rho_m)}{RT} = \alpha(\tau,\delta) \tag{6.34}$$

Table 6.3 Modified BWR EOS parameters for Eq. 6.33, as presented by Leachman [33]

Parameter	Value
G_1	$4.675528393416 \times 10^{-4}$
G_2	$4.289274251454 \times 10^{-2}$
G_3	-0.5164085596504
G_4	2.961790279801
G_5	-30.27194968412
G_6	$1.908100320379 \times 10^{-5}$
G_7	$-1.339776859288 \times 10^{-3}$
G_8	0.3056473115421
G_9	51.61197159532
G_{10}	$1.999981550224 \times 10^{-7}$
G_{11}	$2.896367059356 \times 10^{-4}$
G_{12}	$-2.257803939041 \times 10^{-2}$
G_{13}	$-2.287392761826 \times 10^{-6}$
G_{14}	$2.446261478645 \times 10^{-5}$
G_{15}	$-1.718181601119 \times 10^{-3}$
G_{16}	$-5.465142603459 \times 10^{-7}$
G_{17}	$4.051941401315 \times 10^{-9}$
G_{18}	$1.157595123961 \times 10^{-6}$
G_{19}	$-1.269162728389 \times 10^{-8}$
G_{20}	-49.83023605519
G_{21}	-160.6676092098
G_{22}	-0.192679918531
G_{23}	9.319894638928
G_{24}	$-3.222596554434 \times 10^{-4}$
G_{25}	$1.206839307669 \times 10^{-3}$
G_{26}	$-3.84158819747 \times 10^{-7}$
G_{27}	$-4.036157453608 \times 10^{-5}$
G_{28}	$-1.250868123513 \times 10^{-10}$
G_{29}	$1.976107321888 \times 10^{-9}$
G_{30}	$-2.411883474011 \times 10^{-13}$
G_{31}	$-4.127551498251 \times 10^{-13}$
G_{32}	$8.91797288361 \times 10^{-12}$

where α is the reduced Helmholtz free energy, T is the temperature, ρ_{m} is the molar density, R is the universal gas constant, and τ and δ are given by the following two expressions,

$$\tau = \frac{T_c}{T} \tag{6.35}$$

and,

$$\delta = \frac{\rho_m}{\rho_{cm}} \tag{6.36}$$

where T_c is the critical temperature and ρ_{cm} is the critical molar density. τ is therefore the inverse reduced temperature and δ is the reduced density.

Table 6.4 Parameters for the ideal gas contribution to the reduced Helmholtz free energy in the Leachman et al. EOS [33, 34] for normal hydrogen

k	a_k	b_k
1	−1.4579856475	–
2	1.888076782	–
3	1.616	−16.0205159149
4	−0.4117	−22.6580178006
5	−0.792	−60.0090511389
6	0.758	−74.9434303817
7	1.217	−206.9392065168

The reduced Helmholtz free energy consists of an ideal gas contribution, α^0, and the residual contribution, α^r, so that,

$$\alpha(\tau, \delta) = \alpha^0(\tau, \delta) + \alpha^r(\tau, \delta). \tag{6.37}$$

The ideal gas contribution is given by,

$$\alpha^0 = \ln \delta + 1.5 \ln \tau + a_1 + a_2 \tau + \sum_{k=3}^{N} a_k \ln(1 - \exp(b_k \tau)) \tag{6.38}$$

where the a_k and b_k parameters are given in Table 6.4.

In the Leachman et al. EOS, the residual contribution to the reduced Helmholtz free energy is given by,

$$\begin{aligned}\alpha^r(\tau, \delta) = &\sum_{i=1}^{l} N_i \delta^{d_i} \tau^{t_i} + \sum_{i=l+1}^{m} N_i \delta^{d_i} \tau^{t_i} \exp(-\delta^{p_i}) \\ &+ \sum_{i=m+1}^{n} N_i \delta^{d_i} \tau^{t_i} \exp\left(\phi_i (\delta - D_i)^2 + \beta_i (\tau - \gamma_i)^2\right)\end{aligned} \tag{6.39}$$

where $l = 7$, $m = 9$ and $n = 14$. The parameters N_i, d_i, t_i, p_i, φ_i, β_i, γ_i and D_i for normal hydrogen are given in Tables 6.5 and 6.6.

The pressure and compressibility factors are then given by,

$$P(T, \rho_m) = \rho_m RT \left[1 + \left(\frac{\partial \alpha^r}{\partial \delta}\right)_\tau\right] \tag{6.40}$$

and,

$$Z(T, \rho_m) = \frac{P}{\rho_m RT} = 1 + \left(\frac{\partial \alpha^r}{\partial \delta}\right)_\tau. \tag{6.41}$$

The Leachman et al. EOS is the current state of the art equation for hydrogen. It should therefore provide the highest accuracy description of the compressibility of hydrogen available to date. It has not yet been widely tested for use in hydrogen sorption measurement, but it would seem reasonable to recommend its use due to the higher accuracy it should provide, over a wide measurement temperature and pressure range, in comparison with older expressions.

Table 6.5 Parameters for the residual contribution to the reduced Helmholtz free energy in the Leachman et al. EOS [33, 34] for normal hydrogen

i	N_i	t_i	d_i	p_i
1	−6.93643	0.6844	1	0
2	0.01	1	4	0
3	2.1101	0.989	1	0
4	4.52059	0.489	1	0
5	0.732564	0.803	2	0
6	−1.34086	1.1444	2	0
7	0.130985	1.409	3	0
8	−0.777414	1.754	1	1
9	0.351944	1.311	3	1
10	−0.0211716	4.187	2	–
11	0.0226312	5.646	1	–
12	0.032187	0.791	3	–
13	−0.0231752	7.249	1	–
14	0.055734	2.986	1	–

6.1.2 The Joule–Thomson Effect

The *Joule–Thomson Effect*, which is also known as the *Joule-Kelvin Effect*, occurs when a real gas expands through a throttling device, such as a valve orifice or a porous filter. It results in a temperature change, defined by the Joule–Thomson coefficient, μ_{JT}, which is given by,

$$\mu_{JT} = \left(\frac{\partial T}{\partial P}\right)_H \tag{6.42}$$

where T is the temperature and P is the pressure. The subscript H denotes the constant enthalpy conditions under which the Joule–Thomson effect occurs. μ_{JT}, expressed in units of K Pa^{-1}, is positive below a so-called *inversion temperature*, T_{inv}, at which $\mu_{JT} = 0$, and is negative above this threshold. Practically speaking, this means that a gas cools when undergoing isenthalpic expansion below its inversion temperature and heats when above it. This effect is widely exploited in industrial liquefaction processes [35].

The inversion temperature is dependent on the gas pressure and is above ambient for most gases. A plot of inversion temperature versus pressure is known as the *Joule–Thomson inversion curve*. T_{inv} reaches a maximum at zero pressure and for nitrogen, for example, this is 625 K [36].[8] Hydrogen and helium are the only pure fluids that have an inversion temperature below ambient, with maximum inversion temperatures of 205 K and 40 K, respectively. This means that hydrogen delivered to sorption apparatus at ambient temperatures or higher will undergo a heating effect, but for low temperature adsorption measurements it will cool. Joule–Thomson effects for hydrogen were determined experimentally by Johnston

[8] Joule–Thomson data for a number of pure fluids can be found in Perry and Green [17].

Table 6.6 Parameters for the residual contribution to the reduced Helmholtz free energy in the Leachman et al. EOS [33, 34] for normal hydrogen

i	φ_i	β_i	γ_i	D_i
10	−1.685	−0.171	0.7164	1.506
11	−0.489	−0.2245	1.3444	0.156
12	−0.103	−0.1304	1.4517	1.736
13	−2.506	−0.2785	0.7204	0.67
14	−1.607	−0.3967	1.5445	1.662

et al. [37] at a range of temperatures from 64 K to 288 K and pressures up to 200 atm (20.3 MPa). The results show positive excursions at near ambient temperatures (288 K) of nearly 6 K at the higher pressures, and negative excursions of over 20 K under a number of low temperature, high pressure conditions. It can be seen that the thermal effects of such gas expansion can be relatively severe and so this should be taken into account during a sorption measurement if such expansion is likely to occur in the apparatus. It is necessary to allow a return to thermal equilibrium if heating or cooling effects such as these occur, and it is also important not to misattribute associated experimental artefacts to a sorption process or processes.

6.1.3 Thermal Conductivity

The thermal conductivity of gases varies with temperature, pressure and species. In the case of hydrogen, the thermal conductivity under atmospheric conditions is relatively high (0.1971 W m^{-1} K^{-1}). Although comparable to the thermal conductivity of helium (0.1574 W m^{-1} K^{-1}), it is an order of magnitude greater than many common adsorptives, including nitrogen (0.0275 W m^{-1} K^{-1}), argon (0.0190 W m^{-1} K^{-1}) and carbon dioxide (0.0183 W m^{-1} K^{-1}) [38]. These figures are for $P = 101$ kPa and $T = 273$ K, for comparative purposes, but a survey of the literature on the experimental determination of the thermal conductivity of hydrogen, covering a wide range of temperatures and pressures, was presented recently by Leachman et al. [39] and this article is therefore recommended to interested readers. The thermal conductivity can also be calculated from equations of state, and it is therefore one of the many thermophysical properties that can be calculated using the NIST REFPROP database [30].

With regard to sorption measurement, the thermal conductivity of any gas will affect the measurement of temperature in the gas phase as a function of pressure. At vacuum there is limited heat transfer and so a thermal gradient will exist. In gravimetric measurement, in which the temperature sensor cannot be in direct contact with the sample (see Sect. 6.3.4), it is not always possible to achieve the required sample temperature at vacuum and so a small amount of gas may be necessary to provide some heat transfer between the thermostat and the sample at the start of a measurement. Once this first dose of gas is present any observed gradient is likely to then change as a function of pressure through the isotherm measurement. The same effect will occur to a certain extent in volumetric measurement but this is dependent on the exact positioning of the sample temperature measurement sensor and the size of the sample cell. The key point is that any heat transfer effects, or temperature variations due to these effects, could be greater with hydrogen than other common adsorptives because of its higher thermal conductivity, and so this should be considered during the performance of sorption measurements.

6.1.4 Continuum, Transition and Free Molecule Regimes

The physical behaviour of a gas changes significantly during the transition from high vacuum conditions to elevated pressure. Under high vacuum the gas exists in the free molecule regime. At these pressures, the mean free path of a gas molecule is significantly greater than the length scale of typical laboratory apparatus and the gas molecule transport is dominated by collisions with the walls of the vacuum or pressure system. As the pressure increases, the gas will pass through the transition region, before reaching the continuum (viscous flow) regime. In the continuum regime the gas molecule transport is dominated by intermolecular collisions.

The regime can be determined by the assessment of the Knudsen number, Kn, of the system, which is defined by,

$$Kn = \frac{\lambda}{L} \tag{6.43}$$

where L is the system length scale and λ is the mean free path of the molecule, which is given by,

$$\lambda = \frac{k_B T}{\sqrt{2}\pi d_m^2 P} \tag{6.44}$$

where k_B is the Boltzmann constant and d_m is the molecular diameter.

The boundaries between the different regimes are not that clearly defined but, generally speaking, the free molecule regime refers to $Kn > 1$, which is the point at which the mean free path begins to exceed the length scale of the system. The continuum regime is in the region $Kn < 0.01$, and the transition region is given by $0.01 > Kn > 1.0$. The Knudsen number has a significant effect on the likelihood of errors associated with thermal transpiration, or thermomolecular flow, effects covered in the next section. It also affects the fluid transport mechanisms in porous media discussed in Sect. 3.3.1, but this is not an accuracy issue unless it affects experimental equilibration times (Sect. 6.4.2). This can have significant experimental error implications in many adsorption measurements because of kinetic limitations on pore diffusion, but the rapid diffusion of hydrogen at the temperatures of interest for storage applications (see Sect. 3.3.1) means that it is generally not a consideration for hydrogen adsorption measurement.

6.1.5 Thermal Transpiration (Thermomolecular Flow)

A possible source of error at low pressure is the occurrence of *thermal transpiration*, which is also known as *thermomolecular flow*, whereby a thermal gradient along a tube of a diameter close to, or below, the mean free path of the gas molecule ($Kn > 1$) will result in a pressure gradient along the length of the tube. In practical terms this can occur when measuring the pressure of a system, or the pressure of a sample cell or calibrated volume, which is at a different temperature to the pressure measuring device (Sect. 6.3.3.2). A similar effect can also lead to the disturbance of the microbalance during gravimetric measurements at low pressure in the presence of a temperature gradient (Sect. 6.6.3).

With regard to pressure measurement, there are two practical consequences of thermal transpiration effects. Firstly, pressure measurement data that is subject to this error should be corrected using one of the empirical corrections we describe later; secondly, the internal diameter of apparatus intended for low pressure measurement should be considered during instrument design. The use of large bore tubing will minimise thermal transpiration effects, as well as being favourable for high vacuum operation due to its higher conductance. Broom [2] presented plots of the mean free path for hydrogen at a range of temperatures in the pressure range of interest for sorption measurements, and of the characteristic length scale versus mean free path, indicating the three regimes. These plots show that thermal transpiration effects could start to become significant at the lower end of the low vacuum range (10^2–10^3 Pa) in typical laboratory instrumentation. Some of the empirical corrections that have been applied to account for thermal transpiration effects during pressure measurement will be covered in Sect. 6.3.3.2.

6.1.6 Gas Purity

High purity hydrogen should be used for sorption measurement purposes for a number of reasons. During the measurement of hydrogen adsorption by microporous materials, gas phase impurities can preferentially adsorb on the material (see Sect. 6.2.5). This will potentially affect the actual uptake of hydrogen due to pore blocking or by the occupation of adsorption sites. It will also affect the measurement in the case of gravimetry because the weight increase due to impurity adsorption is potentially indistinguishable from the weight increase due to hydrogen uptake, although the likelihood of impurity adsorption can be assessed from the kinetics of the sorption process (see Sect. 6.2.5). In the case of hydrides, gas phase impurities can inhibit the surface adsorption and dissociation of molecular hydrogen or can react with the surface to form a passivating layer. Although this may depend on the impurity type and the material, the most obvious way to ensure that surface poisoning or passivation, or any other process that could be detrimental to hydrogen absorption, does not occur is to use high purity gas. The minimum purity that should be considered in any hydrogen sorption measurement is 99.999%, but higher purity gas should ideally be used and, for adsorption experiments, 99.9999% purity and the addition of filtering is the ideal. During the controversy over the hydrogen storage properties of carbon nanostructures, for example, Yang [40] showed that the capacities of 20 and 16 wt% reported by Chen et al. [41] for lithium and potassium-doped carbon nanotubes, respectively, were most likely due to moisture contamination in the hydrogen supply. In this case, the reaction of water with the alkali dopants resulted in a false reading in the gravimetric measurements due to the formation of hydroxides, hydrates, and other compounds, such as semiquinones and phenolate. This example clearly illustrates the need to ensure that measurements are performed with a sufficiently pure hydrogen supply.

Practical experience in the Hiden Isochema applications laboratory[9] suggests that sufficient purity gas can be delivered to an instrument for hydrogen adsorption measurement without filtration providing, firstly, that sufficient provisions are made to minimise contamination of the regulator during bottle changes and, secondly, that the gas bottle is connected directly to the instrument without a significant length of gas supply line. Thomas [42], on the other hand, suggests that filtration is obligatory; however, we should add the caveat that if measurements are to be performed without filtration, in the case of hydrogen adsorption, it is essential to confirm that this approach is suitable using microporous materials of known hydrogen storage capacities and to confirm the complete reversibility of the hydrogen adsorption process. Contaminants are likely to introduce some form of hysteresis into the adsorption and desorption isotherms, as well as affecting the overall isotherm shape. In the case of laboratories with an external gas supply, which necessitates a gas supply line between the cylinder and the instrument, filtration will almost certainly be required and is therefore recommended.

6.2 Properties of Materials

In this section we will look at the properties of potential storage materials that can affect the determination of their hydrogen sorption behaviour. We are concerned primarily with the degree to which uncertainty in the properties of a material can lead to uncertainty in the determination of its sorption properties. Many physical or

---

[9] See, for example, Xiang et al. [18] in which the hydrogen adsorption measurements show very good reversibility, which is indicative of pure physisorption and the absence of impurity contamination. These data were measured without hydrogen filtration using 99.9999% purity hydrogen and a so-called "T-purge" regulator supplied by Air Products. Further examples are the highly reversible, variable temperature datasets for hydrogen adsorption by Na-X zeolite shown in Fig. 2.1 (see Sect. 2.1).

chemical properties will not affect the hydrogen adsorption or absorption of hydrogen, and hence its measurement, but some aspects can have a significant impact. Any uncertainty will directly affect the accuracy of the determination of the hydrogen sorption properties of a particular material and could potentially affect the reporting of its hydrogen storage capability. As with many of the points discussed here, the material type will have a significant effect on the importance of each of these issues, and so we will try to identify the material types that are most significantly affected in each case. It is also worth noting that there is considerable interplay between many of the points discussed in this section and the basic characterisation of the host materials.[10]

Firstly, we look at the definition and determination of the sample volume, density and mass. The accurate determination of the density of a real material is challenging and the measured value will differ from the theoretical density due to the presence of impurities and defects. Any error in the knowledge of the real density will lead to an error in one or more of the corrections necessary for high pressure hydrogen sorption measurements. In the case of high density materials these may be negligible but the lower the material density the greater the potential error. Secondly, many hydrogen storage materials are air or moisture sensitive and so we will briefly look at the effects that this property could have on measurement uncertainty. We will then discuss the importance of either knowledge or understanding of the history of the sample and the degree to which different storage conditions, thermal history, hydrogen cycling, and so forth, can affect the properties and hence the measured hydrogen sorption behaviour of a material. We will then look at the importance of sample purity before briefly discussing gaseous impurity gettering.

6.2.1 *Sample Volume, Density and Mass*

Accurate determination of the volume, density and mass of a sample is crucial in all measurements but it can vary in importance depending on the material type. In the gravimetric technique the mass is known to high accuracy from the direct

[10] By this we mean the determination of the microstructural and chemical composition of the material. Prominent examples are the carbon nanotube samples of Dillon et al. [43]. After the initial reports of high temperature hydrogen adsorption, the samples were later shown by Hirscher et al. [44, 45] to be storing hydrogen via the interaction of hydrogen with Ti alloy particles rather than the carbon nanotubes themselves. The alloy particles had been deposited on the nanotubes during a purification process, which was performed using an ultrasonic probe constructed from Ti-6Al-4V alloy, and they subsequently reacted with the hydrogen during the hydrogen loading process. This resulted in the appearance of a high temperature TDS peak during the subsequent desorption experiment, which was incorrectly attributed to the carbon nanotubes. Thorough sample characterisation prior to hydrogen sorption measurement and careful consideration of the effects of any impurities present is therefore clearly essential, as this case demonstrates.

in situ measurement of the sample mass at the start of an experiment.[11] However, the sample volume and hence the density, or vice versa, is unknown and must be determined using another method in order for the buoyancy corrections to be applied to elevated pressure data. The buoyancy corrections increase, for a sample of a given mass, with decreasing sample density (Sect. 6.6.2). In volumetric measurement the degassed sample mass is not known to high accuracy and knowledge of the sample volume is necessary for the application of the dead volume corrections (Sect. 6.5.3). Although the direct determination of the dead volume does not necessarily require knowledge of the sample volume or density for the calculation of the observed uptake, the direct determination of the dead volume in a sample cell of known volume is itself a helium pycnometry measurement and therefore implicitly contains the errors associated with this type of sample density determination. In this section we will look at the different definitions of density, before discussing, firstly, the effects of sample density and volume on sorption measurement accuracy and, secondly, the effects of sample mass.

6.2.1.1 Definitions of Density

There are a number of definitions of the density of a material, with some variation between their use in different fields. Density determination methods and definitions have been covered by a number of standards, and the topic is covered in more depth by Lowell et al. [46]. There are two definitions that are of interest to us, with regard to measurement accuracy, and a further two that are of interest generally in the application of hydrogen storage materials. The most important density definition, in terms of measurement accuracy, is known commonly as the *skeletal density*. This is the ratio of the mass to the volume occupied by the sample, excluding the volume of any open pores. This is the density that should be used in the buoyancy effect (Sect. 6.6.2) and dead volume (Sect. 6.5.3) corrections, in the gravimetric and volumetric techniques, respectively. This may be different to the *true density* of a material, if we define the true density as the ratio of the mass to the volume occupied by the sample, excluding all pores, both open and closed.[12] The true and skeletal densities will be equal if a microporous adsorbent contains no blocked pore regions or, in the hydride case, if the host material is completely non-porous. The second definition, which has come to prominence in recent research into hydrogen storage using MOFs, is the *envelope* or *geometric density*. This is the density calculated from the mass and the volume occupied by the solid, including all internal pore space. This has been widely used for the surface

[11] Practically speaking, in gravimetric measurement a decision must be made as to the exact point at which the weight at vacuum is recorded but the possible variation is small compared to the uncertainty of the sample mass in volumetric measurements (see Sect. 6.2.1.3).

[12] *Closed pores*, in this context, are internal volumes and voids that are inaccessible to hydrogen.

excess-to-absolute adsorption conversion using the total pore volume approximation (see Sect. 3.1.1.3), even though the adsorbed phase volume does not necessarily equal the total pore volume.

Meanwhile, the two definitions of interest practically for hydrogen storage, which are therefore relevant in the context of the comparison of the performance of different materials, are the bulk and the tap density. The *bulk density* is calculated from the volume occupied by the solid, including all internal pore space and the volume of the voids between particles. The *tap density* is similar except it is obtained after the container holding the material is tapped in a specified manner to allow more efficient packing of the bed. This will tend to result in the tap density being greater than the bulk density and is obviously only relevant to powders.[13]

6.2.1.2 Sample Density and Volume

The specific effects of our knowledge of sample density and hence volume, or vice versa, in the separate cases of adsorption and absorption measurement are different. However, the errors introduced by uncertainty in our knowledge of the sample volume for both hydrides and porous adsorbents have an identical origin: uncertainty in either the dead volume of volumetric apparatus or the buoyancy effect corrections in gravimetry. Although the density of a material is a simple concept, its accurate determination for many real materials is not a trivial task. In our introduction above, we have chosen a straightforward definition of skeletal density but what we determine in practice is an *apparent density* [46]. This relates to the skeletal volume observed by our chosen characterisation fluid or species. This is invariably helium for microporous materials, which is probably the best choice but we have to assume, firstly, that there is no adsorption of helium in the micropores at the measurement temperature and pressure and, secondly, that the apparent helium density is equal to the apparent hydrogen density. The determination of the apparent helium density of a microporous material at low temperature is likely to lead to some adsorption [48] and the errors associated with this effect. Therefore, it is recommended that helium density determination is performed at elevated temperatures. A convenient measurement temperature is ambient but some authors have advocated the use of significantly higher temperatures of up to 400°C (673 K) [49][14] to avoid errors associated with the adsorption of helium. It seems, however,

[13] As mentioned in Chap. 2, Jordá-Beneyto et al. [47] advocate the use of activated carbon monoliths for hydrogen storage because microporous activated carbon in this form has a greater bulk density than an activated carbon powder. As well as the tap density for their powder samples, Jordá-Beneyto et al. [47] also use the term *packing density*, which in their case is the bulk density we have defined above determined after the powder has been pressed in a mould with a pressure of 550 kg cm^{-2}.

[14] The upper temperature will obviously be limited by the thermal stability of the material and therefore the highest temperatures will not always be practical. Malbrunot et al. [49] suggest using the degassing temperature of the adsorbent.

possible that higher temperature measurements could involve increased errors due to the thermal gradients that exist in the pycnometry apparatus.

If helium adsorption occurs, the density of a material will be overestimated in a volumetric pycnometry measurement because the helium will appear to fill a larger volume than it would in the absence of adsorption. Therefore, the sample volume will be underestimated, and hence the calculated density overestimated for any given sample mass. This will result in a reduced buoyancy correction for a subsequent gravimetric experiment or too large a dead volume in the volumetric equivalent. In both cases this will reduce the perceived hydrogen uptake by the material. Another possibility is that helium cannot access the same pore volume as hydrogen. If it can access a greater volume, it will have the same effect as helium adsorption and the density will be overestimated further. If it cannot access some of the volume that is accessible to hydrogen then it will have the reverse effect and ultimately the hydrogen uptake will be overestimated. However, the kinetic diameters of hydrogen and helium are 0.283–0.289 and 0.255 nm, respectively, according to values tabulated by Li et al. [50], and so the former is more likely.

In adsorption, this issue is linked to the determination of the *Gibbs Dividing Surface* (GDS), which defines the boundary between the solid and the gas phase. This topic has been addressed by a number of authors because it is of significance in many areas of adsorption measurement. Gumma and Talu [51] presented a method that corrects for helium adsorption, and other approaches have also been proposed (see Gumma and Talu [51] and references therein). Their method involves measuring the weight change of the sample as a function of helium pressure at a number of different temperatures. These data are then used to plot β versus temperature, where β is a parameter dependent on the isosteric heat of adsorption, a second parameter related to the entropy of adsorption, and the volume of the solid. By fitting the experimentally determined values of β, the volume of the solid and hence the location of the GDS can be determined. The method was demonstrated using helium adsorption data measured for a silicalite sample at 12 temperatures in the range 93–515 K, up to approximately 3.5 MPa. Gumma and Talu [51] also advocate the implementation of a standard set of conditions, by IUPAC or the International Adsorption Society, for the determination of the GDS. Such a definition would be of great interest for the study of adsorptive hydrogen storage, and would play a valuable role in the development or definition of hydrogen adsorption measurement guidelines (see Sect. 7.3).

Absorption measurements are not affected by helium adsorption or pore accessibility problems, because the materials of interest are not microporous; although the apparent density of a material will still differ from its theoretical value. In the case of absorption, however, the volume and mass, and hence density, of the sample will change throughout a measurement and so it is important to ensure that this volume dilation does not substantially affect the results of a measurement. This can be viewed as the analogue of the excess-to-absolute conversion problem in adsorption measurement and the necessary corrections are nontrivial. However, it does not create such an issue because, firstly, the lattice expansion is not as significant in relation to the empty sample volume compared to the ratio of adsorbed phase volume to the skeletal volume of a microporous material and, secondly, for single phase samples the lattice expansion associated with hydrogen absorption can be measured using in situ diffraction methods. It is therefore known to a certain extent, unlike the adsorbed phase volume, which can only be assumed. However, for samples containing multiple phases this remains a problem. Another reason, associated with the first point, is that the absorption does not occur exclusively in the sample cell dead space, as it does in the case of adsorption. The hydrogen enters the sample and the overall decrease in the sample cell dead volume is relatively small per hydrogen atom. Furthermore, the host materials in the absorption case tend to have a higher density than adsorbents and so the associated buoyancy effect or dead space corrections are less significant; although this is not the case for many complex hydrides, which can also have low skeletal or apparent densities. Further investigation of this issue for complex hydrides would be very valuable.

6.2.1.3 Sample Mass

The mass of a sample at the beginning of a measurement must be known accurately. In gravimetric measurement this is determined in situ and is used as the reference point for the calculation of hydrogen uptake. For hydrogen absorbing materials this is unlikely to be significantly different to the mass of the sample measured before either mounting or activating the sample. However, in the case of adsorbents it can be substantially different due to the removal of pre-adsorbed species during the degassing process (see Sect. 6.4.1). In some cases, this may only be a few wt% but for adsorbents synthesised using wet chemistry methods it can be a significant fraction of the total mass because they may still contain solvents from the synthesis process. The degassed sample mass must, however, be determined to sufficient accuracy, otherwise the calculated gravimetric hydrogen capacity (wt%) will contain a significant associated error. Different methods can be used for this purpose. One common approach is to degas the sample in a removable sample cell equipped with an isolation valve. The sample mass can then be determined by subtracting the empty cell weight from the weight of the cell containing the degassed sample. However, this method is subject to the additive errors from the subtraction of one relatively large mass from another. For small samples this could add considerable error to the calculated wt% hydrogen uptake. Another approach is to correct the loaded (wet) mass for the percentage loss observed in a parallel gravimetric degassing process. However, the error in this case originates from the likely difference in the vacuum degassing conditions at the sample, and hence the mass loss as a function of time, between different apparatus. If the mass loss during degassing is significant, the uncertainty in the activated sample mass is a major source of error in volumetric measurement. The same problem can also affect independent density determinations, although if the dead volume is determined directly this should not be such an issue.

Another related consideration in gravimetric measurement is the point at which the sample weight at vacuum is recorded. In the case of measurements at low temperatures the sample will not reach the measurement temperature at vacuum because the presence of some gas is required for heat transfer purposes. The choice is somewhat arbitrary but for comparison between samples consistency is the most important consideration. In terms of accuracy, or rather uncertainty, the potential variation in the sample weight at vacuum is another source of error.

6.2.2 *Air and Moisture Sensitivity*

Many hydrogen storage materials are sensitive to air or moisture. In the case of the complex hydrides this sensitivity can result in the decomposition of the sample upon exposure to air. Samples must therefore be loaded into the instrument under an inert atmosphere. In volumetric instrumentation this can be performed using a detachable sample cell that can be sealed sufficiently in a glovebox or by loading the sample in a glovebox attached to the instrument. For gravimetric systems this task is more problematic as the sample must be loaded onto the microbalance pan. One approach is to attach a glovebox to the instrument for this purpose. An alternative is the method used in the Hiden Isochema IGA instruments in which an inert atmosphere sample loader can be temporarily attached to the instrument during the loading of the sample and then removed before starting the measurement. The sample is passed into the inert gas environment of the loader pre-mounted in a sample holder in a small sealed transfer chamber and attached to the microbalance hangdown using a wobble stick.

The important point with regard to measurement accuracy is that the sample should not be affected significantly during the transfer process. In gravimetric measurements the weight can be monitored in situ, which will give an indication of the stability of the sample. In volumetric systems the weight cannot be monitored and so any uncertainty in the loaded sample mass will contribute to measurement error. This point is related to the sample mass discussion in the previous section. In addition, the stability of the sample cannot be monitored in volumetric instruments. If the sample mass is determined after measurement, however, the uncertainty can be determined or eliminated from consideration. Repeated hydrogen cycling with no loss of capacity would also indicate that there is no significant deterioration of the sample during the loading or sample transfer process. An example of a scenario that could result in significant measurement error is the case of ammonia evolution from lithium amide and imide in which a portion of the nitrogen that forms part of the host material would be lost as a result. A similar effect could occur for borohydrides with the evolution of boranes.

Interstitial metal hydrides are often air sensitive in that an oxide layer will form on the surface of an activated sample upon exposure to air. If the sample is to be subsequently hydrogen cycled, this layer must be reduced or fractured sufficiently to allow rehydrogenation. Samples exposed to air for different periods of time are likely to have different surface properties. For comparison purposes, it is therefore crucial that sufficient attention is paid to the surface state of different samples. It is possible that samples treated differently in terms of air exposure following activation or hydrogen cycling will not be returned to the same state, regardless of the treatment used. Further discussion of sample surface oxidation is given in Sect. 6.2.5.

6.2.3 Sample History

The history of a sample, including periods of storage, can have a significant effect on its sorption properties. The history referred to here can include the thermal history, including any annealing that has been performed in the case of hydride-forming compounds, the hydrogen cycling and activation processes, and the exposure of materials to air, but it can also include the synthesis process itself. As discussed in Sect. 3.1.2, the hydrogen cycling of metallic hydrides will result in disproportionation and the creation of significant numbers of lattice defects.[15] Any annealing performed could return the sample to its virgin state, to a lesser or greater degree. Annealing can also be used to increase the compositional homogeneity of a metal host prior to hydrogenation. The significant effects that annealing can have on a sample, whether during sample preparation or for the purpose of regeneration after hydrogen cycling-induced degradation, mean that it is important to record and pay close attention to the effect that this could have on the hydrogen absorption properties.[16] Any previous hydrogen cycling that has been performed on a sample will also have an effect on the subsequent absorption properties, and this should also therefore be recorded. In the case of the activation of metallic hydrides, which normally involves a hydrogen cycling process of some kind, this should also be recorded for the same reasons. Disproportionation, and any other form of compositional or microstructural degradation, or surface modification, that occurs through the activation process could later affect the hydrogen absorption behaviour of a material.

With regard to microporous materials, exposure to air for prolonged periods or previous activation and hydrogen adsorption procedures could potentially affect subsequent adsorption behaviour. In the case of crystalline samples, such as zeolites and MOFs, the synthesis procedure could affect the hydrogen adsorption behaviour of a material. However, this is perhaps more closely related to sample purity, which is discussed in the next section. Prolonged exposure to air can result

---

[15] Measured dislocation densities in some hydrogen-cycled AB_5 intermetallics, for example, can exceed those of heavily deformed (cold worked) metals [52].

[16] The extended cycling study of Wanner et al. [53] included periodic sample regeneration through annealing, which was shown to restore lost storage capacity, and so this is a good example of the effect that an annealing process can have on the hydrogen absorption properties of a material.

in surface oxidation in the case of carbon samples [54–56], and this may alter their adsorption properties. This should therefore be considered as a potential source of uncertainty during reproducibility studies on carbon samples.

In conclusion, close attention should be paid to the synthesis, preparation, activation and thermal history of samples when assessing the accuracy of a measurement performed on any given material. Rigorous microstructural, crystallographic and chemical analysis of samples prior to hydrogen sorption measurement will reduce any uncertainty relating to the history of the sample but, as concluded by Buckley et al. [57] in their study of the effects of the thermal history of $LaNi_5$ on its hydrogenation properties, it is also important to record the history of a sample when data are published.

6.2.4 Sample Purity

The purity of a sample can greatly affect its hydrogen sorption properties. In the interstitial hydrides the purity can be affected by contaminant elements in the host matrix or by the presence of minority phases. The hydrogen absorption properties of alloys or intermetallics are known to strongly depend on the presence of partial substituents and so any uncertainty in the composition will carry over into the absorption measurement. Thorough sample characterisation will ensure sufficient knowledge of the presence of any impurity elements or minority phases. The preparation of high purity intermetallic host materials was discussed in detail by Percheron-Guégan and Welter [58]. Among the sample purity-related issues highlighted are the greater possibility of compositional variation in the intermetallics compared to elemental metals, the effect that this variation may have on the homogeneity of the hydrogen concentration in the hydride, and the more severe effect that impurities can have on the stability of ternary or higher intermetallic hydrides compared to the equivalent effect on the stability of binary hydrides.

With regard to microporous adsorbents, the synthesis method used to produce MOFs has been shown to have a significant effect on the measured uptake [59, 60]. Therefore careful attention should be paid to the importance of the synthesis method as well as the activation procedure (Sect. 6.4.1). This uncertainty could originate from partial framework collapse, pore blockage or the presence of impurity phases. The results of a recent study by Hafizovic et al. [61] suggest that large variations in the reported adsorption properties, and hence hydrogen storage capacity, of MOF-5 (Sect. 2.1.3) can be explained by framework interpenetration and the blocking of pores by $Zn(OH_2)$ species, while Tsao et al. [62] found that MOF-5 samples prepared using different synthesis routes exhibited different mesoporous characteristics (see Sect. 5.3.3). Such a variation in the pore structure will greatly affect the measured hydrogen storage properties of a microporous material. For these materials, in general, the presence of any impurities within the structure could potentially alter the interaction between the internal surface and hydrogen due to changes in the surface chemistry. Their presence could also affect the skeletal density of the material, which could potentially change the measured hydrogen sorption properties (see Sect. 6.2.1). Thorough sample characterisation is therefore crucial and sufficient detail should be included when reporting hydrogen sorption data to allow the effects of any chemical or structural impurities to be assessed.

6.2.5 Gaseous Impurity Gettering

A *getter* is a reactive material that can be used to remove impurities from a vacuum or gaseous environment. In the presence of impurities, any hydrogen storage material can act as a getter to some extent. Impurities can be present in the gas supply, they can outgas from the internal surfaces of the measurement system, or be present due to virtual and real leakage within the system.

Porous materials that are effective hydrogen adsorbents are likely to adsorb contaminants, or impurities, from the hydrogen that is being used to perform the measurement, particularly in view of their large surface area and narrow pores. The type of impurity adsorbed and the extent to which the adsorption is preferential will depend on the adsorbent. This may be difficult to predict and the measurement of multicomponent gas mixture adsorption is also challenging. The best approach is to simply minimise the amount of impurities present. The first essential step is to use high purity hydrogen. Practically, this means hydrogen greater than 99.999% purity but higher purity is preferable (see Sect. 6.1.6). The second step is either filtration of the gas or the minimisation of the chance of contamination being introduced into the gas stream from the gas bottle regulator or the gas delivery lines. As discussed in Sect. 6.1.6, special purge regulators are available commercially that can help ensure the cleanliness of an ultra-high purity or research grade gas supply following a cylinder change.

In gravimetric instrumentation it is possible to observe impurity adsorption by monitoring the mass change of the sample during equilibration. Hydrogen appears to exhibit particularly rapid kinetics in comparison to gases of larger molecular size [63]. A very slow adsorption kinetic, combined with a failure to reach a stable equilibrium uptake, is therefore indicative of impurity contamination [42]. Repeatability and reversibility studies will help confirm whether significant contamination is present. In the absence of contamination, molecular hydrogen physisorption is expected to be fully reversible, with no significant hysteresis, and this can be confirmed by the measurement of both adsorption and desorption isotherms.

With regard to hydrides, activated samples require a reactive surface in order to dissociate molecular hydrogen and absorb it in atomic form. Therefore, it is likely that the surface of a hydride sample will react with impurities in gas phase hydrogen. If a metal hydride sample attracts impurities from the gas phase, there is a chance that this could be misinterpreted as hydrogen uptake. As discussed in Sect. 3.1.3, the presence of some impurity species will result in surface passivation or poisoning, although this is dependent on the material; for example, surface segregation of $LaNi_5$ is known to lead to the binding of O_2 and H_2O impurities as La oxide or hydroxide while metallic Ni continues to dissociate molecular hydrogen [64].

In the context of hydrogen concentration determination techniques for the purpose of accurately correlating metal hydride lattice expansion to the hydrogen-to-metal ratio of the sample, Peisl [65] discussed the effect that gaseous impurities, such as oxygen and nitrogen, can have on gravimetric measurements. As the influence of these impurities, which will result in much greater mass changes than hydrogen, is mainly within the surface layer, Peisl concluded that the accuracy of a gravimetric measurement increases with a decreasing surface area-to-volume ratio. Therefore, this is not an issue for large single crystals but it can be significant for powders. In a recent study of the surface oxidation of a commercial AB_2 Laves Phase alloy, $Ti_{0.96}Zr_{0.04}Mn_{1.43}V_{0.45}Fe_{0.08}$, Schülke et al. [66] found that an oxide layer approximately 15 nm thick was formed on the surface of the material after exposure to air for a prolonged period. The analysis was performed using Secondary Neutral Mass Spectrometry (SNMS) after an exposure period of several months.

As the importance of gaseous impurity gettering is dependent on the surface area-to-volume ratio, particular care should be taken with fine powder samples. As discussed by Lowell et al. [46], particle size and shape can have a considerable effect on the surface area. The ratio of the surface area of a powder of an identical material consisting of cubic and spherical particles is given by,

$$\frac{S_{cube}}{S_{sphere}} = \frac{2\, r_{sphere}}{l_{cube}} \tag{6.45}$$

where S_{cube} and S_{sphere} are the surface areas of the cubic and spherical particles, respectively, r_{sphere} is the sphere radius and l_{cube} is the cube edge length.

The surface area of a powder consisting of uniform spheres is given by,

$$S_{sphere} = \frac{3}{\rho\, r_{sphere}} \tag{6.46}$$

where ρ is the density of the material. Thus, an order of magnitude decrease in the particle radius results in an order of magnitude increase in the surface area; as does a similar decrease in the material density. A sphere has the minimum surface area-to-volume ratio of all geometric forms and any deviation from this will increase the surface area. Furthermore, surface irregularities, which are present in any real material, will also increase the absolute surface area. The available surface area for gaseous impurity gettering will therefore be affected by the particle size, morphology and the surface topography, and the potential significance of the effects of this phenomenon will depend on each of these factors.

In their study of the loss of reversible hydrogen capacity of $LaNi_5$ with hydrogen cycling, Gray et al. [67] discussed the effect that impurity gettering would have on their data of hydrogen uptake as a function of cycle number. They observed a positive shift in the zero point mass of their sample after hydrogen cycling and suggested that this could be due to impurity gettering rather than the trapping of residual hydrogen in their sample. However, they concluded that, in this case, impurity gettering was unlikely to be responsible for the loss in reversible capacity because the isotherm data they measured gravimetrically did not show an overall increase in the sample mass with time. The loss in capacity resulted only in a narrowing of the plateau width rather than any shift to higher, and unphysical, hydrogen-to-metal ratios. Careful consideration of the measured pressure-composition behaviour with both time and cycle number, as demonstrated in this work, can therefore help assess the likelihood of problems with gaseous impurity contamination, and hence gettering.

Generally speaking, gaseous impurity gettering is more of an issue in gravimetric instrumentation because it will directly affect the measurement. The sample is also likely to be exposed to a larger volume of hydrogen gas and therefore, after the adsorption of, or reaction with, the impurities, the supply of contaminant species near to the surface of the sample is more readily replenished. However, gravimetric measurement also allows identification of the occurrence of impurity gettering, as described above. In the case that impurity adsorption or reaction passivates the surface of the sample preventing full hydrogenation, but does not significantly affect measurement accuracy, the gravimetric technique has the advantage that the kinetic profile will help diagnose the problem. In gravimetric measurement the mass signal from an adsorbed or absorbed impurity atom or molecule will be significantly greater than that of hydrogen. In volumetric measurement the impurity adsorption or reaction will affect the pressure to a similar extent to hydrogen and may therefore be more difficult to differentiate from a hydrogen sorption process.

6.3 General Instrumentation Issues

In this section we will look at the accuracy issues associated with the instrumentation used for hydrogen sorption measurement. We begin with the vacuum and pressure capability of any given system, before examining the thermal stability and homogeneity of measurement apparatus, in terms of sample temperature and the overall thermal performance of the instrument. We then cover both pressure and temperature measurement.

6.3.1 Vacuum and Pressure Capability Considerations

The vacuum and pressure handling capability of apparatus is important for several reasons. High vacuum operation with a dry pump system is required to ensure a clean environment, because any contamination can affect sorption measurement accuracy. A difference in the vacuum quality above the sample following activation will result in sorption measurements beginning from different reference points. It will also lead to different degassing conditions, which can be crucial for microporous adsorbents (see Sect. 6.4.1). Meanwhile, leakage and hydrogen permeation at high pressures, as well as being a safety concern, will lead directly to measurement errors, although this mainly applies to volumetric measurement (see Sect. 6.5.5); in addition, leakage can permit the ingress of contaminants. In this section, we will look, firstly, at vacuum pump systems before discussing the implications of hydrogen leakage.

6.3.1.1 Vacuum Pump Systems

High vacuum operation should ensure a clean system. Generally speaking, the lower the vacuum capability, the higher the chance of contamination of the gaseous hydrogen and the subsequent effects that this will have on the hydrogen sorption measurement. The quality of the vacuum achievable in sorption apparatus will depend on a number of factors. These are principally the choice of vacuum pump, the diameter and length of the system tubing, the number of valves, which are likely to reduce the conductance of the system, the quality of the vacuum seals and the tortuosity of the pumping path. However, it will also be affected by the smoothness of the internal system surfaces and the existence of virtual leaks through, for example, poor quality welds. Impurities can be flushed from a system, to a certain extent, by flowing an inert gas or hydrogen, following an initial evacuation, or by repeated system pressurisation and evacuation, but the provision of high vacuum capability is a preferable route to ensuring a clean system. The simultaneous heating and evacuation of a system, a process known as *baking* or *bakeout* in vacuum science [68], will help desorb impurities and dry the apparatus, particularly if moisture contamination is likely. The application of heat rapidly accelerates any thermally activated desorption processes, and this treatment is necessary in UHV systems to achieve ultra-high vacuum conditions.

The choice of vacuum pump is important. A typical and recommended dry high vacuum pump configuration for operation with hydrogen is a turbomolecular pump backed by a scroll or membrane (diaphragm) pump. As mentioned in Chap. 4 (Sects. 4.1.1 and 4.2.1) a dry pump system should be used because the backstreaming of oil vapour can contaminate the system [68]. Oil vapour filters, which are often called *foreline traps*, are available but a contaminated filter will result in a contaminated system and there seems to be little advantage in risking the chance of oil contamination when a membrane or scroll pump will suffice for roughing purposes. The advice to use a dry system is particularly pertinent to the case of hydrogen adsorption measurement because any contamination can potentially adsorb to the microporous adsorbent during the degassing procedure (Sect. 6.4.1). Any preferential adsorption of impurities will affect the accuracy of a measurement either by increasing the sample mass during a gravimetric measurement or by partially blocking the process of hydrogen adsorption in either the volumetric or gravimetric case. Furthermore, microporous materials require high vacuum conditions for degassing and so the use of a turbomolecular pump is essential in this case.

6.3.1.2 Leakage

In Sect. 4.2.3.2 we covered the technical system requirements for high pressure hydrogen sorption measurement. These are necessary for safe operation with high pressure hydrogen but they are also crucial for the performance of high accuracy measurements. The effect of leakage on the accuracy of a measurement relates primarily to the false detection of sorption or desorption in volumetric apparatus due to the misinterpretation of the resultant drop or rise in pressure, respectively. However, internal leakage through valves is also a problem for the maintenance of isobaric conditions in gravimetric instrumentation. Any leakage can be safely detected before an experiment using helium pressure hold tests. Leakage during an experiment will be seen as an approximately linear decrease in pressure with time which is atypical of a sorption process, because kinetic profiles normally show some curvature. This should be seen during the kinetics of sorption for a single isotherm step. If the kinetics of hydrogen sorption are expected to be very slow, as can be the case for some of the complex hydrides, then it can become difficult to differentiate between leakage and the slow sorption process. In this situation, the removal of the sample and extended leak testing with a blank sample cell is probably the best approach, combined with repeatability and reproducibility testing. If, during uptake, the kinetics are not recorded then problems with leakage can potentially be diagnosed from the isotherm shape of a well known or understood sample. Leakage will result in unusual isotherm curvature or unexpected linear sections in terms of hydrogen uptake or release versus pressure. Leakage has been identified as a significant source of error in volumetric measurement by a number of authors (see Sect. 6.5.5).

Although less significant during gravimetric measurements, hydrogen leakage should also be avoided in this case. It has less significant implications for the accuracy of a measurement but internal leakage through mass flow controllers or pressure and vacuum control valves will affect the ability of the measurement system to control the hydrogen pressure. However, this should be relatively obvious from the time-dependent pressure data during a measurement. Leakage in a gravimetric system could also result in the increased supply of contaminants to the sample chamber, which may then be adsorbed by the sample. External leakage will affect the achievable vacuum level and so the pressure above the pump system can be used to check for any leakage, to a certain extent, in addition to pressure hold tests.

6.3.2 Thermal Stability and Homogeneity

The temperature stability of both the measurement apparatus and the sample is important in all techniques. Firstly, we will look at the sample temperature, primarily with regard to the maintenance of stable isothermal conditions and, secondly, we shall emphasise some of the most significant effects of thermal instability in a system and a lack of homogeneity throughout, or in a particular part of, the apparatus.

6.3.2.1 Sample Temperature

A stable and homogeneous sample temperature is a key factor in accurate sorption measurement. Temperature gradients across a sample bed will result in differing sorption characteristics, which will affect the overall accuracy of a measurement. The exo- and endothermic nature of the sorption and desorption processes can result in significant temperature excursions during the characterisation of a material; hence the need for special reactors for the study of the isothermal absorption kinetics of hydrides (Sect. 4.1.3), for example. During normal isotherm determination, it is therefore necessary to carefully monitor the sample temperature to ensure that the sample has returned to thermal equilibrium before an isotherm point (the equilibrium hydrogen uptake) is determined. This is an important consideration during experimental method development and complements the consideration of equilibration times with regard to hydrogen sorption equilibrium, at each isotherm point. A larger sample size is likely to both exacerbate this problem and increase the likely presence of thermal gradients even under 'equilibrium' conditions. Therefore, a strong argument can be made for reducing the sample size for this reason providing the ability of the instrument to determine the uptake of hydrogen by the sample at each point is not compromised significantly by the sample size reduction.

The thermostat, or the method used to control the sample temperature, must be able to maintain a stable temperature. The stability must be as high as possible, with variations of the order of 1 K or higher liable to affect the measurement significantly. This is less important at elevated temperatures but crucial for low temperature measurements. The Japanese Industrial Standard (JIS) on Sieverts' method measurements on hydrogen absorbing alloys [69] specifies a precision of $\pm$ 0.5 K for temperature stability of the thermostat and of the temperature measurement, although higher precision, as well as accuracy, than this is required at low temperatures.

6.3.2.2 Instrumentation

Hot or cold spots, whereby a part of the apparatus is at a different temperature to the majority of the system, can greatly affect the volumetric measurement of adsorption by porous adsorbents. This is particularly crucial in the measurement of subcritical adsorptives because a cold spot can result in bulk condensation within the instrument, which will lead to a false measurement of adsorption. Although bulk condensation will not occur in supercritical hydrogen adsorption measurement, unless perhaps the measurement is close to the critical temperature, it could still lead to false or spurious decreases in pressure that could be misinterpreted as adsorption by the sample. Hot spots will have the opposite effect. For volumetric measurement any temperature inhomogeneity will affect the calculated sorbed quantity due to an increased uncertainty in the $PV = nZRT$ terms of the calculation (see Sect. 6.5.4). Thermal stability is essential in gravimetric measurement in order to achieve an accurate weight reading. Any fluctuations over short or long time scales will affect the accuracy of the measurement regardless of the intrinsic accuracy of the microbalance. In other words, high thermal stability is required for a stable microbalance reading, and hence high gravimetric measurement accuracy (see Sect. 6.6.3).

6.3.3 Pressure Measurement

Pressure measurement is obviously important in both the volumetric and gravimetric techniques. In the volumetric method it is used directly to calculate the hydrogen uptake and therefore must be determined to sufficient accuracy for this purpose. However, the accurate determination of the pressure is also crucial for gravimetric techniques. Significant uncertainty in the pressure will add significant error bars to an isotherm plot and the maintenance of a stable pressure is critical for isobaric measurements. Inaccurate pressure measurement will contribute to uncertainty in the determination of the hydrogen density or the compressibility at the measurement temperature and pressure. This could affect the buoyancy effect corrections in gravimetric measurements (Sect. 6.6.2) and add to the cumulative error in the calculation of the sorbed quantity for volumetric measurements (Sect. 6.5.4).

6.3.3.1 Manometer Accuracy

For volumetric measurement, when selecting a pressure measuring device it is important to ensure that the uncertainty in the pressure measurement is not significant compared to the expected decrease in pressure upon hydrogen sorption, or vice versa. Practically speaking, significant improvements in transducer accuracy involve a concomitant increase in cost: at the time of writing, a transducer with a quoted accuracy of $\pm$ 0.05% is reasonably priced and could be used on relatively low cost laboratory equipment, as well as typical commercial hydrogen sorption instrumentation, but significant increases in accuracy beyond this require substantially higher expenditure. However, this will presumably change as pressure measurement technology continues to improve. To ensure sufficient accuracy over a broad range of pressures, it may be necessary to use different devices to cover different pressure measurement ranges. In low pressure sorption instrumentation used for pore size distribution determination, this can be essential to cover the low

P/P_0 region during which micropore filling occurs [46]. In high pressure hydrogen sorption instrumentation, however, it is not so crucial but it will allow higher accuracy sub-ambient pressure sorption measurements to be performed and it is a prerequisite if the Henry's law region for hydrogen adsorption is being investigated. The higher accuracy of differential pressure transducers is a good argument in favour of the differential volumetric method (see Sect. 4.1.2).

6.3.3.2 Thermal Transpiration Effects

There are a number of empirical corrections that have been applied to account for the effects of thermal transpiration (Sect. 6.1.5) on measured pressures. In the free molecule regime ($Kn > 1$), using the tube diameter as the characteristic length scale, the correction necessary for thermal transpiration effects is given by,

$$\frac{P_t}{P_m} = \left(\frac{T_t}{T_m}\right)^{1/2} \tag{6.47}$$

where P_t is the true equilibrium pressure at the sample temperature, T_t, and P_m is the measured hydrogen pressure at a temperature, T_m. As Kn approaches unity the thermal transpiration effect decreases with increasing pressure, and it becomes zero once the pressure is above the threshold at which the system has fully entered the continuum regime. In the transition region, the pressure dependence must be described and a number of approaches have been used. Knudsen's original derivation gave,

$$\frac{P_t}{P_m} = \left(\frac{T_t}{T_m}\right)^{\chi/2} \tag{6.48}$$

with,

$$\chi = \left(1 + \frac{d}{\lambda}\right)^{-1} \tag{6.49}$$

where d is the tube diameter. Wallbank and McQuillan [70] applied thermal transpiration corrections to Ti–H data at low pressures by representing χ using the following empirical expression,

$$\chi = \left[1 + \left(\frac{P_m}{P_0}\right)^{1/2}\right] \exp\left[-\left(\frac{P_m}{P_0}\right)^{1/2}\right] \tag{6.50}$$

with $P_0 = 3.4$ Pa, a value that was determined empirically. They demonstrate the correction with experimental data measured using significantly different tube diameters of 3.5 and 18 mm, and applied the correction to data below 15 Pa.

Two other corrections, which are of a more complicated form than the correction applied by Wallbank and McQuillan, are due to Liang [71] and Takaishi and Sensui [72]. The Liang correction takes the form,

$$\frac{P_t}{P_m} = \frac{\alpha\varphi^2 X^2 + \beta\varphi X + (T_t/T_m)^{1/2}}{\alpha\varphi^2 X^2 + B\varphi X + 1} \tag{6.51}$$

where $X = P_m d$ with P_m in Torr and d in mm [73]. The value of φ is 1.0 for helium, which is used as the reference, and 1.52 for hydrogen [71]. α and β are empirically determined constants for any given gas.

The Takaishi and Sensui [72] correction is an extension of the Liang expression,

$$\frac{P_t}{P_m} = \frac{A^*(X/T^*)^2 + B^*(X/T^*) + C^*(X/T^*)^{1/2} + (T_t/T_m)^{1/2}}{A^*(X/T)^2 + B^*(X/T^*) + C^*(X/T^*)^{1/2} + 1} \tag{6.52}$$

where $T^* = (T_t + T_m)/2$ and X is defined as above. A^*, B^* and C^* are coefficients obtained empirically.

These corrections are commonly applied in low pressure measurements using UHV systems [74]. They are not relevant at elevated pressures but low pressure hydrogen sorption measurements can be affected at surprisingly large tube diameters, particularly in the presence of large temperature gradients. The Wallbank and McQuillan [70] study shows how high temperature hydrogen absorption measurements can be affected, and low temperature pressure measurement can contain errors of up to 100% if these effects are not properly taken into account [75]. Wilson et al. [74] corrected all pressure measurements below 2 Torr (0.267 kPa) in their low temperature (20–100 K) study of hydrogen and deuterium adsorption by carbon nanotubes.

6.3.4 Temperature Measurement

The sample temperature must be monitored carefully during a sorption measurement. In volumetric techniques this can be performed directly because a temperature sensor can be in contact with the sample. Any error in the measured value in the volumetric case will, however, contribute to the cumulative error in the calculation of the sorbed quantity (see Sect. 6.5.4). In gravimetric measurement the sensor can only be in the vicinity of the sample because, if it is in direct contact, the presence of the sensor will affect the weight measurement. In the gravimetric case, discrepancies between the real sample temperature and the measured value will not significantly affect the determination of the sorbed quantity, but any systematic error in the measured sample temperature will result in a deviation of the isotherm from its true pressure and hydrogen content for that particular temperature. The significance of this error will depend on the sensitivity to temperature of the hydrogen uptake for the material in question. For temperature-programmed measurements, if the analysis of the data includes such information as the TDS peak position, it is essential that the measured temperature accurately represents the real increase experienced by the sample. Otherwise, for integration of the signal for total content determination, sample temperature measurement is not so crucial. In general, the accuracy of the temperature measurement will depend on the sensor type and the regularity of recalibration, and so attention should be paid to both of these aspects. It is likely that sensor readings will drift over time and so regular checks are recommended.

6.4 Experimental Methodology

In this section we will cover two of the general issues related to the development of a suitable experimental method. Firstly, we will look at the degassing of microporous materials and the activation of hydrides and, secondly, we will discuss the importance of ensuring that sufficient equilibration times are allowed during the approach to equilibrium at each isotherm point.

6.4.1 Sample Degassing and Activation

The degassing of porous adsorbents is crucial to the accurate characterisation of their adsorption properties because a material must be degassed[17] before the performance of any adsorption experiment. If this process is not performed thoroughly the adsorption process could be affected significantly. Degassing is the method by which microporous adsorbents are activated in preparation for

[17] In vacuum science, *outgassing* is defined as the evolution of gas from a solid or liquid in a vacuum, *degassing* as the deliberate removal of gas from a solid or a liquid in vacuum as a result of the impact of molecules, electrons, ions, or photons; or by heating and *desorption* as the release of adsorbed atoms and/or molecules, either neutral or ionised, from the surface of a solid or liquid as a result of the impact of molecules, electrons, ions, and photons; or by thermal energy at the temperature of the material [76]. Using these definitions, *degassing* is the most appropriate term in this case, because we are interested in deliberately removing gas, or liquid, either by heating in a vacuum or by bombarding the surface with inert gas molecules. In the context of hydrogen adsorption measurement and apparatus, the term *outgassing* should be reserved for the general phenomenon in which the evolution of potential contaminants occurs from either the sample surface or the internal surfaces of the apparatus under vacuum conditions. This clearly distinguishes between, for example, the different processes of sample degassing by evacuating and heating the sample to deliberately drive off contaminants, and the unavoidable outgassing from the apparatus walls at vacuum that can contaminate the sample due to gettering. *Desorption* is then simply used for the temperature-programmed techniques (thermal desorption spectroscopy or temperature-programmed desorption) in which hydrogen is deliberately desorbed from the sample.

hydrogen adsorption (see Sect. 3.1.4) and involves the evacuation of the sample at elevated temperatures to remove any environmental contaminants or remnants from the synthesis process. This is clearly crucial regardless of the measurement technique and insufficient removal of contaminants will inevitably result in measurement errors. If the measurements are being made at elevated temperatures at which the contaminants could potentially react with hydrogen, then the result of the contamination could be quite unpredictable. On the other hand, if the measurement is performed at low temperatures it is likely that the contaminants will freeze on or in the sample and will result in a reduced uptake of hydrogen due to a reduction in the number of adsorption sites or the accessible pore volume.

In gravimetric apparatus the sample degassing process can be monitored directly by measuring the weight change of the sample as a function of time. In any other apparatus, whether a TDS or volumetric system, it cannot be determined directly, although it is possible to monitor the evolution of gases from the sample using a mass spectrometer if the experimental configuration allows for this possibility. However, in a standard volumetric instrument the only method of monitoring the degassing process is to measure the pressure above the vacuum pump system. In this case, it is necessary to monitor the quality of the vacuum achieved above the vacuum pump until it has reduced to a level low enough to indicate that there is no further significant evolution of pre-adsorbed species from the sample. The effectiveness of this method will depend, to a certain extent, on the quality of the vacuum achievable in a particular piece of apparatus (see Sect. 6.3.1.1). In the case of a system that can achieve a lower vacuum, the vacuum reading should not be relied upon entirely. It is better to err on the side of caution, bearing in mind that any microporous material is likely to require degassing for a number of hours. The effect of degassing time on the subsequent adsorption behaviour of a material can be investigated by determining the reproducibility or repeatability of the measurement following different degassing procedures.

For hydrides, sample activation can be a more involved process than simple degassing (see Sect. 3.1.4). A good understanding of the activation process is necessary before accurate characterisation of the hydrogen sorption properties of a material can be performed. In the case of intermetallic hydrides, the hydrogen absorption and desorption properties change significantly during the first few cycles while the sample is undergoing full activation. It is possible for a hydride sample to be only partially activated if the full procedure is not performed, leading to different portions of the sample exhibiting different absorption behaviour. In this situation, macroscopic hydrogen absorption measurements will only determine the average behaviour of the sample, which will not represent the real behaviour of the fully activated material, and therefore lead to significant inaccuracy in the characterisation process. The changes in the hydrogen absorption and desorption behaviour of a material through the activation process can be determined by measuring isotherms during the initial cycling process. If a material exhibits good cycling stability (see Sect. 3.1.2), its hydrogen absorption and desorption behaviour will remain relatively consistent after full activation has been achieved. A detailed description of the procedure should ideally be given when reporting data.

6.4.2 Equilibration Times

Consideration of the time required for the system to reach equilibrium at each isotherm point during a hydrogen sorption measurement should include the approach to both thermal and sorption equilibrium. If the sorption kinetics are rapid it is possible that the time required will be dictated by the approach to thermal equilibrium rather than the approach to sorption equilibrium. This is more likely to be the case for hydrogen adsorption than it is for hydrogen absorption because the former is typically a rapid process. In Sect. 4.2.1.2, we saw that gravimetric data can be fitted in real time to allow the approach to equilibrium to be carefully monitored and an asymptotic uptake to be predicted. In volumetric measurement, meanwhile, the pressure decay is monitored. In the case of rapid adsorption, with the addition of the pressure drive for volumetric measurement, the pressure decay is likely to be rapid with thermal effects perhaps making the most significant contribution to non-equilibrium conditions, and therefore monitoring the sample temperature, as well as the pressure, during the approach to equilibrium is important.

For hydrogen absorption measurements, the more complex nature of the absorption process (see Sect. 3.3.2) means that equilibration times between different materials can vary greatly, and also that they can vary significantly through the course of a single isotherm measurement for any given material. If sufficient equilibrium is not achieved at each point using either technique, the isotherm shape can be significantly affected. For example, for an intermetallic hydride, the plateau pressure in the absorption isotherm will be higher and the plateau is likely to be shortened in the higher hydrogen content region. The plateau could also exhibit a steeper gradient. In the desorption case, the plateau pressure may appear lower than expected because the pressure will not have risen sufficiently, at each step, to represent technical equilibrium. A substantial exothermic reaction in the case of hydrides means that the sample temperature must be monitored carefully. As mentioned above, in gravimetric measurement the weight change as a function of time is monitored to assess the approach to equilibrium. In theory, in an isobaric experiment, if the uptake is exponential in nature, as expected for a diffusion-limited process, the asymptotic mass will not be reached. A given percentage of the predicted asymptotic mass can therefore be used to define the necessary conditions for the achievement of technical equilibrium. Exponential behaviour will not necessarily be the case, however, and so the reaction may reach completion but nevertheless it is necessary to define the required equilibration conditions (see Sect. 4.2.1.2).

In the volumetric case, during absorption, the system will reach an apparent equilibrium when the hydrogen pressure drops sufficiently and therefore prevents further hydrogen absorption. At this point any further absorption would result in the hydrogen pressure, and hence the chemical potential, dropping below the level required for more hydrogen to enter the sample or for further hydride phase formation to occur. The reaction therefore cannot continue until more hydrogen is supplied to the system, and so there is a significant difference between the hydriding processes in the isochoric manometric and isobaric gravimetric measurement methods. In the isobaric gravimetric case the chemical potential of the gaseous hydrogen is held constant[18] on the surface of the sample. In the mixed phase region, for a homogeneous sample, the reaction will therefore continue until the hydride phase transformation is complete [77]. However, in the case of most real materials, particularly those being considered for hydrogen storage applications, the presence of significant hetereogeneity means that the plateau is unlikely to be flat and stepwise hydriding will be required to allow complete conversion of the sample from the α to the β phase.

For measurements using Sieverts' method on hydrogen absorbing alloys, the Japanese Industrial Standard [69] suggests that no pressure change over the period of 1 min indicates that equilibrium has been achieved. However, it includes the caveat that this is dependent on the sample and if different degrees of hysteresis are seen in the absorption and desorption isotherms with an extension of the equilibrium period at each point, then the equilibrium time should be extended.

6.5 Volumetric Measurement

In this section we consider some of the issues specific to volumetric measurement. We first look at the effect of the presence of thermal gradients, which must be taken into account unless measurements are made under fully isothermal conditions. We then cover the ratio of the chosen sample size to the overall system volume. This can affect the ability of a system to detect hydrogen sorption for a given transducer accuracy, for example, or it can exaggerate effects from errors in either the compressibility description or our knowledge of the dead volume or sample volume and density. We then cover the dead volume calibration or corrections, the accumulative errors inherent in any multipoint volumetric measurement, and the consequences of hydrogen leakage.

6.5.1 Thermal Gradients

Although thermal gradients are a practical problem in gravimetric as well as volumetric measurement, the uncontrolled gradient between the sample cell and delivery manifold is of particular importance in the volumetric case. In Chap. 4 we looked at the calculation of the hydrogen uptake in the presence of a gradient between the sample cell and the system (Eqs. 4.5 and 4.6), using the parameter f,

---

[18] This is not strictly true because the heat generated by the reaction will change the chemical potential of the gas phase hydrogen.

which is the decimal fraction that defines a hypothetical dividing line between the two temperature zones of the apparatus. Any uncertainty in this value will contribute to the uncertainty in the measurement. The minimisation of the volume between the two temperature zones, in order to reduce the proportion of the overall volume that is uncontrolled, will therefore help reduce this potential source of error. Thermal gradients within either the system or the sample cell will also contribute an error. The problem of hot and cold spots was introduced in Sect. 6.3.2.2. These two phenomena, in which regions of the apparatus are at significantly higher or lower temperatures than the assumed or measured average temperature, will be detrimental to measurement accuracy because the accurate calculation of the sorbed quantity relies on the assumption of isothermal conditions. Cold regions will produce a decrease in the overall pressure that could be misinterpreted as sorption, and warm regions will have the opposite effect. Any uncertainty in the overall average system temperature will be transferred to the calculated uptake. In addition to the elimination of thermal gradients, the system temperature, or the temperature of the housing for the dosing volume, should be resistant to external ambient temperature fluctuations and ideally maintained constant.

6.5.2 Sample Size to System Volume Ratio

The ratio of the sample size to the system volume is an important aspect of the design and operation of a volumetric instrument, and it affects measurement accuracy. Let us consider the case of a small sample mounted in a volumetric system of large internal volume. When the hydrogen is dosed into the system by opening valve C in Fig. 4.1, the pressure will drop by at least the amount expected due to the difference between the volumes V_1 and $(V_1 + V_2)$. If the sample is too small the amount of hydrogen uptake will not be detected by the additional pressure drop, ΔP, even if the pressure is measured with high accuracy. A straightforward conclusion is, therefore, that for a system of a given volume, a sufficient amount of sample must be used, taking into account the thermal stability of the instrument (Sect. 6.3.2.2), the accuracy with which the pressure can be determined (Sect. 6.3.3) and the accuracy of the description of the compressibility (Sect. 6.1.1). Conversely, if a system is being designed for samples of a given size, then its total volume must be such that the expected hydrogen uptake can be determined accurately. The JIS on Sieverts' method measurements on hydrogen absorbing alloys [69] suggests that a minimum of 1 g per 100 cm^3 of system volume should be used, with no upper limit specified.

However, the ratio of V_1 and V_2 is also important and also the density of the material. Blach and Gray [78] examined the effect of the sample size-to-system volume, as well as the dosing volume-to-sample cell ratio, on measurements performed on low density hydrogen storage materials. They considered the determination of hydrogen uptake using the indirect dead volume determination method and therefore used an independently determined sample density. They concluded that both the dosing volume and empty sample cell volume should be considerably larger than the volume occupied by the sample, by a factor of at least 100, and that both volumes should be roughly equal. They also note that, because of the sensitivity of these measurements to sample density, accurate measurement with a relatively high density hydride sample, such as $LaNi_5$, does not necessarily mean that the instrument will be accurate for low density samples. Therefore, the use of high density hydride samples to check apparatus subsequently used for the measurement of hydrogen sorption by low density hydrogen storage materials should be avoided and a low density reference material used instead (see Sect. 7.2).

6.5.3 Dead Volume Corrections

The dead volume corrections account for the difference between the total sample cell volume and the total cell volume minus the volume occupied by the sample. They are the analogue of the buoyancy effect corrections for gravimetric measurement [79], because they also account for the presence of the sample in the chamber. As for the buoyancy effect corrections, their significance increases with increasing hydrogen pressure, as well as decreasing sample density, and so any error associated with the correction will increase in absolute terms with the pressure. As we saw in Sect. 4.1.1, the dead volume determination necessary for the manometric (volumetric) method can take the direct or indirect form. Using the direct method, the dead volume of the sample cell is determined using a gas that is assumed to be non-interacting, which is typically helium. The indirect approach, meanwhile, uses a dead volume calculated by subtracting an independently determined sample volume (see Sect. 6.2.1.2) from the measured volume of the empty sample cell. As we saw in Eq. 4.4, the expression for the total number of moles at the mth sorption point, n_m, is given by,

$$n_m = \sum_{j=1}^{m} \left[\frac{P_{f,j-1,T}\left(V_{cell} - m_s/\rho_s\right)}{Z_{f,j-1,T}RT} + \frac{P_{i,j,T}V_1}{Z_{i,j,T}RT} - \frac{P_{f,j,T}\left(V_1 + \left(V_{cell} - m_s/\rho_s\right)\right)}{Z_{f,j,T}RT} \right] \tag{6.53}$$

where V_{cell} is the empty sample cell volume, V_1 is the calibrated dosing cell volume, m_s is the sample mass, ρ_s is the sample density, $P_{i,j,T}$ and $P_{f,j,T}$ are the initial and final pressures at isotherm point j, $Z_{i,j,T}$ and $Z_{f,j,T}$ are the compressibilities of hydrogen at pressures, $P_{i,j,T}$ and $P_{f,j,T}$, T is the measurement temperature and R is the universal gas constant.

In this expression it can be seen that the two terms $P_{f,j\text{-}1,T}m_s/Z_{f,j-1,T}RT\rho_s$ and $P_{f,j,T}m_s/Z_{f,j,T}RT\rho_s$ both account for the presence of the sample in the cell and increase with increasing pressure, as well as decreasing sample density. Both terms are of the same approximate form as the sample buoyancy correction term

$m_s g(\rho_H/\rho_s)$ in Eq. 4.10, ignoring the conversion to a force in the latter, because the P/ZRT contribution to the dead space correction terms from Eq. 6.53 represents the density of gaseous hydrogen, ρ_H, shown in the buoyancy correction. The magnitude of both terms therefore depends on the ratio ρ_H/ρ_s. The key difference for multipoint manometric isotherm determination is that the corrections for the dead volume accumulate error through the course of a measurement, whereas the buoyancy effect corrections do not because, in the latter case, every isotherm point is referenced to the empty sample mass at vacuum (see Sect. 6.6.2).

6.5.4 Accumulative Errors

The expression for the number of moles of hydrogen absorbed by a material measured volumetrically, at the mth point on the isotherm, assuming direct dead space volume determination, is given by,

$$n_m = \sum_{j=1}^{m} \left[\begin{array}{l} \left(\dfrac{fP_{f,j-1,T_{sys}} V_2}{Z_{f,j-1,T_{sys}} RT_{sys}} + \dfrac{(1-f)P_{f,j-1,T_{sample}} V_2}{Z_{f,j-1,T_{sample}} RT_{sample}} \right) + \\ \dfrac{P_{i,j,T_{sys}} V_1}{Z_{i,j,T_{sys}} RT_{sys}} - \left(\dfrac{P_{f,j,T_{sys}} (V_1 + fV_2)}{Z_{f,j,T_{sys}} RT_{sys}} + \dfrac{(1-f)P_{f,j,T_{sample}} V_2}{Z_{f,j,T_{ssmple}} RT_{sample}} \right) \end{array} \right] \quad (6.54)$$

where the various parameters have the same definitions as those for Eq. 6.53, except we have non-isothermal conditions and therefore T_{sys} is the system temperature and T_{sample} is the sample temperature, with f being the decimal fraction of the sample cell volume that remains at T_{sys}. Also, V_2 is the sample cell dead volume determined directly using a non-interacting gas such as helium. It can be seen that the determination of n_m is an iterative calculation and any errors in the pressures, volumes, temperatures and gas compressibilities used in the above will accumulate through the measurement. This is in contrast to the gravimetric technique in which the sorbed quantity is calculated using a non-iterative function referenced to the empty sample mass measured at vacuum. This is a disadvantage of the volumetric technique and contributes to the lower accuracy of volumetric measurements in comparison to gravimetry. The accumulative errors also mean that measuring a smaller number of isotherm points will reduce the total accumulative error in volumetric measurement, and so this should be considered during the development of the experimental method.

6.5.5 Leakage

Leakage can be a significant practical problem in volumetric measurement, and it has been identified as such by a number of authors [14, 42, 80, 81]. As discussed in Sect. 6.3.1.2, any measurable leakage could be misinterpreted as either sorption or desorption depending on its direction. Any suspected leakage should be investigated through pressure hold tests, but the measurement of pressure versus uptake for an empty sample cell and the measurement of hydrogen uptake by well understood materials can also help identify leakage problems. Kiyobayashi et al. [80] measured the hydrogen leak rate of their system over a 10–20 h period to use as a reference for comparison with the pressure reduction observed during adsorption. Such a measurement is recommended during the characterisation of the performance of an instrument. The measurement of uptake for an empty cell in the absence of leakage should, meanwhile, yield an approximately zero uptake as a function of pressure. Any significant deviation from zero could indicate leakage problems, particularly if it increases with pressure; although it is important to monitor the pressure as a function of time for each point to confirm that leakage is the issue. The measurement of hydrogen sorption isotherms for well understood materials should also help confirm leakage problems. Significant leakage will result in unusual curvature or unexpected linear sections in the hydrogen sorption isotherms (see Sect. 6.3.1.2) and so any such deviation from the expected isotherm shape could be indicative of such problems, particularly if the deviation increases with increasing pressure.

6.6 Gravimetric Measurement

In this section we consider the issues specific to gravimetric measurement, and begin with the choice of sample size, which is important when performing measurements using a balance of a particular resolution, stability or accuracy. We then look at the buoyancy effect (Archimedes' Principle) corrections that are essential for any pressure dependent gravimetric measurement to account for the upthrust experienced by the balance due to the increasing gas density at elevated pressure. Finally we consider potential problems caused by the disturbance of the balance.

6.6.1 Sample Size Considerations

Sample size considerations in gravimetric measurement largely depend on the balance resolution, long term stability and maximum capacity. The compensating beam balance used in the Hiden Isochema IGA instruments has a resolution of ± 0.1 μg and a long term stability[19] of ± 1 μg, with a maximum capacity of

[19] The long term stability is determined primarily by the thermal stability of the balance enclosure, although it is also limited to a certain extent by the balance resolution.

200 mg. These values, combined with the expected hydrogen uptake of a material, govern the feasible sample size range for the performance of an accurate hydrogen sorption measurement. The isotherm points, from one measured hydrogen composition to the next, should be sufficiently separated in terms of weight to allow accurate determination of the associated change in the balance reading. The sample size cannot be specifically defined without taking into account the magnitude of the buoyancy effect corrections, which are inversely proportional to the sample density; however, generally speaking, the minimum sample size in the case of the balance characteristics described above, will be of the order of tens of milligrams. If the balance accuracy is lower then the sample size should be increased accordingly to achieve equivalent measurement accuracy.

6.6.2 Buoyancy Effect Corrections

The basic calculation of the hydrogen uptake in a gravimetric measurement was covered in Sect. 4.2.1.1. The buoyancy effect corrections are necessary in gravimetry to account for the upthrust on the microbalance originating from the presence of the sample in gas phase hydrogen of differing density in the microbalance chamber. The upthrust is dependent on the mass of hydrogen displaced by the sample, according to Archimedes' Principle, and is denoted by $f_{buoyancy}$ in Eq. 4.8. As it is dependent on the mass of gaseous hydrogen displaced it will increase with increasing hydrogen pressure for any given sample size. In Eq. 4.10 we gave the buoyancy correction necessary for a hypothetical gravimetric experimental configuration in which there is a contribution from each of the four other components in the gas phase,

$$f_{buoyancy} = \left(m_{sh}g\left(\frac{\rho_H}{\rho_{hd}}\right) + m_{sp}g\left(\frac{\rho_H}{\rho_{sp}}\right) + m_s g\left(\frac{\rho_H}{\rho_s}\right)\right) - \left(m_{ch}g\left(\frac{\rho_H}{\rho_{hd}}\right) + m_{cw}g\left(\frac{\rho_H}{\rho_{cw}}\right)\right) \quad (6.55)$$

where the first term on the right hand side of the equation represents the buoyancy effect from the sample side of the balance and the second term represents the buoyancy effect from the counterweight side. A hangdown of density, ρ_{hd}, is suspended from both arms, contributing a mass, m_{sh}, on the sample side and a mass, m_{ch}, on the counterweight side. The sample of mass, m_s, and density, ρ_s, is mounted in a sample pan of mass, m_{sp}, and density, ρ_{sp}, and a counterweight of mass, m_{cw}, and density, ρ_{cw}, is mounted on the counterweight side. g is the acceleration due to gravity. In this expression, m_s is taken as the mass of the sample at vacuum and stays constant regardless of any interaction with the gas in the microbalance chamber. In addition, in adsorption measurement, the conversion

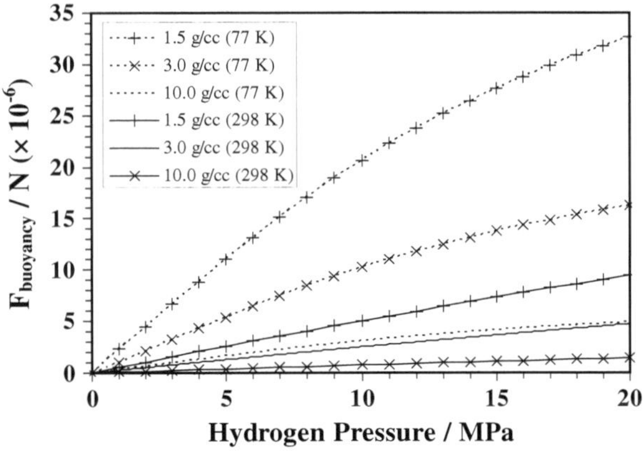

Fig. 6.1 The buoyancy contribution of 100 mg samples of three different densities at 77 and 298 K at pressures up to 20.0 MPa

from an excess to an absolute adsorbed quantity should account for the buoyancy contribution of the adsorbed phase (see Sect. 3.1.1.3).

It can be seen that these corrections increase with increasing pressure. For this reason, it can be argued that beyond a certain threshold the magnitude of the corrections necessary in gravimetry begins to outweigh the benefits that make gravimetric measurements intrinsically more accurate than volumetric measurements in the lower pressure range. However, the level of any such pressure is dependent on the particular experimental configuration, as well as the temperature of operation and the density of the sample. Therefore, if such a threshold exists, it will be a rather complex function of the measurement temperature and the sample density, for any given experimental configuration.

As an example, however, let us consider the buoyancy contribution of 100 mg samples of three different material densities, 1.5 g cm^{-3}, 3 g cm^{-3} and 10 g cm^{-3}. Figure 6.1 shows the buoyancy contribution from these 100 mg samples at 77 K and 298 K in the pressure range up to 20.0 MPa. The hydrogen density was calculated using the NIST REFPROP database [30]. The wide range of corrections can be seen with the correction for the low density sample at low temperature unsurprisingly being considerably larger than the high density sample at ambient temperature for the equivalent hydrogen pressure. Furthermore, Fig. 6.2 shows a practical example of the application of buoyancy effect corrections to hydrogen adsorption data measured gravimetrically for an activated carbon sample at 77 K. The figure shows the uncorrected weight data, expressed as a wt% uptake, excess adsorption data, which is corrected for the buoyancy of the sample, and absolute adsorption data, which is corrected for both the buoyancy of the sample and the buoyancy of the adsorbed phase. The conversion of the excess to absolute adsorption uses the assumption that the density of the adsorbed phase is equal to the density of liquid hydrogen (see Sect. 3.1.1.3). This figure shows that the corrections at low temperatures can be significant even at pressures below 2.0 MPa and therefore must be applied with care.

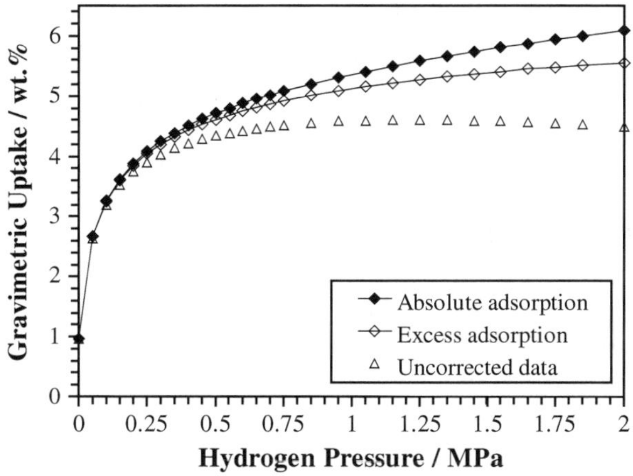

Fig. 6.2 Hydrogen adsorption by an activated carbon at 77 K measured using the gravimetric technique. The uncorrected data show the raw weight data from the microbalance reading without buoyancy corrections applied, the excess adsorption data show the raw data corrected for buoyancy, but not accounting for the buoyancy of the adsorbed phase, and the absolute adsorption data show the raw data corrected for buoyancy of both the sample and the adsorbed phase assuming that the density of the latter is equal to the liquid phase hydrogen density. Data plotted from 'Definitions of the Hydrogen Storage Capacity of Materials', 2008, Hiden Isochema Technical Note 201, with permission from Hiden Isochema Ltd

6.6.3 Disturbance of the Balance

All microbalance-based gravimetric measurements are susceptible to errors originating from the disturbance of the balance. These can arise from a number of sources. One obvious example is external vibration. This can be eliminated by appropriate mounting of the balance. Another is thermal instability of the balance enclosure, as covered briefly in Sect. 6.3.2.2. A further two sources relating to the measurement temperature and pressure are thermal transpiration effects, which occur at low pressure, and convection currents, which can affect measurements at elevated pressures. Both of these effects occur when there is a significant temperature gradient between the sample and the balance chamber. In the former case, it is useful to assess the likely occurrence of thermal transpiration effects by considering the Knudsen number, *Kn*, of the system for any particular measured pressure and system dimension (see Sect. 6.1.4), or to ensure their minimisation during instrument design. The effects are, however, complex and depend on a number of factors, including the geometry of the experimental set-up, the density and geometry of the sample, the hangdown and the gas type [82–85]. In the case of convection currents, baffles can help dampen the disturbance. Empty pan measurements can help determine the likely disturbance of the balance due to each of these effects and to assess the effectiveness of baffles in eliminating instability problems associated with convection currents. A stable empty balance pan reading in the temperature and pressure range of interest should help eliminate these potential problems from consideration.

6.7 Thermal Desorption Measurement

In this section we consider the issues specific to the thermal desorption measurements covered in Sect. 4.3. These include sample size considerations, the temperature ramp rate at which the desorption is performed and the method used to calibrate the hydrogen desorption signal.

6.7.1 Sample Size Considerations

Generally speaking, smaller sample sizes can be used for thermal desorption spectroscopy, performed using a mass spectrometer, compared to volumetric or gravimetric measurement. One of the key errors associated with the measurement of desorption from a small sample is the accuracy with which the sample mass can be determined (Sect. 6.2.1.3). If a total desorbed quantity, in terms of wt%, is calculated this will obviously be affected by the accuracy with which either the empty or the hydrogenated sample mass is known. Another important aspect of temperature-programmed desorption measurement is heat transfer. This should be considered in conjunction with the temperature ramp rate, which is addressed in the next section, but the smaller the sample the easier it is to maintain a relatively homogeneous sample temperature, taking into account the endothermic nature of the hydrogen desorption process.[20] In general, the problem of heat transfer in larger samples is more likely to affect the shape of thermal desorption spectra, and hence activation energy determination (see Sect. 3.3.2.1), than the calculation of total desorbed quantities determined using this method. This is because the activation energy determination methods typically depend on the measurement of peak position as a function of heating rate. If heat transfer problems affect the measured peak positions, the result will be an associated error in the calculated activation energy. Total desorbed quantity, or hydrogen content, determination, meanwhile, relies only on the integration of the complete thermal desorption spectrum.

[20] In the case of the hydrides of interest for hydrogen storage, the desorption process is usually endothermic.

6.7.2 Temperature Ramp Rate

The temperature ramp rates in TDS studies of hydrides should be relatively slow (1–20 K min^{-1}, for example), in comparison with the rapid rates typically used in surface science, because the desorption process for a hydride is slow in comparison with desorption from a surface. The chosen ramp rate should reflect the sample size to a certain extent because if the ramp rate is too high there is little chance of the sample temperature following the applied rate and considerable inhomogeneity could be introduced as a result. The endothermic nature of the desorption process should also be considered because the associated cooling effect will oppose the heating rate. In the case of the determination of the total desorbed quantity, the effect of ramp rate on the calculated quantity should be checked to ensure reasonable agreement between measurements performed under different conditions.

6.7.3 Signal Calibration Method

The hydrogen signal from a thermal desorption experiment must be calibrated if quantification of the desorbed quantity is required. Calibration is not necessary for TDS measurements intended for activation energy determination, as described in Sect. 4.3.1, because in this case the peak desorption temperature is used rather than the total desorbed quantity. In the case of TGA, the weight reading gives a direct measurement of the desorbed quantity of hydrogen, assuming that the weight loss is all due to hydrogen desorption.[21] We will therefore now consider the calibration of non-TGA TDS measurements, which tend to fall into one of two categories.

The first approach has been used by Fernández et al. [86] and von Zeppelin et al. [87], and involves the use of a calibrant sample. This method involves desorbing a known quantity of hydrogen from a well understood material and measuring the resultant mass spectrometer signal, which can then be used for calibration purposes. Fernández et al. [86] chose $TiH_{1.92}$ as a calibration material, whereas von Zeppelin et al. [87] used three different calibration materials, namely a hydrogenated PdGd alloy, TiH_2 and CaH_2. They found that the first two of these were better suited to the task and that CaH_2 should be avoided due to its hygroscopic nature. The latter was examined because it was chosen as a calibrant sample by Dillon et al. [43] in previous carbon nanotube work, discussed briefly in Sect. 2.1.1.2 and Sect. 6.2. Von Zeppelin et al. [87] estimate an error of $\pm$ 5% in their quantitative calculations of desorbed hydrogen. They used a series of TiH_2 samples that were in the mass range (1–6) mg, but the PdGd alloy sample was significantly larger ($\sim$ 1 g) due to its lower gravimetric hydrogen content. In later

[21] If there is a possibility of the evolution of other species during the desorption or decomposition process, this can be confirmed by coupled quadrupole mass spectrometry.

work on low temperature thermal desorption spectroscopy from microporous materials, Panella et al. [88, 89] used $PdH_{0.7}$ as a calibrant material.

The second approach involves measuring the mass spectrometer response to the flow of a calibrant gas mixture at a series of flow rates [90]. The accuracy of this method is partially dependent on both the uncertainty in the pre-mixed calibrant gas mixture and the accuracy of the mass flow controllers used for both the desorption measurement and the calibration routines.

6.8 Summary

In this chapter we have covered a number of the experimental issues relating to the performance of accurate hydrogen sorption measurements. We began with a number of the properties of hydrogen, including coverage of some of the equations of state that can be used to represent the pressure-density relationship of hydrogen as a function of temperature. We then addressed some of the other relevant properties, such as the Joule–Thomson effect, thermal conductivity and thermal transpiration, as well as considering the pressures and length scales at which hydrogen exists in the free molecule, transition and continuum regimes. We have emphasised the importance of the gas purity, as well as discussing the properties of a material that can affect the accuracy of hydrogen sorption measurements, such as its density, sensitivity to air or moisture contamination and the purity of the crystal or compound. Various general aspects of sorption instrumentation have been discussed including vacuum and pressure measurement and compatibility, and temperature measurement and stability. Two of the main considerations when developing an experimental method, the sample degassing and activation procedure, and the equilibration time at each isotherm point, were then addressed. The following three sections covered some of the issues that apply specifically to each of the laboratory gas sorption measurement techniques covered in Chap. 4. We began with issues relevant to the gravimetric measurement technique, which include sample size considerations, the buoyancy effect corrections, and disturbance of the balance. We then considered the issues relating specifically to volumetric measurement. These include thermal gradients, the ratio of the sample size to the internal system volume, the dead space volume corrections, the accumulative errors inherent in the multipoint manometric isotherm determination method, and hydrogen leakage. We concluded the chapter with considerations for thermal desorption techniques, which include sample size, the choice of temperature ramp rate and the TDS signal calibration method. All of the aspects discussed in this chapter that are relevant to the particular material being studied and the measurement type being used should be considered when performing hydrogen sorption measurements on potential hydrogen storage materials.

References

1. Broom DP, Moretto P (2007) Accuracy in hydrogen sorption measurements. J Alloy Compd 446–447:687–691
2. Broom DP (2007) The accuracy of hydrogen sorption measurements on potential storage materials. Int J Hydrogen Energy 32:4871–4888
3. Broom DP (2008) Hydrogen sorption measurements on potential storage materials: experimental methods and measurement accuracy. EUR 23242 EN. Office for Official Publications of the European Communities, Luxembourg
4. McCarty RD (1975) Hydrogen technical survey—thermophysical properties. Report number: NASA-SP-3089. NASA Scientific and Technical Information Office, Washington DC
5. McCarty RD, Hord J, Roder HM (1981) Selected properties of hydrogen (engineering design data). NBS monograph 168. National Bureau of Standards, Washington DC
6. Pitzer KS (1955) The volumetric and thermodynamic properties of fluids. I. Theoretical basis and virial coefficients. J Am Chem Soc 77(13):3427–3433
7. Pitzer KS, Lippmann DZ, Curl RF, Huggins CM, Petersen DE (1955) The volumetric and thermodynamic properties of fluids. II. Compressibility factor, vapor pressure and entropy of vaporization. J Am Chem Soc 77(13):3433–3440
8. Span R, Wagner W, Lemmon EW, Jacobsen RT (2001) Multiparameter equations of state—recent trends and future challenges. Fluid Phase Equilib 183–184:1–20
9. Zhou L, Zhou Y (2001) Determination of compressibility factor and fugacity coefficient of hydrogen in studies of adsorptive storage. Int J Hydrogen Energy 26:597–601
10. Zhang C, Lu X, Gu A (2004) How to accurately determine the uptake of hydrogen in carbonaceous materials. Int J Hydrogen Energy 29:1271–1276
11. Zhou W, Wu H, Hartman MR, Yildirim T (2007) Hydrogen and methane adsorption in metal-organic frameworks: a high-pressure volumetric study. J Phys Chem C 111(44):16131–16137
12. Nasrifar K (2010) Comparative study of eleven equations of state in predicting the thermodynamic properties of hydrogen. Int J Hydrogen Energy 35:3802–3811
13. Kumar KH, Starling KE (1982) The most general density-cubic equation of state: application to pure nonpolar fluids. Ind Eng Chem Fundam 21:255–262
14. Tibbetts GG, Meisner GP, Olk CH (2001) Hydrogen storage capacity of carbon nanotubes, filaments, and vapor-grown fibers. Carbon 39:2291–2301
15. Redlich O, Kwong JNS (1949) On the thermodynamics of solutions. V. An equation of state. Fugacities of gaseous solutions. Chem Rev 44(1):233–244
16. Soave G (1972) Equilibrium constants from a modified Redlich–Kwong equation of state. Chem Eng Sci 27:1197–1203
17. Perry RH, Green DW (1997) Perry's chemical engineers' handbook, 7th edn. McGraw-Hill, New York
18. Xiang Z, Lan J, Cao D, Shao X, Wang W, Broom DP (2009) Hydrogen storage in mesoporous coordination frameworks: experiment and molecular simulation. J Phys Chem C 113:15106–15109
19. Lin X, Telepeni I, Blake AJ, Dailly A, Brown CM, Simmons JM, Zoppi M, Walker GS, Thomas KM, Mays TJ, Hubberstey P, Champness NR, Schröder M (2009) High capacity hydrogen adsorption in Cu(II) tetracarboxylate framework materials: the role of pore size, ligand functionalization, and exposed metal sites. J Am Chem Soc 131(6):2159–2171
20. Peng D-Y, Robinson DB (1976) A new two-constant equation of state. Ind Eng Chem Fundam 15(1):59–64
21. Ansón A, Benham M, Jagiello J, Callejas MA, Benito AM, Maser WK, Züttel A, Sudan P, Martínez MT (2004) Hydrogen adsorption on a single-walled carbon nanotube material: a comparative study of three different adsorption techniques. Nanotechnology 15:1503–1508
22. Jagiello J, Ansón A, Martínez MT (2006) DFT-based prediction of high-pressure H_2 adsorption on porous carbons at ambient temperatures from low-pressure adsorption data measured at 77 K. J Phys Chem B 110:4531–4534

23. Frost H, Düren T, Snurr RQ (2006) Effects of surface area, free volume, and heat of adsorption on hydrogen uptake in metal-organic frameworks. J Phys Chem B 110:9565–9570
24. Hemmes H, Driessen A, Griessen R (1986) Thermodynamic properties of hydrogen at pressures up to 1 Mbar and temperatures between 100 and 1,000 K. J Phys C: Solid State Phys 19:3571–3585
25. McLennan KG, Gray EM (2004) An equation of state for deuterium gas to 1,000 bar. Meas Sci Technol 15:211–215
26. Beattie JA, Bridgeman OC (1928) A new equation of state for fluids. Proc Am Acad Arts Sci 63(5):229–308
27. Rao YVC (2004) Introduction to thermodynamics, 2nd edn. Universities Press, Hyderabad
28. Paskevicius M, Sheppard DA, Buckley CE (2010) Thermodynamic changes in mechanochemically synthesized magnesium hydride nanoparticles. J Am Chem Soc 132:5077–5083
29. Stampfer JF, Holley CE, Suttle JF (1960) The magnesium-hydrogen system. J Am Chem Soc 82(14):3504–3508
30. Lemmon EW, Huber ML, McLinden MO (2007) NIST Standard Reference Database 23: Reference Fluid Thermodynamic and Transport Properties-REFPROP, Version 8.0, National Institute of Standards and Technology, Standard Reference Data Program, Gaithersburg
31. Benedict M, Webb GB, Rubin LC (1940) An empirical equation for thermodynamic properties of light hydrocarbons and their mixtures. I. Methane, ethane, propane and *n*-butane. J Chem Phys 8:334–345
32. Starling KE (1973) Fluid thermodynamic properties for light petroleum systems. Gulf Publishing Co, Houston
33. Leachman JW (2007) Fundamental equations of state for parahydrogen, normal hydrogen, and orthohydrogen. MSc thesis, University of Idaho
34. Leachman JW, Jacobsen RT, Penoncello SG, Lemmon EW (2009) Fundamental equations of state for parahydrogen, normal hydrogen, and orthohydrogen. J Phys Chem Ref Data 38(3):721–748
35. Latimer RE, Mostello RA (1964) Thermodynamic comparison of large-scale liquefaction of air, hydrogen, and helium. AIChE J 10(3):407–415
36. Wisniak J (1999) The Joule–Thomson coefficient for pure gases and their mixtures. Chem Educ 4:51–57
37. Johnston HL, Bezman II, Hood CB (1946) Joule–Thomson effects in hydrogen at liquid air and at room temperatures. J Am Chem Soc 68(11):2367–2373
38. Bruno TJ, Svoronos PDN (1989) CRC handbook of basic tables for chemical analysis. CRC Press, Boca Raton
39. Leachman JW, Jacobsen RT, Penoncello SG, Huber ML (2007) Current status of transport properties of hydrogen. Int J Thermophys 28:773–795
40. Yang RT (2000) Hydrogen storage by alkali-doped carbon nanotubes–revisited. Carbon 38:623–626
41. Chen P, Wu X, Lin J, Tan KL (1999) High H_2 uptake by alkali-doped carbon nanotubes under ambient pressure and moderate temperatures. Science 285:91–93
42. Thomas KM (2007) Hydrogen adsorption and storage on porous materials. Catal Today 120:389–398
43. Dillon AC, Gennett T, Alleman JL, Jones KM, Parilla PA, Heben MJ (2000) Carbon nanotube materials for hydrogen storage. In: Proceedings of the 2000 Hydrogen Program Review NRL/CP-570-28890. http://www1.eere.energy.gov/hydrogenandfuelcells/pdfs/28890kkk.pdf. Accessed 13 Sept 2009
44. Hirscher M, Becher M, Haluska M, Dettlaff-Weglikowska U, Quintel A, Duesberg GS, Choi Y-M, Downes P, Hulman M, Roth S, Stepanek I, Bernier P (2001) Hydrogen storage in sonicated carbon materials. Appl Phys A 72:129–132
45. Züttel A, Orimo S (2002) Hydrogen in nanostructured, carbon-related, and metallic materials. MRS Bull 27(9):705–711
46. Lowell S, Shields JE, Thomas MA, Thommes M (2004) Characterization of porous solids and powders: surface area, pore size and density. Springer, Dordrecht

47. Jordá-Beneyto M, Lozano-Castelló D, Suárez-García F, Cazorla-Amorós D, Linares-Solano Á (2008) Advanced activated carbon monoliths and activated carbons for hydrogen storage. Microporous Mesoporous Mater 112:235–242
48. Neimark AV, Ravikovitch PI (1997) Calibration of pore volume in adsorption experiments and theoretical models. Langmuir 13:5148–5160
49. Malbrunot P, Vidal D, Vermesse J, Chahine R, Bose TK (1997) Adsorbent helium density measurement and its effect on adsorption isotherms at high pressure. Langmuir 13:539–544
50. Li J-R, Kuppler RJ, Zhou H-C (2009) Selective gas adsorption and separation in metal-organic frameworks. Chem Soc Rev 38:1477–1504
51. Gumma S, Talu O (2003) Gibbs dividing surface and helium adsorption. Adsorption 9:17–28
52. Černý R, Joubert J-M, Latroche M, Percheron-Guégan A, Yvon K (2000) Anisotropic diffraction peak broadening and dislocation substructure in hydrogen-cycled $LaNi_5$ and substitutional derivatives. J Appl Crystallogr 33:997–1005 (Erratum: Černý R et al (2002) J Appl Crystallogr 35:288)
53. Wanner M, Friedlmeier G, Hoffmann G, Groll M (1997) Thermodynamic and structural changes of various intermetallic compounds during extended cycling in closed systems. J Alloy Compd 253–254:692–697
54. Verma SK, Walker PL (1992) Carbon molecular sieves with stable hydrophobic surfaces. Carbon 30:837–844
55. Menendez I, Fuertes AB (2001) Aging of carbon membranes under different environments. Carbon 39:733–740
56. Strelko V, Malik DJ, Streat M (2002) Characterisation of the surface of oxidised carbon adsorbents. Carbon 40:95–104
57. Buckley CE, Gray EM, Kisi EH (1995) Stability of the hydrogen absorption and desorption plateaux in $LaNi_5$-H. Part 4: thermal history effects. J Alloy Compd 231:460–466
58. Percheron-Guégan A, Welter JM (1988) Preparation of intermetallics and hydrides. In: Schlapbach L (ed) Topics in applied physics vol. 63: hydrogen in intermetallic compounds I. Electronic, thermodynamic and crystallographic properties, preparation. Springer-Verlag, Berlin
59. Férey G (2008) Hybrid porous solids: past, present, future. Chem Soc Rev 37:191–214
60. Thomas KM (2009) Adsorption and desorption of hydrogen on metal-organic framework materials for storage applications: comparison with other nanoporous materials. Dalton Trans 1487–1505
61. Hafizovic J, Bjørgen M, Olsbye U, Dietzel PDC, Bordiga S, Prestipino C, Lamberti C, Lillerud KP (2007) The inconsistency in adsorption properties and powder XRD data of MOF-5 is rationalized by framework interpenetration and the presence of organic and inorganic species in the nanocavities. J Am Chem Soc 129(12):3612–3620
62. Tsao C-S, Yu M-S, Chung T-Y, Wu H-C, Wang C-Y, Chang K-S, Chen H-L (2007) Characterization of pore structure in metal-organic framework by small-angle X-ray scattering. J Am Chem Soc 129:15997–16004
63. Zhao XB, Xiao B, Fletcher AJ, Thomas KM (2005) Hydrogen adsorption on functionalized nanoporous activated carbons. J Phys Chem B 109:8880–8888
64. Siegmann HC, Schlapbach L, Brundle CR (1978) Self-restoring of the active surface in the hydrogen sponge $LaNi_5$. Phys Rev Lett 40(14):972–975
65. Peisl J (1978) Lattice strains due to hydrogen in metals. In: Alefeld G, Völkl J (eds) Topics in applied physics vol. 28: hydrogen in metals I. Basic properties. Springer-Verlag, Berlin
66. Schülke M, Paulus H, Lammers M, Kiss G, Réti F, Müller K-H (2008) Influence of surface contaminations on the hydrogen storage behaviour of metal hydride alloys. Anal Bioanal Chem 390:1495–1505
67. Gray EM, Blach TP, Buckley CE (1999) Stability of the hydrogen absorption and desorption plateaux in $LaNi_5$-H. Part 5: H capacity. J Alloy Compd 293–295:57–61
68. Chambers A, Fitch RK, Halliday BS (1998) Basic vacuum technology, 2nd edn. Institute of Physics Publishing, Bristol

69. Japanese Industrial Standards Committee (2007) Method for measurement of pressure-composition-temperature (PCT) relations of hydrogen absorbing alloys. JIS H 7201:2007 (E)
70. Wallbank AD, McQuillan AD (1975) Thermal transpiration correction of hydrogen equilibrium pressure measurements in metal/hydrogen solution. J Chem Soc Faraday Trans 1(71):685–689
71. Liang SC (1953) On the calculation of thermal transpiration. J Phys Chem 57:910–911
72. Takaishi T, Sensui Y (1963) Thermal transpiration effect of hydrogen, rare gases and methane. Trans Faraday Soc 59:2503–2514
73. York DC, Chambers A, Chew AD (2000) Thermal transpiration of helium and nitrogen in 50-μm bore silica capillaries. Vacuum 59:910–918
74. Wilson T, Tyburski A, DePies MR, Vilches OE, Becquet D, Bienfait M (2002) Adsorption of H_2 and D_2 on carbon nanotube bundles. J Low Temp Phys 126(1–2):403–408
75. Rouquerol F, Rouquerol J, Sing K (1999) Adsorption by powders and porous solids: principles, methodology and applications. Academic Press, London
76. Redhead PA (2002) Recommended practices for measuring and reporting outgassing data. J Vac Sci Technol A 20(5):1667–1675
77. Benham MJ, Ross DK (1989) Experimental determination of absorption-desorption isotherms by computer-controlled gravimetric analysis. Z Phys Chem NF 163:S25–S32
78. Blach TP, Gray EM (2007) Sieverts apparatus and methodology for accurate determination of hydrogen uptake by light-atom hosts. J Alloy Compd 446–447:692–697
79. Fuller EL, Poulis JA, Czanderna AW, Robens E (1979) Volumetric and gravimetric methods of determining monolayer capacities. Thermochim Acta 29:315–318
80. Kiyobayashi T, Hiroyuki T, Takeshita T, Tanaka H, Takeichi N, Züttel A, Schlapbach L, Kuriyama N (2002) Hydrogen adsorption in carbonaceous materials—how to determine the storage capacity accurately. J Alloy Compd 330–332:666–669
81. Blackman JM, Patrick JW, Snape CE (2006) An accurate volumetric differential pressure method for the determination of hydrogen storage capacity at high pressures in carbon materials. Carbon 44:918–927
82. Poulis JA, Thomas JM (1963) Disturbances arising from thermal transpiration in microbalance experiments. J Sci Instrum 40:95–100
83. Poulis JA, Pelupessy B, Massen CH, Thomas JM (1964) Longitudinal Knudsen forces. J Sci Instrum 41:295–301
84. Poulis JA, Massen CH, Thomas JM (1966) Longitudinal Knudsen forces II. J Sci Instrum 43:234–237
85. Behrndt KH, Massen CH, Poulis JA, Steensland T (1966) Longitudinal thermomolecular flow at intermediate pressures—a comparison. In: Behrndt KH (ed) Vacuum microbalance techniques, vol 5. Plenum Press, New York
86. Fernández JF, Cuevas F, Sánchez C (2000) Simultaneous differential scanning calorimetry and thermal desorption spectroscopy measurements for the study of the decomposition of metal hydrides. J Alloy Compd 298:244–253
87. von Zeppelin F, Haluška M, Hirscher M (2003) Thermal desorption spectroscopy as a quantitative tool to determine the hydrogen content in solids. Thermochim Acta 404:251–258
88. Panella B, Hirscher M, Ludescher B (2007) Low-temperature thermal-desorption mass spectroscopy applied to investigate the hydrogen adsorption on porous materials. Microporous Mesoporous Mater 103(1–3):230–234
89. Panella B, Hönes K, Müller U, Trukhan N, Schubert M, Pütter H, Hirscher M (2008) Desorption studies of hydrogen in metal-organic frameworks. Angew Chem Int Ed 47:2138–2142
90. Zlotea C, Sahlberg M, Özbilen S, Moretto P, Andersson Y (2008) Hydrogen desorption studies of the $Mg_{24}Y_5$–H system: formation of Mg tubes, kinetics and cycling effects. Acta Mater 56:2421–2428

Chapter 7
Concluding Remarks

One of the main topics addressed by this book is the accurate characterisation of the hydrogen storage properties of potential hydrogen storage materials. An effective approach to the investigation of the accuracy of any particular experimental technique or measurement method is to perform an interlaboratory study, in which measurements are made on a chosen sample by a number of different laboratories and the results are later compared. This type of exercise can help certify reference materials, which can be used subsequently by individual laboratories to assess the performance of a piece of apparatus, or the suitability of a related experimental method, for the task in hand. If reference materials have been subject to interlaboratory investigation, an uncertainty can be assigned to the particular property of interest and an instrument can then be tested by performing a measurement on a sample of the material to ensure that the results obtained fall within the range defined by the maximum permissible measurement error. In the case of hydrogen sorption measurements for hydrogen storage purposes, the most basic measurement parameter is the hydrogen uptake at a particular temperature and pressure, although as we have seen the uptake as a function of pressure at any given temperature is also of crucial importance. If an interlaboratory study has been performed under tightly controlled experimental conditions that have been shown to result in reproducible data, the chosen method could also be used to form the basis of guidelines that would help other laboratories perform sufficiently accurate measurements on closely related materials. In this concluding chapter we therefore firstly discuss the results of a recent interlaboratory study on hydrogen adsorption measurement before moving on to discuss potential reference materials. We then define some provisional measurement guidelines, which provide both a tentative guide to the most important considerations for the performance of high accuracy hydrogen sorption measurements as well as a practical and useful summary of many of the topics discussed in the previous chapter. We conclude with a discussion of the importance of future research in this area.

D. P. Broom, *Hydrogen Storage Materials*, Green Energy and Technology,
DOI: 10.1007/978-0-85729-221-6_7,

7.1 Interlaboratory Studies

An effective way to investigate or demonstrate the reproducibility of a measurement technique, or to compare the results of different techniques, is to conduct an *interlaboratory study* [1], also known as a *Round Robin exercise*. Interlaboratory studies involve the measurement of a sample or samples in a number of different laboratories, and this type of exercise has been performed in many different research fields. Recent work related to our area of interest include interlaboratory studies of the characterisation of the pore structure characteristics of porous materials using gas adsorption [2] and the measurement of carbon dioxide adsorption by coals and activated carbon [3–5]. However, an interlaboratory study of hydrogen adsorption measurement was recently reported by Zlotea et al. [6] and, due to its direct relevance to the subject of this book, we shall now discuss the results. In the study, the hydrogen adsorption and desorption characteristics of a microporous carbon, namely Takeda CMS 4A, at both 77 K and ambient temperature, were determined by 14 participating European laboratories. Figs. 7.1 and 7.2 show the data obtained at both 77 K and ambient temperature (293–301 K). It can be seen that there is a considerable spread in the data, particularly at the higher hydrogen pressures, for both measurement temperatures.

The data measured at 77 K show reasonable agreement at low pressures but a considerable spread at elevated pressures. It is notable, however, that dataset 10 appears to show a particularly large uptake even at lower pressures with a very steep rise up to the upper measurement pressure of just below 1.0 MPa and so it is not only the highest pressure measurements that vary. With regard to the highest pressures, only datasets 1–3 include data above 3.0 MPa. Dataset 1 shows a significantly higher uptake at these elevated pressures, reaching approximately 2.6 wt% at 8 MPa, whereas dataset 3 shows less than 1.5 wt% uptake at a similar pressure. Zlotea et al. [6] attribute the large uptake seen in dataset 1, which was determined volumetrically, to possible hydrogen leakage or impurity contamination

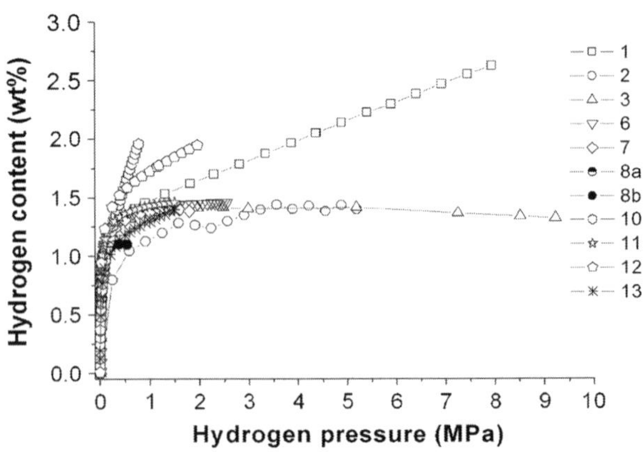

Fig. 7.1 Hydrogen adsorption isotherms for Takeda CMS 4A measured at 77 K by 11 participants in the interlaboratory study reported by Zlotea et al. [6]. Reproduced with permission from Zlotea et al. [6]. Copyright 2009 International Association for Hydrogen Energy

due to the 99.99% purity of the hydrogen used in the measurement. The low hydrogen purity in this case is likely to affect the measurement accuracy (see Sect. 6.1.6), but there are other potential sources of error that could also have contributed to the deviation seen in this data. Firstly, the recorded sample density for this dataset was 2.72 g cm^{-3}, which is rather high for a microporous carbon. This suggests that there could be an error in the calibration of the dead volume or the dosing volume of the apparatus (see Sect. 6.5.3). Secondly, the equation of state used is recorded only as 'real' and so an error in the compressibility description could also be the cause of the higher measured capacity, although this depends on the chosen expression (see Sect. 6.1.1). Regardless of the exact reason for the larger overall uptake, the shape of the isotherm looks unusually linear for hydrogen adsorption at 77 K by a microporous material, which further supports the argument that this dataset is erroneous. Dataset 3 on the other hand, which shows more curvature, was reported with a more physically reasonable density of 1.72 g cm^{-3}. The difference in the position of the Gibbs dividing surface, as defined by material densities of 1.72 and 2.72 g cm^{-3}, is very significant and could account for the majority of the difference in the observed uptakes. According to Blach and Gray [7], the ratio of the dosing volume to sample cell volume may have a significant effect on the calculated hydrogen uptake by a low density material in volumetric (manometric) apparatus (see Sect. 6.5.2). The authors suggest that an inappropriate choice of internal volumes can increase problems associated with uncertainty in the volume, and hence density, or vice versa, of the sample. Therefore, a problem relating to the discrepancy in the densities used or found, which could be exacerbated by the choice of internal volumes, would seem to be a possible origin of the discrepancy between these datasets; however, this possibility is difficult to assess because the internal volumes were not reported in the study.

According to Zlotea et al. [6], when considering the data below 2.5 MPa, datasets 3, 6, 7 and 11 show similar Type I behaviour with a plateau of around 1.4 ± 0.1 wt% at elevated pressures. Datasets 3 and 6 were measured volumetrically, whereas datasets 7 and 11 were obtained using the gravimetric method. They conclude that the data reported by these participants can be considered as the reference behaviour for this carbon because each of these datasets were reported with reasonable density values, high vacuum degassing conditions, and high purity hydrogen.[1] In addition, 1.4 wt% is consistent with a value of 1.45 wt% reported for this carbon in the literature [8]. The authors also argue that the porous structure characteristics exhibited by the carbon, including a micropore volume of 0.17 cm^3 g^{-1} and an

[1] The authors also mention the satisfactory agreement between equilibrium times (see Sect. 6.4.2) but these were only recorded for datasets 7 and 11. It is also interesting to note that these two datasets used only the ideal gas law to represent the pressure-density relationship for hydrogen. The description of the hydrogen compressibility is, however, of lower importance for gravimetric measurements in the lower pressure regime. In comparison to the spread in the other data, this is therefore unlikely to be a particularly important factor.

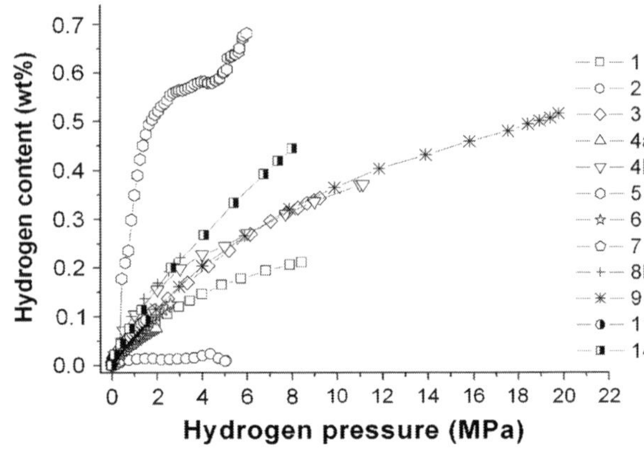

Fig. 7.2 Hydrogen adsorption isotherms for Takeda CMS 4A measured at ambient temperature (293–301 K) by 12 participants in the interlaboratory study reported by Zlotea et al. [6]. Reproduced with permission from Zlotea et al. [6].) Copyright 2009 International Association for Hydrogen Energy

"apparent surface area" of 423 $m^2\ g^{-1}$,[2] are consistent with a relatively modest hydrogen uptake of 1.4 wt%. The other datasets were grouped into (1) those that exhibit Type I behaviour but an underestimated adsorption capacity in relation to the reference value, (2) those that show Type I behaviour but overestimate the capacity, and (3) those that do not show Type I behaviour. Each dataset in the third group also indicates a lower capacity than the reference value. Datasets 8a and 8b are in the first group. The second group consists of datasets 1, 10 and 12, and the third group includes datasets 2 and 13. It is interesting to note that in the pressure range 3–5 MPa dataset 3, which is in the reference group, is in good agreement with dataset 2, which is in the third group. Although it is difficult to draw definite conclusions from these results, they clearly demonstrate the potential variability in the measurement of hydrogen adsorption due to differences in either the experimental conditions or the real behaviour of the material. Some of the potential errors are known to become more significant with increasing pressure, and decreasing temperature, and this would appear to be borne out by the data presented here. These results clearly illustrate the need to carefully consider many of the issues discussed in the previous chapter, particularly at elevated pressures.

With regard to the ambient temperature measurements, the results of which are shown in Fig. 7.2, two datasets clearly appear to be erroneous. Firstly, dataset 5 appears to show a step, which in sub or near-critical temperature adsorption is indicative of either monolayer formation (or rather a monolayer to multilayer transition), or the existence of two distinct pore size regimes (e.g., mixed micro-porous and mesoporous characteristics, or a bimodal pore size distribution). However, Takeda CMS 4A is known to be predominantly microporous and the behaviour indicative of a monolayer to multilayer transition does not occur in

[2] Note that values in the range 313–422 $m^2\ g^{-1}$ were determined for the nitrogen BET surface area of this carbon in the recent interlaboratory study reported by Silvestre-Albero et al. [2], and so the value quoted by Zlotea et al. [6] is at the upper end of this range.

microporous materials because adsorption is dominated by pore-filling rather than Langmuir-type monolayer formation. In addition, regardless of the pore structure, it is also highly unlikely to occur at such a high temperature ($T/T_c \approx 9$). Secondly, dataset 2 indicates negligible hydrogen uptake and appears close to an empty sample cell measurement. This also seems unlikely for a microporous material with such a high surface area or micropore volume because some adsorption is expected to occur at elevated pressures; this seems particularly true in view of the greater amount of adsorption observed by the other participants. Furthermore, the adsorbed quantity appears to decrease at approximately 4.5 MPa, which is unexpected at such a high temperature.[3] Zlotea et al. [6] suggest that the large uptake indicated by dataset 5 could be the result of an incorrect dead volume measurement, noting that the density of the material was reported as 0.2 g cm^{-3}. This is a remarkably low material density and clearly indicates a problem with this measurement. Dataset 2, meanwhile, was measured in a volumetric system in which the sample volume was approximately 1% of the "reaction chamber volume". It is possible that the internal system volume was too large, in this case, to accurately measure the hydrogen uptake for the chosen sample size, but it is also notable that the temperature stability is reported as only 2 K at ambient temperature. Thermal stability at least an order of magnitude better than this should be achievable using a waterbath to maintain near ambient temperatures. If we therefore exclude the two outlying datasets[4] (2 and 5) from the 300 K measurements, the spread is significantly reduced, but there is still a notable variation in the measured uptake at any particular chosen pressure. At 8.0 MPa, for example, datasets 1 and 14 show uptakes of approximately 0.2 and 0.45 wt%, respectively. In their analysis, Zlotea et al. [6] again separate the datasets into different groups. One group consists of the outlying datasets 2 and 5, while the other three consist of datasets 3, 7, 9 and 11, which are considered to have produced reference behaviour consistent with the literature value of 0.1 wt% at 2.0 MPa reported by Schmitz et al. [8], and two groups that show uptakes either below or above the assumed reference value.

The authors conclude that the maximum hydrogen capacity of the material at 77 K and 1 MPa is 1.4 ± 0.1 wt% and at ambient temperature it is 0.07 ± 0.01 wt% and 0.12 ± 0.01 wt% at 1.0 and 2.0 MPa, respectively, based on the behaviour determined by the two groups of isotherms that were assumed to represent reference behaviour at each temperature. A statistical analysis of the data, meanwhile, resulted in calculated relative standard deviations of 14 and 36% for the 77 K and ambient temperature measurements, respectively. It is not entirely

[3] The peak observed in the excess adsorption tends to move to higher pressures with increasing temperature. See, for example, the data of Zhou et al. [9] for MOF-5 in which the peak moves from below 1.0 MPa at 40 K to approximately 4.0 MPa at 77 K.

[4] In the statistical analysis of data, *outliers*, which are data or observations that appear inconsistent with the remainder of the dataset, should be included if they are the result of random fluctuations in the measurement. However, if they are due to human error or another fault in the measurement process, the cause should be investigated and the value corrected, or the data excluded from consideration [10].

clear whether the spread in the data discussed above is due primarily to intrinsic uncertainties in the hydrogen adsorption behaviour of the carbon sample or to errors in the sorption measurements. If it is the former, this could be due potentially to modification of the surface chemistry of the various batches of sample due to exposure to different environmental conditions prior to measurement (see Sect. 6.4.1). However, it seems more likely to be due to measurement error because the experimental conditions, in each case, show significant variation between the participating laboratories. As indicated in the discussion above, the sample density is reported to range from 0.2 to 2.72 g cm^{-3}, while the degassing pressures range from 3×10^{-5} to 60 Pa (see Sects. 6.3.1.1 and 6.4.1) and the recorded equilibration times vary between 1.6 and 30 min (see Sect. 6.4.2). The actual spread of the latter, however, may be greater because equilibration times were not reported by half of the participants. Furthermore, 4 of the 14 laboratories used the ideal gas equation to represent the pressure-density relationship for hydrogen and eight of the participants have not stated the hydrogen purity used. It therefore seems reasonable to conclude that the spread in the data is most likely due to the variation in the experimental conditions used by the participants, rather than the intrinsic uncertainty in the hydrogen uptake behaviour of the sample. If this is the case, then it is a good demonstration of the need to carefully address many of the points discussed in Chap. 6. If, on the other hand, the variation is due to intrinsic uncertainty in the adsorption behaviour of the carbon, the results perhaps provide good evidence against the selection of this particular carbon for use as a reference material. However, further investigation under more tightly controlled experimental conditions would be required before any firm conclusions could be drawn.

7.2 Reference Materials

To characterise and corroborate both sorption instrument performance and experimental methodology, it is necessary, or at least useful, to have reference materials [11] available, for which the hydrogen sorption behaviour is known or understood to a sufficient extent; preferably with a predetermined and hence known uncertainty.

With regard to microporous materials, Zlotea et al. [6] conclude in their study that Takeda CMS 4A would be suitable for 'benchmarking', while Furukawa et al. [12] have previously proposed MOF-177 as a suitable 'benchmark material'. Although MOF-177 has not yet been the subject of a published interlaboratory study, Furukawa et al. [12] present volumetrically and gravimetrically determined hydrogen adsorption data that appear to be in good agreement with each other. However, this material shows a high hydrogen uptake and it could be argued that in order to benchmark other materials, characterise the performance of an instrument or check experimental protocols, it is better to use a material with a more modest hydrogen capacity, because lower measurement sensitivity is required to achieve high measurement accuracy in the case of materials with a high adsorptive capacity. Furthermore, the hydrogen uptake of metal–organic frameworks is known to depend on their purity [13–15] (see Sect. 6.2.4). In the author's opinion a more suitable choice would be one of the synthetic zeolites, such as Na-X, because these are stable materials with more modest hydrogen adsorption capacities than MOF-177, and they are crystalline, unlike Takeda CMS 4A. The latter means that they are easier to characterise in terms of their crystallography and pore structure than amorphous carbons. This suggestion would need to be tested in an interlaboratory study, but this is also the case for the other candidates before they could become reference materials because Takeda CMS 4A would need to exhibit more reproducible data than those presented by Zlotea et al. [6] and the hydrogen uptake of MOF-177 reported by Furukawa et al. [12] would have to be confirmed independently by a number of different laboratories.

Another, perhaps obvious, suggestion would be to use a microporous material that has already been certified for surface area, pore volume or pore size distribution determination. A good example is the recently certified BAM-P109,[5] which is a porous carbon material with a micropore diameter of approximately 0.8 nm and a certified BET Specific Surface Area (SSA) of 1,396 $m^2\ g^{-1}$. During the certification process, 22 laboratories using 23 different instruments found mean BET SSAs of between 1,354 and 1,445 $m^2\ g^{-1}$. The measurements were performed with nitrogen at 77 K over a minimum relative pressure range from $P/P_0 = 0.01$ to $P/P_0 = 0.1$. The results of an interlaboratory study for high pressure hydrogen adsorption measurement on this material, performed under suitably controlled experimental conditions, would be very interesting to see, in order to determine whether the disparity in results broadens significantly with increasing hydrogen pressure, and whether it could serve as a feasible reference material for high pressure hydrogen adsorption measurement. If we assume that the range of BET SSA values reported represents a micropore capacity that correlates well with the amount of hydrogen adsorbed at high pressure (see Sect. 5.2.4), then a series of accurate hydrogen adsorption measurements should result in a significantly reduced variation in the data compared with the spread seen in the data presented by Zlotea et al. [6].

For hydrides it is challenging to produce a reference material because the hydrogen absorption and desorption properties of hydride compounds depend on both the activation procedure and the history of the sample (see Sects. 6.2.3 and 6.4.1). The activation of a hydride-forming material can be viewed as a hydride synthesis process, in which different synthesis routes will potentially lead to different hydride materials due to the complex microstructural effects and modifications that accompany hydride phase formation. However, $LaNi_5$ and Pd have both been used extensively as test materials for volumetric (Sieverts') and

[5] http://www.rm-certificates.bam.de/de/rm-certificates_media/rm_cert_porous_materials/bam_p109e.pdf, Accessed 19 August 2010.

gravimetric apparatus. Furthermore, Ti alloy standard samples are available from NIST,[6] under the reference numbers SRM 2452, SRM 2453 and SRM 2454, although the hydrogen contents of these certified materials are very low.[7] Each of these standard reference materials are Ti–V-Al alloys, which have had their hydrogen contents determined using either cold-neutron Prompt-Gamma Activation Analysis (PGAA) and volumetric measurement (SRM 2452 and SRM 2454) or by these methods and Neutron Incoherent Scattering (NIS) analysis (SRM 2453). By virtue of their low hydrogen content, these standard reference materials do not appear to make particularly good materials for the testing of hydrogen storage material characterisation equipment, although they could be used effectively to test TDS apparatus because the hydrogen could be desorbed in a single desorption scan.[8] Both $LaNi_5$ and Pd, on the other hand, would seem to make reasonable choices as test materials for hydrogen absorption measurement apparatus, although it is worth noting that they may be inappropriate to test apparatus for measurements on low density hydrides [7]. In addition, the dependence of the plateau pressures, and the amount of uptake at any given pressure, on the sample history, purity, activation procedure, and the amount of residual hydrogen trapped in the sample should also be taken into account.

To conclude this section, we will emphasise the need to perform test measurements on reference materials that are similar to those that are to be characterised. The practice of using a high density hydride-forming compound to test apparatus subsequently used for measurements on a low density adsorbent is not uncommon [6, 16–19]; however, as mentioned above, Blach and Gray [7] note that an accurate measurement with a high density material, such as $LaNi_5$, in manometric apparatus does not necessarily mean that a system will provide high accuracy measurements on low density samples. It is also evident that the measurement of hydrogen adsorption should be referenced against test measurements on another adsorbent, rather than on a hydrogen absorbing material, because the accurate determination of the excess adsorbed quantity is so critically dependent on the location of the Gibbs dividing surface and hence on the accuracy of its determination. A measurement performed on a hydride-forming compound will not test the ability of the apparatus to determine the location of the Gibbs dividing surface with sufficient accuracy, or test the sensitivity of the instrument to a realistic uncertainty in the determined position of the dividing surface, or the

[6] http://www.nist.gov/ts/msd/srm/, Accessed 19 Aug 2010.

[7] The hydrogen contents of SRM 2452, SRM 2453 and SRM 2454 are 62.5 ± 1.6, 114 ± 5 and 211 ± 4 mg kg^{-1}, respectively.

[8] There would be an uncertainty in this characterisation process, however, because it is not clear whether some residual trapped hydrogen would remain in the material following the thermal desorption process. This is likely to depend, to a certain extent, on the temperature range over which the desorption is performed.

skeletal density of the sample, during subsequent measurements. Therefore, any further work on the development of reference materials of different densities and types (hydrides and porous adsorbents) would clearly be very valuable in helping to ensure that appropriate materials are available for the testing of apparatus for different measurement types.

7.3 Measurement Guidelines

One possible consequence of the interlaboratory studies discussed in Sect. 7.1, and other research in the area, is the development of guidelines and standards for the measurement of hydrogen sorption. IUPAC guidelines currently exist for adsorption measurement [20], while BET measurements are covered by a British standard [21]. There are many other standards available for particle characterisation techniques, a number of which cover associated adsorption measurement methods [22], and Japanese Industrial Standards (JIS) also exist for metal hydride characterisation [23–26]. An extension of one or more of these standards or guidelines to hydrogen storage materials, in general, should be possible with further research. The challenge, however, is to define procedures that ensure high accuracy but do not potentially restrict the application of the methods to new materials or limit the use of different apparatus that may still be suitable for the task.

A useful precursor to standardisation, which is a lengthy and complicated process, would be the establishment of a set of accepted guidelines for measurement that, at least, provides a checklist for the minimum amount of experimental information that should be provided when presenting hydrogen sorption data, but also includes measures that should help ensure avoidance of the most common pitfalls encountered during hydrogen sorption measurement. The experimental conditions reported by Zlotea et al. [6] were quite varied, and a good starting point would therefore seem to be a more tightly controlled set of such conditions, expanded to cover different material types. The author has previously proposed tentative checklists for volumetric and gravimetric measurements of both hydrogen absorption and adsorption [27], and a modified version of these is presented below, in the context of the relevant sections of Chap. 6. Thermal desorption spectroscopy has not been included, although the guidelines could easily be extended to such measurements in the future. The following guidelines are worth bearing in mind during the study of potential hydrogen storage materials and would also seem to provide a useful and practical summary of the discussion presented in the previous chapter. The guidelines presented previously were separated into four different checklists, with each point in each list corresponding to a different section in the report. Here we combine the guidelines, due to the significant amount of overlap between many of the points addressed previously. The guidelines consist of five sections: (1) gas supply and instrumentation, (2) sample considerations, (3) methodology, (4) data reduction and (5) data reporting.

7.3.1 Gas Supply and Instrumentation

The minimum purity of the hydrogen should be 99.9999%, or 99.999% with filtration (see Sects. 6.1.6 and 6.2.5). The temperature of the instrumentation should be maintained constant with a variation of less than 1 K, but preferably better (see Sect. 6.3.2.2), with no hot or cold spots present. The sample environment should be controlled, using a cryostat, coolant bath, or furnace that can maintain the sample temperature with a variation of less than 1 K (see Sect. 6.3.2.1). Pressure should be measured to high accuracy with particular attention paid to volumetric instrumentation. The uncertainty should be substantially lower than the differences in pressure before and after a sorption or desorption step; for example, better than 0.1% of the range (see Sect. 6.3.3.1). Different devices should be used, where necessary, to cover different pressure measurement ranges. The accuracy of the weight measurement in gravimetric systems must be substantially lower than the expected weight change due to hydrogen sorption or desorption at each isotherm point (see Sects. 6.3.2.2 and 6.6.1). For volumetric measurement, the system volume must be low enough, for a given sample size, for the pressure measuring device to be able to detect the required or expected drop in pressure due to hydrogen sorption or desorption (see Sect. 6.5.2). The sample temperature should be monitored throughout a measurement by a sensor in contact with, or close to, the sample (see Sects. 6.3.2.1 and 6.3.4). The temperature sensor should be calibrated regularly and be accurate to within ± 1 K. The achievable accuracy will depend on the sensor type, which should be stated (see Sect. 6.3.4). All measuring devices should be regularly calibrated, including the pressure and temperature measuring systems in volumetric and gravimetric instruments (see Sect. 6.3.4), and the microbalance in gravimetric apparatus (see Sect. 6.6.1). The volumes of a manometric system, including the sample cell dead volume (see Sect. 6.5.3), must also be calibrated with an uncertainty better than the volume change that would significantly alter the calculated quantity of hydrogen present in the system. The required level of uncertainty will depend on the accuracy of the pressure and temperature measuring devices, and the thermal stability of the instrument. Leakage should be minimised, with particular attention paid to this issue for volumetric systems. The leakage rate of an instrument should be determined using pressure hold tests (see Sect. 6.5.5). A dry pump system should be used to avoid oil vapour contamination (see Sect. 6.3.1.1). High vacuum capability is the ideal for all measurement types, but is a prerequisite for thorough degassing of porous materials (see Sects. 6.3.1.1 and 6.4.1). System design should minimise the tortuosity of the pumping path and other factors that will reduce the achievable vacuum in the sample cell (see Sect. 6.3.1.1). The performance of the instrument should be tested with a suitable reference material; the chosen material or materials should be similar to the material under investigation, in terms of density and the proposed sorption mechanism (see Sect. 6.5.2 and 7.2).

7.3.2 Sample Considerations

An appropriate sample size in relation to the system volume and pressure measurement accuracy should be chosen for volumetric measurements (see Sect. 6.5.2). For gravimetric measurement it should be selected according to the resolution, long term stability and the capacity of the microbalance (see Sect. 6.6.1). For volumetric measurement, the sample should be large enough to allow the accurate determination of the sample mass in an external balance; for example, greater than 50 mg. Any pretreatment, such as annealing in the case of hydrides, should be considered before the characterisation of a material. The history of a sample should be known and recorded (see Sect. 6.2.3). A hydride sample should be fully activated (see Sect. 6.4.1).

7.3.3 Methodology

For porous materials, the degassing conditions for the sample must be considered carefully. High vacuum ($<$ 10 mPa) conditions are required and the process must be performed for a sufficient period of time (see Sects. 6.3.1.1 and 6.4.1). The temperature must be high enough to remove any environmental contaminants or remnants from the synthesis process (see Sect. 6.4.1) without affecting the stability and porous structure of the material, and the process should be monitored effectively by either measuring the pressure decrease above the vacuum pump in volumetric apparatus or the sample weight loss in a gravimetric system (see Sect. 6.4.1). For volumetric measurement, the dead volume should be determined either directly using helium at ambient temperature or above, or indirectly using a physically reasonable value for the apparent or skeletal density of the material (see Sects. 6.2.1.2 and 6.5.3). For gravimetric measurement, the weight reading should be corrected for buoyancy as a function of hydrogen pressure using the density and mass of each component in the microbalance chamber and a physically reasonable value for the apparent or skeletal density of the material (see Sects. 6.2.1.2 and 6.6.2). The accuracy of both the dead volume calibration in the volumetric technique and the buoyancy correction in the gravimetric technique is crucial for adsorption measurement because they both define the position of the Gibbs dividing surface or the apparent volume and density of the sample assumed in the measurement (see Sect. 6.2.1.2). The errors that accumulate in multi-point volumetric isotherm determination mean that the greater the number of points in a single isotherm measurement, the greater the accumulated error. The number of points in a volumetric isotherm measurement should therefore be minimised, providing sufficient information regarding the shape of the isotherm can be obtained (see Sect. 6.5.4). In volumetric measurement, the pressure should be monitored at each isotherm point, to ensure sorption equilibrium has been achieved. Similarly, the weight should be monitored in the gravimetric case.

The equilibration times should be long enough to ensure both sufficient sorption and thermal equilibrium has been achieved at each isotherm point. Thermal effects from the hydrogen absorption and desorption reactions for hydrides must be carefully monitored to ensure thermal equilibrium has been achieved at each point. The equilibration times used for each point should therefore be considered carefully and subsequently recorded (see Sect. 6.4.2), or the method used to monitor the approach to equilibrium stated.

7.3.4 Data Reduction

The pressure-density relationship for hydrogen must be represented accurately because use of the ideal gas law at elevated pressure, during the calculation of hydrogen uptake, will result in an error. The effects are significantly more severe for volumetric measurement and so particular attention must be paid to this issue in the volumetric case. If an equation of state is used directly it should be sufficiently accurate in the temperature and pressure range of the measurement (see Sect. 6.1.1) and the expression used should be stated. If the compressibility values are tabulated for use in the data reduction process, they should be defined in small steps of temperature and pressure to ensure errors are not introduced as a result. If measurements are made at low pressures, the measured values should be corrected for thermal transpiration, or thermomolecular flow, effects using the expressions of Liang [28] or Takaishi and Sensui [29], or an appropriate alternative (see Sect. 6.3.3.2). The pressure at which the corrections are required will depend on the system dimension (see Sects. 6.1.4 and 6.1.5) but they are likely to apply only at pressures below 100 Pa. For volumetric measurements, the method used to account for the thermal gradient between the dosing volume and the sample cell in the calculation of the sorbed quantity should be considered carefully and stated (see Sects. 4.1.1 and 6.5.1). For adsorption measurement, if the excess is converted to the absolute adsorption, a physically reasonable assumption regarding the adsorbed phase volume or density must be used during the data reduction (see Sect. 3.1.1.3). If the results are reported as the excess quantity, this should be clearly stated.

7.3.5 Data Reporting

Based on the guidelines detailed above, the following information regarding the experimental conditions should be provided to allow a reasonable assessment of the quality of published hydrogen sorption data. Measurements performed using variations on either microbalance-based gravimetric techniques or the isochoric manometric (Sieverts') method can be accommodated by simply replacing some of the points with the equivalent piece of relevant information. So, for example, the flowing volumetric method may require a statement regarding the accuracy of the flow meter or mass flow controller used in the measurement, rather than the accuracy of the volume calibration, although the internal volume of the system is still important and should therefore be reported. The proposed list of relevant information is, as follows:

1. The purity of hydrogen gas supply, and any filtration or contamination reduction strategies employed.
2. The temperature stability of the apparatus and the sample.
3. The pressure measurement accuracy or uncertainty.
4. The long term stability and the uncertainty of the weight reading for gravimetry.
5. The volume of the dosing volume, sample cell and any additional components for volumetry.
6. The sample temperature measurement accuracy or uncertainty.
7. The leak rate of the empty system.
8. The vacuum pump configuration and base pressure capability.
9. The sample mass.
10. Information regarding the sample synthesis and history, including the activation procedure.
11. A description of the degassing procedure for porous materials, including the period, temperature and base pressure.
12. The dead volume calibration method used for volumetry.
13. The buoyancy effect corrections applied for gravimetry.
14. The equilibration times, or the method used to assess the approach to equilibrium at each isotherm point.
15. The hydrogen compressibility description used, including the equation of state or the source of tabulated compressibility factors.
16. The thermal transpiration effect correction applied, if required.
17. The method used to account for the thermal gradient between the dosing volume and the sample cell in volumetry.
18. For adsorption measurement, the reported quantity (excess or absolute adsorption) should be stated, with the conversion to absolute adsorption defined, where applicable.
19. The reference or benchmark material or materials used to assess the instrument performance prior to the reported measurements.

Although it is probably unrealistic to expect every report of hydrogen sorption to either strictly follow the above guidelines or include such detailed information, the author believes that the above description, including the data reporting list, represents a reasonable description of best practice for hydrogen sorption measurements on potential hydrogen storage materials.

It should be noted that a US DOE-funded report on best practices for the characterisation of hydrogen storage materials has recently been compiled [30] but, at the time of writing, only a draft version of the report is available and the authors do not attempt to either define specific guidelines of the form described above or summarise their extensive discussion of hydrogen sorption measurement-related issues. Zlotea et al. [6], however, presented guidelines for hydrogen adsorption measurement in their report, which are largely the same as those presented above, although there is some deviation. The differences include the potential use of Pd or $LaNi_5$ as a reference material, which is a practice that, as mentioned above, should be avoided for adsorption measurements. They also do not include minimum temperature or pressure measurement accuracies, or the minimum required hydrogen purity, although the authors recommend that this information should be recorded when reporting data.

7.4 Future Research

As we saw in the opening chapter, hydrogen storage technology is of critical importance to the implementation of a future hydrogen economy and it is therefore clearly crucial that the search for a suitable material continues apace. The rapid application-oriented development of new materials in any research field relies on the accurate characterisation of the relevant physical and chemical properties of the materials in question, and hydrogen storage is no exception. Inaccuracies, erroneous data and the ensuing controversies are an unwelcome distraction that can consume both research time and funding. Although some discussion of the properties of a newly discovered or developed material is inevitable during the scientific process, it is undoubtedly beneficial to reduce the repetition of avoidable methodological mistakes that can lead to erroneous characterisation data or conclusions. Any attempts to both improve our understanding of measurement accuracy and to help ensure that erroneous characterisation is minimised in the future will therefore have an important role to play in the development of new hydrogen storage media. The willingness of major funding bodies, including the US DOE and the European Commission, to support such research in recent years is a testament to the practical significance of this issue.

In addition to the performance of interlaboratory studies, as described in Sect. 7.1, it should be possible to further investigate hydrogen sorption measurement accuracy through research into the hydrogen sorption properties of proposed reference materials using existing measurement methods and complementary techniques. In the author's opinion, this work should be continued in parallel with the development of new hydrogen storage materials and the investigation of the hydrogen storage mechanisms involved. Furthermore, it is likely that conclusions drawn from work on hydrogen adsorption measurement accuracy will contribute to our understanding of supercritical gas adsorption in other fields and could therefore have benefits beyond the targeted search for new hydrogen storage media.[9] Our fundamental understanding of sorption processes and the interaction of hydrogen, and other gases, with matter relies on the performance of accurate

[9] Supercritical gas adsorption at elevated pressure is of significant interest in a range of other application areas, including gas separation, methane storage and carbon dioxide capture.

experimental measurements. In view of the controversy over the storage properties of carbon nanostructures (see Sects. 2.1.1.2 and 6.2), the initially erroneous data reported for MOF-5 and the subsequent variation in the hydrogen storage capacities measured for different MOF-5 samples [13], the disagreement regarding the storage capacity of THF-stabilised clathrates (see Sect. 2.4), and the variation in the data presented by Zlotea et al. [6] for an activated carbon, it seems likely that some isolated studies of the hydrogen sorption properties of materials reported in the literature may contain erroneous or inaccurate data. Gaining a better understanding of the relative importance of the various error sources in high pressure hydrogen sorption measurements, in general, should help reduce the probability of such erroneous or inaccurate data being reported in the future. As a consequence, it could help reduce the likelihood of further controversies developing over the hydrogen storage capabilities of particular materials and, furthermore, it would greatly aid and accelerate our understanding of the interaction of hydrogen with matter. Work in this area is therefore of both fundamental and technological importance in the field of hydrogen storage material research, and is crucial to the rapid and effective development of the technology required for the implementation of a future hydrogen economy.

7.5 Summary

In this concluding chapter, we have discussed some of the issues surrounding the accurate characterisation of potential hydrogen storage materials. Firstly, we discussed interlaboratory studies, which are an effective way of investigating the reproducibility of a measurement method or technique. We then discussed reference materials that can be used for the corroboration of experimental hydrogen sorption measurement methodology and for the characterisation of the performance of sorption measurement instrumentation. Although low hydrogen content hydride materials are available from NIST, no high hydrogen content standard reference materials exist that are appropriate for the testing of hydrogen sorption measurement apparatus and so the development of such materials is important to progress in the field of hydrogen storage material research. We have presented tentative guidelines for the performance of hydrogen sorption measurements and a checklist for the minimum amount of information that should be provided when reporting hydrogen sorption data. The guidelines are referenced to the relevant sections in Chap. 6 and therefore represent an effective summary of the topics discussed in that chapter. We concluded by emphasising the importance of future research into hydrogen sorption measurement accuracy, which could greatly aid our understanding of the interaction of hydrogen with matter and help reduce the variation in the reported hydrogen sorption properties of new materials in the future. Any such progress would make a significant contribution to the search for new solutions to the hydrogen storage problem. It is hoped that this book will also make a positive contribution towards these goals.

References

1. ASTM International (1999) Standard practice for conducting an interlaboratory study to determine the precision of a test method. E 691–699
2. Silvestre-Albero J, Sepúlveda-Escribano A, Rodríguez-Reinoso F, Kouvelos V, Pilatos G, Kanellopoulos NK, Krutyeva M, Grinberg F, Kaerger J, Spjelkavik AI, Stöcker M, Ferreira A, Brouwer S, Kapteijn F, Weitkamp J, Sklari SD, Zaspalis VT, Jones DJ, de Menorval LC, Lindheimer M, Caffarelli P, Borsella E, Tomlinson AAG, Linders MJG, Tempelman JL, Bal EA (2009) Characterisation measurements of common reference nanoporous materials by gas adsorption (Round Robin tests). In: Kaskel S, Llewellyn P, Rodríguez-Reinoso F, Seaton NA (eds) Characterisation of porous solids VIII: proceedings of the 8th international symposium on the characterisation of porous solids. RSC Publishing, Cambridge
3. Goodman AL, Busch A, Duffy GJ, Fitzgerald JE, Gasem KAM, Gensterblum Y, Krooss BM, Levy J, Ozdemir E, Pan Z, Robinson RL Jr, Schroeder K, Sudibandriyo M, White CM (2004) An inter-laboratory comparison of CO_2 isotherms measured on Argonne premium coal samples. Energy Fuels 18:1175–1182
4. Goodman AL, Busch A, Bustin RM, Chikatamarla L, Day S, Duffy GJ, Fitzgerald JE, Gasem KAM, Gensterblum Y, Hartman C, Jing C, Krooss BM, Mohammed S, Pratt T, Robinson RL Jr, Romanov V, Sakurovs R, Schroeder K, White CM (2007) Inter-laboratory comparison II: CO_2 isotherms measured on moisture-equilibrated Argonne premium coals at 55°C and up to 15 MPa. Int J Coal Geol 72:153–164
5. Gensterblum Y, van Hemert P, Billemont P, Busch A, Charriére D, Li D, Krooss BM, de Weireld G, Prinz D, Wolf K-HAA (2009) European inter-laboratory comparison of high pressure CO_2 sorption isotherms. I: activated carbon. Carbon 47(13):2958–2969
6. Zlotea C, Moretto P, Steriotis T (2009) A Round Robin characterisation of the hydrogen sorption properties of a carbon based material. Int J Hydrogen Energy 34(7):3044–3057
7. Blach TP, Gray EM (2007) Sieverts apparatus and methodology for accurate determination of hydrogen uptake by light-atom hosts. J Alloy Compd 446–447:692–697
8. Schmitz B, Müller U, Trukhan N, Schubert M, Férey G, Hirscher M (2008) Heat of adsorption for hydrogen in microporous high-surface-area materials. ChemPhysChem 9:2181–2184
9. Zhou W, Wu H, Hartman MR, Yildirim T (2007) Hydrogen and methane adsorption in metal-organic frameworks: a high-pressure volumetric study. J Phys Chem C 111(44):16131–16137
10. Ellison SLR, Barwick VJ, Duguid Farrant TJ (2009) Practical statistics for the analytical scientist: a bench guide. RSC Publishing, Cambridge
11. International Laboratory Accreditation Cooperation (2005) Guidelines for the selection and use of reference materials. ILAC-G9:2005
12. Furukawa H, Miller MA, Yaghi OM (2007) Independent verification of the saturation hydrogen uptake in MOF-177 and establishment of a benchmark for hydrogen adsorption in metal-organic frameworks. J Mater Chem 17:3197–3204
13. Hafizovic J, Bjørgen M, Olsbye U, Dietzel PDC, Bordiga S, Prestipino C, Lamberti C, Lillerud KP (2007) The inconsistency in adsorption properties and powder XRD data of MOF-5 is rationalized by framework interpenetration and the presence of organic and inorganic species in the nanocavities. J Am Chem Soc 129(12):3612–3620
14. Férey G (2008) Hybrid porous solids: past, present, future. Chem Soc Rev 37:191–214
15. Thomas KM (2009) Adsorption and desorption of hydrogen on metal-organic framework materials for storage applications: comparison with other nanoporous materials. Dalton Trans 1487–1505
16. Tibbetts GG, Meisner GP, Olk CH (2001) Hydrogen storage capacity of carbon nanotubes, filaments, and vapor-grown fibers. Carbon 39:2291–2301
17. Poirier E, Chahine R, Tessier A, Bose TK (2005) Gravimetric and volumetric approaches adapted for hydrogen sorption measurements with *in situ* conditioning on small sorbent samples. Rev Sci Instrum 76:055101

18. Blackman JM, Patrick JW, Snape CE (2006) An accurate volumetric differential pressure method for the determination of hydrogen storage capacity at high pressures in carbon materials. Carbon 44:918–927
19. Mishra A, Banerjee S, Mohapatra SK, Graeve OA, Misra M (2008) Synthesis of carbon nanotube-TiO_2 nanotubular material for reversible hydrogen storage. Nanotechnology 19:445607
20. Sing KSW, Everett DH, Haul RAW, Moscou L, Pierotti RA, Rouquérol J, Siemieniewska T (1985) Reporting physisorption data for gas/solid systems with special reference to the determination of surface area and porosity. Pure Appl Chem 57(4):603–619
21. British Standards Institution (1996) Determination of the specific surface area of powders - part 1: BET method of gas adsorption for solids (including porous materials). BS 4359-1:1996 (ISO 9277:1995)
22. Dąbrowski A, Robens E, Klobes P, Meyer K, Podkościelny P (2003) Standardization of methods for characterizing the surface geometry of solids. Part Part Syst Charact 20:311–322
23. Japanese Industrial Standards Committee (2007) Glossary of terms used in hydrogen absorbing alloys. JIS H 7003:2007 (E)
24. Japanese Industrial Standards Committee (2007) Method for measurement of pressure-composition-temperature (PCT) relations of hydrogen absorbing alloys. JIS H 7201:2007 (E)
25. Japanese Industrial Standards Committee (2007) Method for measurement of hydrogen absorption/desorption reaction rate of hydrogen absorbing alloys. JIS H 7202:2007 (E)
26. Japanese Industrial Standards Committee (2007) Method for measurement of hydrogen absorption/desorption cycle characteristic of hydrogen absorbing alloys. JIS H 7203:2007 (E)
27. Broom DP (2008) Hydrogen sorption measurements on potential storage materials: experimental methods and measurement accuracy. EUR 23242 EN. Office for Official Publications of the European Communities, Luxembourg
28. Liang SC (1953) On the calculation of thermal transpiration. J Phys Chem 57:910–911
29. Takaishi T, Sensui Y (1963) Thermal transpiration effect of hydrogen, rare gases and methane. Trans Faraday Soc 59:2503–2514
30. Gross KJ, Carrington KR, Barcelo S, Karkamkar A, Purewal J, Parilla P (2010) Recommended best practices for the characterization of storage properties of hydrogen storage materials, V2-79, US DOE hydrogen program document. http://www1.eere.energy.gov/hydrogenandfuelcells/pdfs/best_practices_hydrogen_storage.pdf. Accessed 4 Jul 2010

Index